Recent Advances in Satellite Aeronautical Communications Modeling

Andrii Mikhailovich Grekhov
National Aviation University, Ukraine

A volume in the Advances in Wireless Technologies and Telecommunication (AWTT) Book Series

IGI Global
DISSEMINATOR OF KNOWLEDGE

Published in the United States of America by
IGI Global
Engineering Science Reference (an imprint of IGI Global)
701 E. Chocolate Avenue
Hershey PA, USA 17033
Tel: 717-533-8845
Fax: 717-533-8661
E-mail: cust@igi-global.com
Web site: http://www.igi-global.com

Library of Congress Cataloging-in-Publication Data

Names: Grekhov, Andrii Mikhailovich, 1951- author.
Title: Recent advances in satellite aeronautical communications modeling / by Andrii Mikhailovich Grekhov.
Description: Hershey, PA : Engineering Science Reference, [2019] | Includes bibliographical references.
Identifiers: LCCN 2018049686| ISBN 9781522582144 (h/c) | ISBN 9781522582151 (eISBN)
Subjects: LCSH: Aeronautics--Communication systems. | Artificial satellites in telecommunication.
Classification: LCC TL692 .G74 2019 | DDC 621.3841/56011--dc23 LC record available at https://lccn.loc.gov/2018049686

This book is published in the IGI Global book series Advances in Wireless Technologies and Telecommunication (AWTT) (ISSN: 2327-3305; eISSN: 2327-3313)

British Cataloguing in Publication Data
A Cataloguing in Publication record for this book is available from the British Library.

Advances in Wireless Technologies and Telecommunication (AWTT) Book Series

ISSN:2327-3305
EISSN:2327-3313

Editor-in-Chief: Xiaoge Xu, University of Nottingham Ningbo China, China

MISSION

The wireless computing industry is constantly evolving, redesigning the ways in which individuals share information. Wireless technology and telecommunication remain one of the most important technologies in business organizations. The utilization of these technologies has enhanced business efficiency by enabling dynamic resources in all aspects of society.

The **Advances in Wireless Technologies and Telecommunication Book Series** aims to provide researchers and academic communities with quality research on the concepts and developments in the wireless technology fields. Developers, engineers, students, research strategists, and IT managers will find this series useful to gain insight into next generation wireless technologies and telecommunication.

COVERAGE

- Cellular Networks
- Wireless Broadband
- Broadcasting
- Mobile Technology
- Network Management
- Mobile Web Services
- Grid Communications
- Mobile Communications
- Radio Communication
- Digital Communication

IGI Global is currently accepting manuscripts for publication within this series. To submit a proposal for a volume in this series, please contact our Acquisition Editors at Acquisitions@igi-global.com or visit: http://www.igi-global.com/publish/.

The Advances in Wireless Technologies and Telecommunication (AWTT) Book Series (ISSN 2327-3305) is published by IGI Global, 701 E. Chocolate Avenue, Hershey, PA 17033-1240, USA, www.igi-global.com. This series is composed of titles available for purchase individually; each title is edited to be contextually exclusive from any other title within the series. For pricing and ordering information please visit http://www.igi-global.com/book-series/advances-wireless-technologies-telecommunication/73684. Postmaster: Send all address changes to above address.

Titles in this Series

For a list of additional titles in this series, please visit:
https://www.igi-global.com/book-series/advances-wireless-technologies-telecommunication/73684

Paving the Way for 5G Through the Convergence of Wireless Systems
Ramona Trestian (Middlesex University, UK) and Gabriel-Miro Muntean (Dublin City University, Ireland)
Information Science Reference • ©2019 • 350pp • H/C (ISBN: 9781522575702) • US $195.00

Enabling Technologies and Architectures for Next-Generation Networking Capabilities
Mahmoud Elkhodr (Central Queensland University, Australia)
Information Science Reference • ©2019 • 384pp • H/C (ISBN: 9781522560234) • US $195.00

Mobile Devices and Smart Gadgets in Human Rights
Sajid Umair (National University of Sciences and Technology (NUST), Pakistan) and Muhammad Yousaf Shah (Ministry of Human Rights, Pakistan)
Information Science Reference • ©2019 • 304pp • H/C (ISBN: 9781522569398) • US $195.00

Emerging Automation Techniques for the Future Internet
Mohamed Boucadair (Orange, France) and Christian Jacquenet (Orange, France)
Information Science Reference • ©2019 • 351pp • H/C (ISBN: 9781522571469) • US $195.00

Emerging Capabilities and Applications of Wireless Power Transfer
Alicia Triviño-Cabrera (University of Málaga, Spain) and José A. Aguado (University of Málaga, Spain)
Information Science Reference • ©2019 • 383pp • H/C (ISBN: 9781522558705) • US $195.00

Sensing Techniques for Next Generation Cognitive Radio Networks
Ashish Bagwari (Uttrakhand Technical University, India) Jyotshana Bagwari (Uttrakhand Technical University, India) and Geetam Singh Tomar (THDC Institute of Hydropower Engineering and Technology, India)
Information Science Reference • ©2019 • 356pp • H/C (ISBN: 9781522553540) • US $215.00

Mobile Applications and Solutions for Social Inclusion
Sara Paiva (Instituto Politécnico de Viana do Castelo, Portugal)
Information Science Reference • ©2018 • 354pp • H/C (ISBN: 9781522552703) • US $195.00

For an entire list of titles in this series, please visit:
https://www.igi-global.com/book-series/advances-wireless-technologies-telecommunication/73684

IGI Global
DISSEMINATOR OF KNOWLEDGE

701 East Chocolate Avenue, Hershey, PA 17033, USA
Tel: 717-533-8845 x100 • Fax: 717-533-8661
E-Mail: cust@igi-global.com • www.igi-global.com

To my wife and children and in the memory of my parents

Table of Contents

Preface

In less than half a century, computers have become an indispensable attribute of modern society. The power of computer technology is increasing and computers will become ubiquitous means of solving intellectual problems. In aviation computerization is entering a new phase and the role of computers is increasing in all areas of aviation. In fact, computers are used at all stages from the design of the glider, engines, hydraulic systems, communication, navigation and surveillance systems to air traffic control, predicting conflict situations and preventing aircraft collisions in the air. Some employees make hypotheses and build models, while others are engaged in quantitative analysis of models in order to test the hypotheses underlying the models. The technical arsenal is constantly expanding, and first of all the modeling methods.

The extent of computer penetration into all spheres of human mental activity has no analogue. Computerization is obviously the most impressive scientific and technical achievement in history. The incentive for the development of modeling methods was the need for an accurate solution of problems with a large number of degrees of freedom. The tasks of aviation are also characterized by a huge number of interacting degrees of freedom. The environment for the functioning of aviation is formed by a set of moving objects, mobile and fixed obstacles, exclusion zones, and also contains communication, navigation and surveillance facilities. The global air traffic control system operates in an air navigation environment, which undergoes global transformation at both the structural and functional levels. Here computers can help not only to make precise calculations, but to identify new dependencies.

Numerical analysis is intended to complement practical results. Codes and algorithms are able to predict the behavior of systems in critical situations and help prevent them. The book is devoted to new research of satellite air communication channels and development of modeling methods.

The process of intellectual progress depends on the emergence of new ideas and finding new ways to "tie" old ideas. In recent decades, a significant step forward in improving the entire aviation structure has been made. Transformation of the functional level of the ATC system is characterized by the introduction of new concepts and modes, such as CNS/ATM, ADS-B, Free Flight, PBN, etc.

Modern development of aviation is characterized by the following tendencies.

The first trend is due to the intensification of air traffic and the expansion of aircraft operating ranges.

The second trend is determined by fundamentally new requirements and capabilities created by the World Transport System. New organizational and computer technologies for GPS management declare flexible, coordinated, rather than regulated use of airspace, taking into account all users in the CNS/ATM environment. The transition from a centralized air traffic control command system to a distributed system allows air traffic participants to choose their flight routes based on criteria of efficiency and economy. Therefore, the formation of the CNS/ATM environment is primarily due to the need to ensure the implementation of the Free Flight concept.

The third trend is the intensification of the development and use of aeronautical robotics systems operating without the participation of the pilot - RPAS (Remotely Piloted Air Systems), UAV (Unmanned Air Vehicle), UAS (Unmanned Aircraft Systems).

THE CHALLENGES

Communication, navigation, surveillance and air traffic control facilities, based on traditional principles, have a number of significant limitations. Surveillance means, which are primary and secondary ground-based radars, have a range of action limited by line of sight. In this case, there are difficulties in creating the necessary radar field, especially at low altitudes. The mean-square error in measuring the aircraft position at the maximum range is hundreds of meters, which does not meet the modern requirements of the ATC for new separation standards. For these reasons, ICAO has proposed the concept of CNS/ATM, which calls for the introduction of satellite navigation systems. In September 1991, at the Tenth Air Navigation Conference, ICAO member states approved the concept of CNS/ATM. A Global Air Navigation Plan for CNS/ATM Systems was developed (Global Plan, Doc 9750), which is a strategic document for management in the implementation of CNS/ATM systems. The existing aeronautical system has such major shortcomings in communications, navigation and surveillance, as well as air traffic services.

Communication

Currently, the dominant type of air-ground communication between the aircraft crew and the air traffic controller is voice communication. The use of transceivers in the very high frequency range (VHF) provides radio communication directly between the pilot and ATC unit within the line of sight. For communication in the FIR (Flight Information Region) outside VHF coverage area, radio stations operating in the high frequency range (HF) are used. Neighboring ATC agencies interact with each other on leased telephone channels (voice frequency), which provides direct voice communication between controllers in the process of negotiation of flight conditions and the implementation of the procedure for receiving and transmitting control. The ATC, other aviation authorities and many airlines are interconnected by AFTN (Aeronautical Fixed Telecommunication Networkxed), which provides the transmission of message-oriented messages (telegrams), and in some cases, lines of the ICAO common data exchange network (CIDIN- Common ICAO Data Interchange Network).

The main drawback of the existing air-ground communication subsystem is that the information exchange between the aircraft and ATC is mainly conducted through voice communication channels without the organization of automated data exchange between on-board and ground equipment. At the same time, the bandwidth of such channels is limited by the speed of pronunciation of voice messages, the language features of each person, and the need for repetition of messages in the event of unfavorable signal transmission conditions or interference effects. As the volume of air traffic increases, the voice communication channels are increasingly overloaded, which necessitates the allocation of additional channels. For the same reason, during the flight along the route, the crew is often forced to change the tuning frequency, which leads to an increase in the workload for communication.

Other disadvantages of the existing subsystem are related to the range in the VHF band limited by the line of sight. Elimination of this shortcoming can be achieved through the use of a territorially-distributed network of VHF stations (terrestrial repeaters) connected with ATC by leased lines of communication. However, as a result of this, the costs of organizing and operating such a communication network are significantly increased. In addition, in some cases the installation of terrestrial repeaters may be difficult or impossible. Communication in the HF range is subject to wave propagation anomalies, interference and signal attenuation. Due to these physical limitations, aerial radio communication in this range is carried out with the help of specially trained radio operators.

In the terrestrial segment of the communication subsystem, when using the AFTN network, only the low-speed transmission using teletypes is provided at the final stage of message delivery. In this case, some message switching centers still work in manual mode. All this delays the exchange of aeronautical information and leads to a decrease in the quality of ATM.

Navigation

Navigation over the land is mainly carried out by non-directional radio beacons within the structure of the routes covered by the zone of VOR operation (Very High Frequency Omni directional Range) and DME (Distance Measuring Equipment). When using omni-directional beacons of HF range due to wave propagation conditions, the same interference occurs as in the operation of HF radio communications. Therefore, the navigation accuracy in this case and the coverage area are limited. It is not always possible to ensure regional navigation in the entire required volume of airspace due to geographic or economic constraints, since these facilities operate in the line of sight. In addition, the very principle of air navigation, based on marking certain ground points and linking airways to the location of navigation aids at these points, limits the choice of routes of movement and causes the appearance of overly congested areas of airspace.

Surveillance

The use of ATC procedures in a specific airspace depends to a large extent on surveillance methods. As a rule, primary and secondary radars are used for surveillance in the continental and coastal regions, and in the oceanic and remote regions reports for this purpose are transmitted through the voice communication channels according to established rules. The main disadvantage of the surveillance subsystem is related to the limited range of primary and secondary radars.

Air Traffic Services (ATS)

The purpose of the ATS is to ensure that aircraft operators comply with the established schedule for the departure and arrival of aircraft and the selection of the most preferred flight profiles with minimum restrictions while maintaining the required level of safety. Therefore, the limitations of existing ATS systems are directly dependent on the shortcomings of the CNS elements.

Due to these shortcomings, in some sections of existing airways, it is sometimes impossible to obtain information about the actual position of the aircraft in real time and the predicted flight trajectory in the short and long term, resulting in the use of procedural ATC methods. The use of procedural ATC methods does not allow selecting the most effective flight profiles and making full use of the system's throughput, since in this case the flights usually have to be planned with increased intervals between the aircraft and with the passage of intermediate control points (compulsory reporting points). This also limits the ability to change the permitted flight profile. As a result, the potential of modern airborne systems is not fully realized, and ATS can not always be provided on an efficient and cost-effective basis.

In addition, the lack of digital air-ground data exchange systems does not allow to fully automate the processing of ATS-related information. Due to this uneven development of ATS, existing ATM systems do not allow the most efficient use of airspace. To eliminate these shortcomings, it is necessary to ensure, as soon as possible, the coordinated implementation of the automated interaction between the onboard and ground equipment and the elements of the system. In general, this will allow airspace users to fly on the most preferred trajectory for them and be more free in their choice.

SEARCHING FOR A SOLUTION

The main task of the future CNS is to ensure the fulfillment of the main goal of ATM concept, which is to satisfy the users' needs in the most preferred flight trajectories. The future CNS system is based on the idea of high-precision positioning of aircraft and the organization of effective automatic interaction of airborne and ground equipment to ensure safe air traffic in the whole world air space along the chosen flight routes.

Development of Communication Subsystem

The required level of efficiency, capacity and flexibility of the future ANS system can be achieved only with the use of digital means of data transmission. Therefore, improved data transmission and a global coverage area will characterize the new communication subsystem.

Although the need for voice communication will continue, nevertheless, the possibility of data transmission between all mobile and fixed communication services users in combination with the use of interworking will allow creating a homogeneous data transmission network in the conditions of applying various technical and administrative solutions. ATN (Aeronautical Telecommunications Network) will become an infrastructure for providing such information exchange of civil aviation on a global scale. ATN will provide interoperability of terrestrial data transmission networks, subnetworks of air-ground data transmission and airborne data networks by adopting common interface services and protocols based on the International Organization for Standardization (ISO) reference model of open systems interconnection.

Voice Air Communication

Voice communication of the VHF band will remain the main type of communication with aircraft for a long time. However, the use of digital data exchange channels will be extended and applied to the transmission of most routine air-ground messages, depending on operational requirements. In this case, voice communication will continue to be available for the transmission of non-routine and emergency messages.

Satellite Voice Communication

Satellite voice communication is likely to be limited to areas where there is no VHF band communication field, but it will not replace VHF voice communication until significant benefits are achieved in terms of productivity and cost. Satellite voice communication can be used to transmit non-routine and emergency messages or duplication where the data channel is the main means of communication.

Voice Terrestrial Communication

Despite the fact that an increasing amount of information will be transmitted through data channels, voice communication will be used, possibly, further, especially for the transmission of non-routine and emergency messages.

Air-Ground Data Transmission

Data exchange with the aircraft is expected to be performed using the following mobile subnets of ATN:

- VDL (Very High Frequency Digital Link);
- Mode S data link (Selective-Address);
- AMSS (Aeronautical Mobile Satellite Service);
- HF (Data transmission lines).

Currently, there are four versions of VDL (modes 1 to 4).

VDL mode 1 is a data link for the widely used communication system for addressing and transmitting ACARS (Aircraft Communications Addressing and Reporting System) messages, developed and commissioned in the late 1970s by ARINC (Aeronautical Radio Incorporated) to provide data exchange between aircraft and their operating agencies. The features of this line are due to the use of existing analog VHF radio equipment for data transmission in the byte-oriented format and consist of applying a two-step modulation with amplitude modulation of carrier and minimum frequency shift keying (AM-MSK-Minimal Shift Keying) as well as multiple access with carrier control (CSMA-Carrier Selecting Multiple Access). The data transfer rate in VDL mode 1 is 2400 bps.

VDL mode 2 is standardized by ICAO and involves the use of digital radio techniques with a set of protocols for various operational application processes. The use of a modem with 8-position relative phase modulation (D8PSK-Differential 8 Phase Shift Keying) provides a nominal transfer rate of 31,5 kbps. However, the use of CSMA still leads to the occurrence of nondeterministic message delays.

VDL mode 3 is based on the use of TDMA (Time Division Multiple Access) and will be a complex system of digital voice communication and data transmission, improving the use of VHF radio spectrum by providing four separate radio channels on the same carrier. Channel synchronization and channel access require the presence of ground stations.

VDL mode 4 summarizes the main advantages of previous versions. VDL mode 4 is self-synchronizing, which makes it possible to use it for autonomous organization of data transfer between aircraft. VDL version 4 is the most efficient digital data link for use in ATN network supporting all known ATM applications:

- ADS-B (Automatic Dependent Surveillance-Broadcasting);
- CPDLC (Controller Pilot Digital Link Communication), controlled by the pilot;
- FIS (Flight Information Service), TIS (Traffic Information Service), meteorological data and other broadcasting information;
- GNSS (Global Navigation Satellite System);
- SMGCS (Surveillance Moving Ground Control System).

Development of Navigation Subsystem

The central element in the ICAO navigation concept is the use of the GNSS system, which is currently able to provide navigation, routing requirements, and will meet all the requirements of precision approaches.

Development of Surveillance Subsystem

For surveillance purposes, according to the CNS/ATM concept, the SSR in A/C/S modes and ADS-B is used.

SSR Mode S

SSR will be used in the continental airspace with a high density of air traffic and in areas of aerodromes. It is envisaged that the accuracy and quality of the position information of the aircraft received in the A/C modes will be improved by the use of monopulse detection methods and/or the use of antennas with a large vertical opening. Advanced SSR Mode S (Selective) will enable the use of selective address data lines, more reliable identification of aircraft and the transfer of additional (flight) information. This will significantly improve the efficiency of surveillance and flight safety. Mode S Extended Squitter uses a squitter signal at the downlink frequency (1090 MHz), containing a set of broadcast messages, for the operation of ADS-B.

ADS-B

ADS-B is a method of monitoring, according to which the aircraft automatically provides information from the on-board navigation systems and positioning systems, including the aircraft identification index, its position in four dimensions and, if necessary, additional data, over the data transmission line. Based on the broadcasting principle of message transmission, ADS-B makes it possible to implement new applications, including the collection on board the aircraft information about the air situation (traffic) and resolving conflict situations.

Currently, ADS-B is designed only for use in line of sight.

The history of these studies dates back to 2010 when we expressed the idea of creating the "Aviation Internet" and a flight data recorder in real time and began to develop a new approach to the use of ADS-B surveillance systems based on the Iridium satellite communication system.

On May 12, 2010, we submitted an application for funding our studies and, in accordance with the Order of the Ministry of Education and Science of Ukraine, No. 1177 of 30.11.2010, received funding (Research work No. 721-DB11 "Development of the methodology for the modernization of ADS-B surveillance systems on base of low-orbit satellite systems", scientific supervisor of the research work - A.M. Grekhov). The financing was started on February 2, 2011, completed on December 31, 2013 and our study consisted in modeling the parameters of communication channels, depending on the transmission conditions, modulation type, error-control channel coding, and data transmission rate. After the completion of this project, research continues to the present.

ORGANIZATION OF THE BOOK

This book is devoted to the modeling of satellite communication channels for aircraft and RPAS/UAV using the Matlab Simulink and NetCraker software. The book is organized into six chapters. A brief description of each of the chapters follows:

Chapter 1 is devoted to the modeling of aircraft data transmission via low-orbit satellites. Original satellite channel models were designed. BER dependencies on the type of signal modulation, signal power, antenna diameters and nonlinearity of a high power amplifier were investigated. Effectiveness of error detection and correction was analyzed using classic linear block and convolutional codes. Free Space Path Loss, AWGN and radio frequency satellite channels were considered. MIMO 2 × 1 and 3 × 2 fading uplink/ downlink channels with antenna diversity were analyzed.

Chapter 2 deals with satellite channels based on IEEE 802.11a standard. A method for parameters estimation of satellite OFDM communication channel was proposed. Channel parameters were received for the Rayleigh and Rician fading, different types of Doppler spectrum, the gain of multipath channels, and the delay time of message flow.

Chapter 3 considers satellite channels based on IEEE 802.16 standard. Dependencies of a SNR in ground receiver on a SNR in downlink for different types of RPAS amplifier nonlinearity were obtained. The influence of the transmitter nonlinearity for different types of fading in the channel was studied. The possibility of the nonlinearity correction using pre-distortion was revealed. The impact of space-time diversity (MISO 2 × 1) for different types of fading in the channels was investigated.

Chapter 4 considers the modeling of ADS-B messages transmission via low-orbit satellite constellation Iridium. Models of satellite communication channel were built using NetCracker Professional 4.1 software. Influence of aircraft and satellites amount on average link utilization and message travel time was studied for channels with intersatellite link and bent-pipe architecture. The effect of communication channel "saturation" during simultaneous data transmission from many aircraft was investigated. Influence of a protocol type, a size of transaction, time between transactions and a channel latency on a traffic was studied. A method for an estimation the traffic loss was proposed and dependencies of the data loss coefficient on the size of transactions were received.

Chapter 5 deals with calculations of the microstrip antenna and the linear phased array using Antenna Magus Software. The analysis of antennas characteristics at different emitter numbers and different types of amplitude distribution was carried out. Calculations of electric field intensity and directional patterns for collinear antennas were provided. The sample of the antenna is described, which is used in operating system for reception of ADS-B signals from airborne transponders.

Chapter 6 describes the system created for ADS-B messages receiving and original software for modeling of real-time TCAS operation.

Andrii Mikhailovich Grekhov
National Aviation University, Ukraine

Acknowledgment

I would like to thank Professor V.P. Kharchenko, Assistant Professor Y.M. Barabanov and each one of the coauthors for their contributions. I want to acknowledge my students who helped me in the calculations and plotting.

Andrii Mikhailovich Grekhov
National Aviation University, Ukraine

Introduction

The proposed book should be read by anyone who is interested in digital satellite communications of aircraft and remotely piloted air systems with ground services. The solution of many functional tasks increasing the efficiency of the aviation transport system as a whole is carried out with the help of civil aviation communication systems. These systems are an integral part of automated air traffic control systems, used in the production, technological and commercial activities of airlines and their services. Aviation aeronautical (mobile) radio communication is the only means of communication between air traffic control center with aircraft crews and between aircraft crews in flight. Technical means of communication are designed to transmit and receive messages and data through the channels of aeronautical mobile and fixed communication services. The achievement of high safety, regularity and economy of flights is largely ensured by the continuous and reliable radio communication of aircraft crews with ground centers during all phases of flights.

Modern systems and means of aeronautical radio communication are continuously being improved. For their correct use the engineering and technical staff should know the principles of the organization of aeronautical radiocommunication, the specifics of the operation of digital data transmission systems, and the specifics of aerospace radio communication systems. Increasing requirements to reliability, throughput and range of aeronautical satellite transmission systems require knowledge of the fundamentals of the theory and technology of radio communication systems from the specialists, without which the operation of existing and development of new technical means by the personnel of aviation enterprises is hampered.

This book is a window into the world of a variety of models of satellite communication channels that help to understand the basic dependencies, the behavior of such systems and make it possible to predict their behavior.

Andrii Mikhailovich Grekhov
National Aviation University, Ukraine

Chapter 1
Modeling of Aircraft and RPAS Data Transmission via Satellites

ABSTRACT

This chapter is devoted to the modeling of aircraft data transmission via low-orbit satellites. Satellite communication channel models were designed, which allow to investigate BER dependencies on the type of signal modulation, information transfer rate, signal power, antenna diameters, and nonlinearity of a high power amplifier. Impact of a modulation type (BPSK, QPSK, 8PSK, 16QAM), E_b/N_0, satellite transponder amplifier gain without and with coding on a BER was investigated. Effectiveness of error detection and correction was analyzed using classic linear block and convolutional codes. Free space path loss, AWGN, and radio frequency satellite channels were considered. MIMO 2×1 and 3×2 fading uplink/downlink channels with antenna diversity were analyzed. Results were compared with AWGN uplink/downlink channels. On the base of these models, channels integrity was investigated.

INTRODUCTION

Aeronautical Telecommunication Network

The Aeronautical Telecommunication Network (ATN) has been designed to provide data communications services to Air Traffic Service (ATS) provider organizations for ATS communication, aeronautical operational control, administrative communication, and aeronautical passenger communication

DOI: 10.4018/978-1-5225-8214-4.ch001

(Doc. 9880-AN/466, 2005; Doc. 9896, 2011). The ATN comprises application entities and communication services, which allow ground-ground, air-to-ground and air-to-air data subnetworks to interoperate using satellite subnetwork (Roddy, 2006).

Satellite communication systems are used for sending data from aircraft to a ground stations accessible to Aeronautical Operation Control (AOC), Air Traffic Management (ATM) and Air Traffic Control (ATC). Wide application of satellites in aviation is connected with the possibility of communication with a considerable amount of aircraft irrespective of a distance, with independence of expenses on a distance between planes, with insignificant influence of atmosphere and sites of land stations on reliability of communication.

A satellite tracking of aircraft is a technology available to aircraft operators that has huge benefits and relatively low costs. An aircraft can report its position via an Aircraft-Satellite-Ground Station data link. Automatic Dependent Surveillance – Broadcasting (ADS-B) – is a technology, which allows pilots and air traffic controllers to track aircrafts with high accuracy (EUROCONTROL, 2013; ADS-B, 2014). ADS-B can make flight safe and allows using of air space more effective. An aircraft equipped with ADS-B avionics transmit their exact position in space by means of digital communication channels. The digital code which contains this information is updated several times per second and transmitted by aircraft on discrete frequencies. ADS-B systems based on Low Earth Orbit (LEO) satellites are of special interest (Osborne, & Xie, 2009).

It was found (Aircraft-to-Satellite Links, 2003) that the most critical factors for all systems using aircraft are the transmitting antenna and the data rate that the communications system is required to carry. The maximum rate for any system in use today is set by the transmit power available from the ground (or aircraft) transmitter and the sensitivity of the satellite receiver. Therefore, it is necessary to determine the amount of data that must be transmitted and the time period in which the transmission must occur. Problems connected with the performance of satellite aeronautical communication channel are very important. Even small degradation of communication channel parameters influences a rate of data transmission, a time delay and a coverage. These factors at once impact on safety of flights and operational expenses. It is necessary to have the system parameters optimized before implementation (Elbert, & Elanix, 2009) and when things go wrong, a simulation model can be used to track down the offending element. The imitation modeling also may be useful for pre-testing any corrective action before attempting it either in space or on the ground. At the same time the role of a satellite transponder as a part of the microwave repeater and antenna system is significant. A transponder is used to amplify carriers on the downlink side. Each transponder is amplified by either a traveling wave tube amplifier or a solid state power amplifier.

Iridium System

Previous Iridium system included 66 LEO satellites at an altitude of 785 km and equally divided into 6 orbital planes. The orbits are circular with an inclination angle of 86.4° degrees (Irid743ium, 2007). Each satellite communicates with the Airborne Earth Station (AES). Each satellite uses three phased-array antennas for the user links, each of which contains an array of transmit/receive modules. These arrays are designed to provide user-link service by communicating within the 1616-1626.5 MHz band. The gateway connects the Iridium satellite network to ground communication networks, such as the terrestrial Public Switched Telephone Networks (PSTNs) and Public Switched Data Networks (PSDNs), and communicates via ground-based antennas with the gateway feeder link antennas on the satellite. The gateway can also serve as a gateway to the ATN for forwarding ATN messages from the aircraft to the required ATC or AOC unit or vice versa.

Channels are implemented in the Iridium Satellite Network using a hybrid Time Division Multiple Access/Frequency Division Multiple Access (TDMA/FDMA) architecture based on Time Division Duplex (TDD) using a 90-millisecond frame.

All L-Band uplink and downlink transmissions used in the Iridium Satellite Network employ variations of 25 Kilosymbols-per-second (ksps) Quadrature Phase Shift Keying (QPSK) modulation and are implemented with 40% square root raised cosine pulse shaping. The variations of QPSK used include differential encoding (DE-QPSK) and Binary Phase Shift Keying (BPSK), which is treated as a special case of QPSK.

Iridium system employs Forward Error Correction Coding (FECC) in the form of convolutional encoding with Viterbi decoding (Viterbi, 1971). Iridium uses a rate 3/4, constraint length 7, (r=3/4; K=7) convolutional code on both transmission and reception. The supportable transmission rates for voice (data) are 4.8 kbps (2.4 kbps).

Iridium NEXT is the second generation of communication satellites of the US company Iridium Communications Inc. (Iridium NEXT, 2018), which is intended to replace the current satellite constellation of Iridium spacecraft. The total cost of the project (development, manufacture, launch, insurance, modernization of ground infrastructure) is $ 2.9 billion.

Customer: United States of America Iridium Communications Inc. *Manufacturer*: France Thales Alenia Space, United States Orbital ATK. *Operator*: United States of America Iridium Communications Inc. *Tasks*: voice and data communication. *Target Equipment*: ADS-B monitoring of air flights, AIS monitoring of maritime transport, transponders of L-band, Ka-band. *Launch vehicle*: Falcon 9.

Specifications: ELiTeBus Platform, weight 860 kg (at start-up), dimensions 3,1 × 2,4 × 1,5 m (at start-up), power 2200 W, power supplies solar batteries, accumulators, active life up to 15 years.

Orbital Elements: the working orbits of the Iridium NEXT satellites are at an altitude of 780 km with an inclination of 86.4 ° and a period of revolution of about 100 minutes, in six orbital planes of 11 apparatus each. As they launch, they will gradually replace the first-generation satellites. The old satellites will be taken from orbit and then burned in the atmosphere. Spare satellites will be located in orbit 667 km high. A total of 81 devices will be manufactured, 75 of them will be launched into space (66 active and 9 orbital reserve satellites), and the remaining 6 will remain on the ground as spare. For maneuvering and holding the orbit 8 hydrazine engines with a thrust of 1 N each are used, the fuel tank holds 141 kg of hydrazine. The expected lifetime of satellites is 15 years. The new generation of satellites is fully compatible with the first-generation satellites. For communication between users of the system, the main satellite antenna generates 48 L-band radios covering a territory of 4700 km in diameter on the Earth's surface. Like the first-generation satellites, the Iridium NEXT will support cross-connection between neighboring satellites using 4 Ka-band transceivers. Another 2 transceivers will communicate with ground stations.

Additional Payload: each Iridium NEXT satellite can carry third-party additional equipment weighing up to 54 kg and an average power consumption of up to 90 watts. The equipment is placed in a special container AppStar produced by Harris Corporations, dimensions 30 × 40 × 70 cm, while the company provides transmission of received data via the Iridium network.

History: on June 2, 2010, Thales Alenia Space was selected as the main developer and manufacturer for the new generation of Iridium satellites. The contract amounted to $ 2.1 billion for the creation of 81 vehicles. Since some of the satellite technologies can not be exported, the final assembly and testing was planned in the USA (previously negotiations were held with Ball Aerospace & Technologies Corp.).

June 16, 2010 was announced a contract for $ 492 million with the company SpaceX. In addition to the launch of satellites, the contract provides for the creation of a special dispenser for the simultaneous launch of several devices. Tentatively, the first launch was scheduled for 2015, the completion of the satellite grouping was to be completed by the end of 2017. Initially, there were 8 launches of the Falcon 9 booster rocket for 9 satellites each.

On January 27, 2011, a contract was signed with Orbital Sciences Corporation (now Orbital ATK). The company will integrate and test all Iridium NEXT satellites at its Gilbert, Arizona plant.

On June 20, 2011, an agreement was concluded with Kosmotras, which launched the launch of the Dnepr carrier rocket.

On April 24, 2013, Harris Corporations was selected responsible for providing the additional payload function for the Iridium NEXT satellites. It was also announced that Aireon, a joint venture between Iridium and the Canadian agency Nav Canada, will deploy ADS-B equipment (Automatic Dependent Surveillance-Broadcast) on each device to provide global monitoring of air flights. The company plans to sell this information to commercial airlines, as well as to government agencies, including the Federal Aviation Administration of the United States of America.

In March 2015, Iridium reported that due to problems with the software, the launch of the first satellites has been postponed for several months. It was expected that the first two vehicles would be launched by the Dnepr launch vehicle in October 2015, and then 7 launches of the Falcon 9 launch vehicle from the SpaceX launch complex at the Vandenberg airbase, 10 satellites each. Under the contract with insurance companies, after the first launch, 2 satellites must undergo a 4-month testing in orbit before the start of the main grouping.

In June 2015, Harris Corp. reported the launch of a partnership with the Canadian company exactEarth [en], which involves the deployment of terminals of the AIS system (English Automatic Identification System) on 58 Iridium NEXT satellites intended for monitoring the movement of maritime transport.

On October 29, 2015, it was announced that when testing the first two satellites, problems were identified with the Ka-band signal reception/ transmission module, which could reduce signal performance when sent to ground stations. Thales Alenia Space reported that the problem will be eliminated, but the launch of the first satellites was postponed.

Remotely Piloted Air System (RPAS)

RPASs are a new way of using flying machines. RPASs are distinguished from manned aircraft by the data link (UAV, 2012) connecting the remote pilot station with the remotely piloted aircraft and used for Command and Control (C2) of RPAS and as a relay for communications between the ATC operator and the remote pilot. The combination of these two functions is termed C3 – Command, Control and ATC Communications. The concept of RPAS Required Communication Performance Methodology for the Command, Control and Communication Link is given in (UAS, 2012). Basic requirements for RPAS data rate are stated in the NATO standards (STANAG 4607/AEDP-7, 2010; 4609/AEDP-8, 2009; 7023/AEDP-9, 2009). RPAS technologies are constantly improved and expand the amount of memory, onboard data processing capabilities, information storage and communication. For widespread use in commercial, military, civil, agricultural and environmental purposes, RPASs should be able effectively communicate with each other and with existing network infrastructures (Gupta, Jain, & Vaszkun, 2015; US Department, 2013; Jawhara, Mohamed, Al-Jaroodi, Agrawald, & Zhange, 2017; Hanscom,& Bedford, 2013; Motlagh, Taleb, & Arouk, 2016; Vikranth, 2017; Szabolcsi, 2015; Tanil, Warty, & Obiedat, 2013). In the review (Motlagh, Taleb, & Arouk, 2016) RPAS's classification, possible cluster architectures, RPAS-based services, obstacle detection techniques, RPAS's networks, RPAS equipment, data collection methods, communication technologies, processing of collected data, use of clouds for computational loading of a RPAS resource are considered. The cluster (or swarm) has several RPASs working synchronously to solve a single task (Vikranth, 2017). Collaborative mission planning for RPAS cluster to optimize relay distance is considered in paper (Tanil, Warty, & Obiedat, 2013).

A promising area in the use of RPASs is their grouping. The principle of the organization of the swarm is taken from the world of insects. In the case of drones, after merging them into a swarm, each drone is controlled by its own automation, and the behavior of the swarm can be controlled by a program with Artificial Intelligence elements or one (several) operators.

Advantages of using RPAS group consist in reduction of the total cost of drones, distribution of the payload across several sides (the opportunity to save on the total cost of the payload), reduction of losses from accidents, improved positioning accuracy of each RPAS due to mutual positioning, improved results due to different angles of view of different UAVs.

The shortcomings and limitations of the use of RPAS groups are high computing capabilities required on board the drone for interaction of drones in flight as part of a group and preliminary processing of collected information in real time, required new types of management software.

On RPAS/UAV board, as a rule, more than two communication complexes are installed: half duplex/ duplex, equipment that transmits control commands/ telemetry and a simplex system that transmits payload information. The complex of control command transmission is intended for low-speed command transmission from the ground station to RPAS/UAV and low-speed telemetry transmission from RPAS to the ground station. The equipment transmitting payload information is intended for one-way high-speed transmission from RPAS to the ground station. The direct communication between RPAS and the ground station in microwave ranges is assumed only within the line of sight. To increase the reliability of the RPAS complex, several transmitters/ receivers of different range and wavelengths are installed on the board. For long-haul flights, telemetry data transmission can be carried out using a satellite communication system (Iridium, Globalstar, etc.).

Channel Coding

In aeronautical communications systems there is a restriction on the transmitted power. For example, in data transmission systems via satellites, the increase in power is very expensive. Error control codes are an excellent means of reducing the required power, because with their help it is possible to correctly recover the attenuated messages. Data transmission in airborne systems is sensitive even to a very small fraction of errors, since a single error can disrupt the calculation program. Coding and controlling errors becomes very important in these applications. For some storage media, the use of error control codes allows for more dense bit packing.

Another type of communication system is a multi-user system and a time-sharing system in which each user is pre-assigned certain time windows (intervals) in which it is allowed to transmit. Long binary messages are divided into packets, and one packet is transmitted to the allocated time window. Due to a violation of synchronization or service discipline, some packages may be lost. Appropriate error control codes protect against such losses.

There are large data flows between subsystems in aviation digital systems. Digital autopilots, digital flight control systems and digital radar signal processing systems are all systems that contain large arrays of digital data.

These data must be transmitted through complex systems with data transmission and time-sharing buses. In any case, an important role is played by error control coding methods, since they allow to guarantee the corresponding characteristics of the communication channel. The history of coding, which controls errors, began in 1948 with the publication of the famous article by Claude Shannon. From the Shannon's theory of information follows the important conclusion that building too good channels is wasteful. It is more economical to use coding. In fact, Shannon's paper states that signal power, channel noise and bandwidth limit only the transmission speed, and not its accuracy.

The first block codes were introduced in 1950, when Hamming described a class of block codes correcting single errors. Almost all encoding schemes used in practice are based on linear codes. Linear codes differ from the non-linear closure of the code set with respect to some linear operator. Linear tree codes are usually called convolutional codes, since the encoding operation can be regarded as a discrete convolution of the input sequence with the impulse response of the encoder.

Multiple-Input Multiple-Output (MIMO)

MIMO was initially represented only as a diversity technology (one transmit and N receive antennas). The processing principle was simple: in two receiving branches, the signal-to-noise ratio was compared, and in accordance with the estimation of this value, weighting factors were assigned to each branch of processing, which played a role in the decision that was transferred: 0 or 1.

In 1997, Alamouti offers Space-Time Block Coding (STBC). After this MIMO becomes very relevant because to increase the frequency-energy efficiency of communication systems began all the more difficult and difficult. And then space-time trellis coding, spatial multiplexing, and a large number of decoding algorithms were proposed.

The radio wave fades passing the path from the transmitter to the receiver. At the same time, the extent to which it will lose energy depends on whether there is a direct visibility between a transmitter and a receiver. If it is, the main cause of the loss is the path loss environment. If there is no line of sight, then, when faced with various obstacles, the wave goes to the destination in several ways (multipath) and accordingly each beam passes a different distance. At reception, all these rays can fold in antiphase, which further reduces the intensity of the signal and causes the signal level to constantly "float". This was called fading, which can be described by different laws. In the presence of a constant component (the presence of line of sight), the distribution of Rice is used, and in its absence, the Rayleigh distribution.

MIMO is distribution of several data flows over a single channel, followed by passing them through a pair or more antennas before reaching the receiving device. This allows significantly improve the signal throughput without the band extension.

The digital stream in the radio channel selectively fades when broadcasting radio waves. To solve this problem, MIMO antenna was created, capable of broadcasting information over several channels with a slight delay. Information is pre-coded, and then restored on the receiving side. As a result, not only the speed of data distribution increases, but the quality of the signal also significantly improves.

One approach of increasing the data transfer rate for WiFi standard 802.11 and for WiMAX standard 802.16 is the use of wireless systems with multiple antennas. MIMO technology plays an important role in the implementation of WiFi standard 802.11n.

An important tool for increasing the physical data rate in wireless networks is the spreading of the bandwidth of the spectral channels. Due to the use of a wider bandwidth of the channel with Orthogonal Frequency Division Multiplexing (OFDM), data transmission is performed with maximum performance. OFDM is a digital modulation that has proven itself as a tool for implementing bi-directional high-speed wireless data transmission in WiMAX/WiFi networks. By combining MIMO architecture with a wider channel bandwidth, a very powerful and cost-effective approach is obtained to increase the physical transmission rate.

Radio Frequency Satellite Channel

Communication systems can be classified into two groups depending on the range of frequencies they use to transmit information. These communication systems are classified into BASEBAND or PASSBAND system. Baseband transmission sends the information signal as it is without modulation (without frequency shifting) while passband transmission shifts the signal to be transmitted to a higher frequency and then transmits it. At the receiver the signal is shifted back to its original frequency.

Almost all sources of information generate baseband signals. Baseband signals are those that have frequencies relatively close to zero such as the human voice (20 Hz – 5 kHz) and the video signal from a TV camera (0 Hz – 5.5 MHz).

In digital passband transmission via radio frequency satellite link, the incoming data stream is modulated onto a carrier with fixed frequency and then transmitted over a band-pass channel.

Radio-frequency satellite communication channels transmit telemetry data or payload data from the satellite to the ground and commands from the ground to the satellite. On both sides of these lines there are modems (modulator/demodulator), one on the satellite and one on the ground station.

Satellite radio communication channels have the following features:

- Large distances between the on-board transmitter, satellite and ground station. The power radiated by the antenna decreases inversely with the square of the distance. The signals coming from the satellite at great distances are very weak.
- The Doppler's effect can take place in the communication channel. Satellites in low Earth orbit have high speeds when they fly over a terrestrial antenna. This leads to a Doppler shift, distorting the frequency and phase of the signal.
- Satellite channel is asymmetric. The antenna of the terrestrial side is capable of generating high transmit power, while the satellite side has a relatively small antenna aperture for receiving this power. Conversely, the satellite side has limited transmit power, while the terrestrial antenna can be quite large.
- Data is continuously transmitted over the satellite communication channel. Unlike network messages, when packets are sent only when there is data to send, each satellite link always contains some data to maintain synchronization.
- Satellite signals are designed for the absebce of transmission errors, since they do not support the retransmission of lost or corrupted data due to a long delay.

SIMULATION OF IRIDIUM SATELLITE LINK OPERATION

In 2010, when our studies began, there were no published data on the calculation of satellite channels parameters and the modeling of their operation in critical regimes. It was only possible to find the demonstration model *RF Satellite Link* in MATLAB Simulink.

Since our goal was to develop models based on the standards of mobile communication IEEE 802.11 and 802.16, taking into account adaptive modulation, OFDM and MIMO, the first task was the development of simple models, which we consider in this chapter (Kharchenko, Barabanov, & Grekhov, 2012a; 2012b; Kharchenko, Barabanov, Grekhov, Bogunenko, & Rudnyk, 2013a; 2013b; Kharchenko, Wang Bo, Grekhov, & Kostynska, 2014; Kharchenko, Wang Bo, Grekhov, & Stahovskyi, 2014).

Aviation telecommunication is a dynamically developing industry, taking into account the evolution of the characteristics and principles of telecommunications between satellite systems and aircraft. Application of satellite communication in aviation is connected with the possibility of servicing a large number of aircraft regardless of the distance, independence of communication costs from the distance between planes, the insignificant influence of the atmosphere and the location of ground stations on the reliability of communication. The principle of aviation satellite systems operation is based on the use of satellite retransmitters, through which the connection between the aircraft and ground stations is carried out (Elbert & Elanix, 2009).

Questions related to the satellite aviation channel and its operation are very important. A small link parameters degradation affects the data rate and coverage area. It is important to know how you can continuously maintain the parameters of the communication channel optimal. That is why it is actual the development of realistic models for satellite aviation links and studying with their help ways of critical situations correction.

Mobile communication systems have small antennas, which makes it difficult to receive a signal. In order to receive signals with sufficient power, the satellites are located in the geostationary orbit (the Inmarsat system - https://www.inmarsat.com/), or many satellites are located in inclined and polar orbits (Iridium - https://www.iridiumnext.com and Globalstar - https://www.n2yo.com systems). In the latter case, the required power of the transmitter is not so high, and the cost of bringing the satellite to an orbit is relatively low. However, this approach requires not only a large number of satellites, but also an extensive network of ground stations.

Satellite communications with moving objects (Roddy, 2006), interaction of satellite communication systems with computer networks and radio systems (Aircraft-to-Satellite Links, 2003), low-orbital satellite systems (Osborne & Xie, 2009), simulation of satellite transponders (Elbert & Elanix, 2009) are important for the development of this industry.

Another feature of satellite communication systems is the need to operate in a relatively low signal/noise ratio caused by several factors: a significant distance from the receiver to the transmitter and the limited power of the satellite transmitter. This causes errors and requires a detailed theoretical study in order to obtain the dependencies of the channel parameters on the characteristics of the traffic.

The Iridium communication system (Manual, 2007) includes subscriber equipment, satellite and terrestrial networks and allows the transmission of data and voice messages. Information is retransmitted from one satellite to another until it reaches the satellite above aircraft and then transmitted back to the ground.

Operating frequency range is 1616-1626.5 MHz. The transmission of information in the L-band is carried out using QPSK and implemented through 40% quadratization in the formation of cosine-wave pulses. Different types of QPSK modulation include DE-QPSK and two-positional phase manipulation BPSK. The data transfer rate is 50 kb/s.

The aim of this study is to create a model of the "Aircraft-Satellite-Ground Station" communication channel using the MATLAB Simulink software, which will allow to investigate the dependence of the Binary Error Rate (BER) on the type of signal modulation, the transmission rate, signal power, antennas diameters, and high power amplifier nonlinearities during ADS-B messages transmission (Kharchenko, Barabanov, & Grekhov, 2012a; 2012b).

Satellite Communication Channel

The original model *FSPL_Sat_FSPL_16QAM* of aircraft satellite link was designed using MATLAB Simulink demonstration example *RF Satellite Link*. This model contains blocks simulating the work of the channel (Figure 1): Aircraft Transmitter, Uplink Path, Satellite Transponder, Downlink Path, and Ground Station Receiver.

Aircraft Transmitter contains Random Integer Generator, Baseband Modulator (BPSK, QPSK, 8PSK, 16QAM), Raised Cosine Transmit Filter, High Power Amplifier (HPA) with Memoryless Nonlinearity, and Transmitter Dash Antenna Gain.

Uplink Path and Downlink Path channels include Free Space Path Loss (FSPL), Phase/Frequency Offset with Doppler and Phase Error.

Satellite Transponder enhances and mirror the signal from aircraft to the ground station.

Figure 1. Model FSPL_Sat_FSPL_16QAM of aircraft satellite link

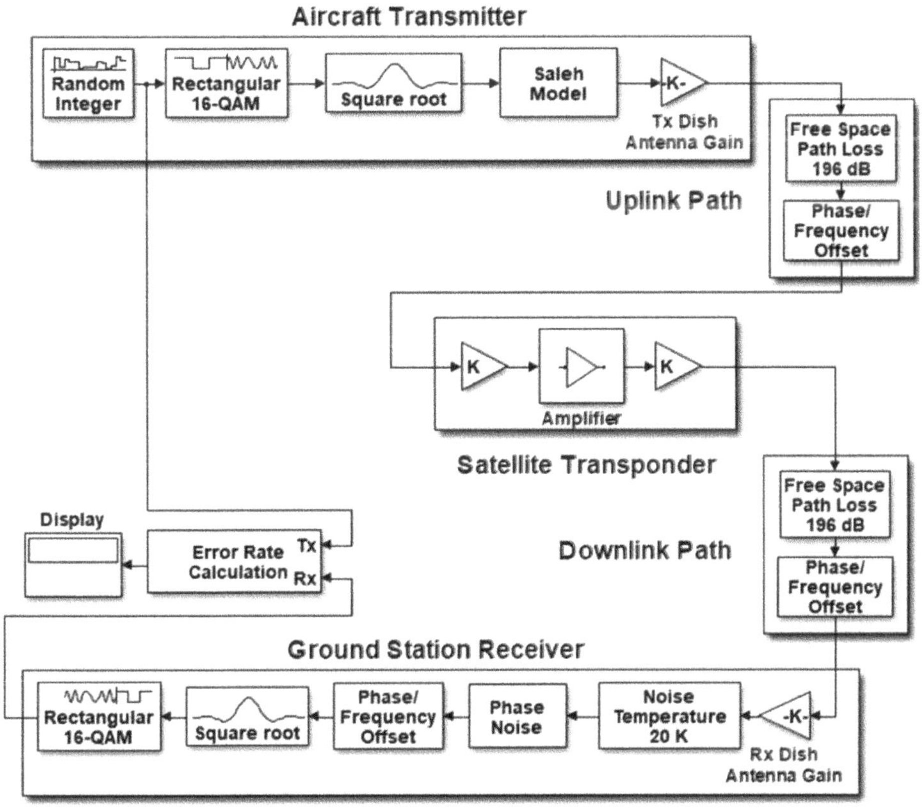

Ground Station Receiver consists of Receiver Dish Antenna Gain, Rcceiver Thermal Noise block, Phase Noise block, Raised Cosine Receiver Filter, and Baseband Demodulator.

During modeling the channel operation, such tools for analyzing simulation results were used as window-indicators, which can be made active for signal changes analysis. BER display shows the number of characters transmitted, the number of errors and the frequency of error bits. The BER calculation is updated every 5000 characters and allows to analyze the effect of changes in a model without restarting the model.

High Power Amplifier block applies memoryless nonlinearity to complex baseband signal and provides five different methods for modeling the nonlinearity. In this paragraph results for Saleh model with standard AM/AM and AM/PM parameters are given. A HPA backoff level is used to determine how close the satellite high power amplifier is driven to saturation.

Simulation of Communication Channel Operation

In simulating the operation of “Aircraft-Satellite-Ground Station” communication channel, the parameters of the low-orbital satellite constellation Iridium were chosen: the satellite’s height - 780 km, the operating frequency - 1616 MHz, the signal modulation types - BPSK, QPSK. For comparison, different types of modulations DQPSK, 8PSK, and 16QAM were also investigated.

During the simulation, the diameters of the transmitting and receiving antennas changed. The first element in the vector $[d_1, d_2]$ is the diameter of the transmitting antenna and is used to calculate a gain of the transmitter antenna amplifier; the second element is the diameter of the receiving antenna and is used to calculate a gain of the receiver antenna amplifier.

The research was carried out for three values of the effective noise temperatures of the receiver: 0 K (no noise), 20 K (very low noise level) and 290 K (typical noise level).

The following selected backoff is used to set the input and output gain of the Memoryless Nonlinearity block: 30 dB - the average input power is 30 dB below the input power that causes amplifier saturation (in this case, AM/AM and AM/PM conversion is negligible); 7 dB - moderate nonlinearity; and 1 dB - severe nonlinearity.

The following experiments were carried out using the model *FSPL_Sat_FSPL_16QAM* for studying the signal passing through the communication channel:

- The losses were symmetrically altered in the channels “up” and “down”;
- The noise temperature of the ground receiver (0 K, 20 K, 290 K) changed;
- Both symmetrically and asymmetrically diameters of the transmitting and receiving antennas changed, which increased or reduced the power of the received signal;
- The gain of the satellite transponder was changed.

1. Dependencies of a BER on free space path loss. Free space path loss is the loss in signal strength that results from a line-of-sight path through free space, does not include the gain of the antennas used at the transmitter

and receiver, is proportional to the square of the distance between the transmitter and receiver, and also proportional to the square of Iridium operational frequency. A dependence of a BER on free space path loss for noise temperature 290 K is shown in Figure 2. Loss values were changed simultaneously in uplink and downlink channels for equal antennas diameters: $d_1 = d_2 = 0.4$ m (satellite transponder antennas diameters – 1.4 m), negligible HPA's nonlinearity (backoff level 30 dB), and satellite transponder gain equals to 0 dB.

2. *Dependencies of a BER on the signal power* was simulated by a change in the average power of the symbols in a signal constellation, referenced to 1 ohm (Watts). Signal modulations 16QAM and 32QAM were considered (Figure 3). The simulation results showed a significant decrease in a BER coefficient with an increase in signal power.

Figure 2. Dependencies of a BER for different modulation schemes on free space path loss in the uplink and the downlink: satellite and ground receivers noise temperatures are 290 K, HPA's backoff level is 30 dB, phase and frequency offsets are equal to zero, $d_1 = d_2 = 0.4$ m, satellite transponder antennas diameters 1.4 m

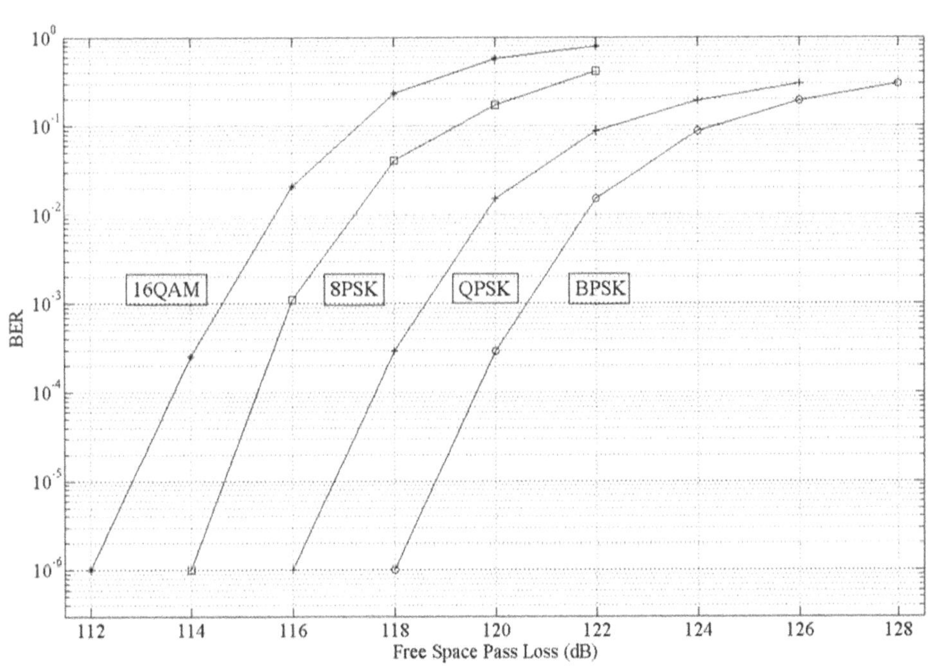

Figure 3. Dependencies of a BER for different modulation schemes on average signal power: free space path loss in the uplink and the downlink 120 dB, satellite and ground receivers noise temperatures are 290 K, HPA's backoff level is 30 dB, phase and frequency offsets are equal to zero, $d_1 = d_2 = 0.4$ m, satellite transponder antennas diameters 1.4 m

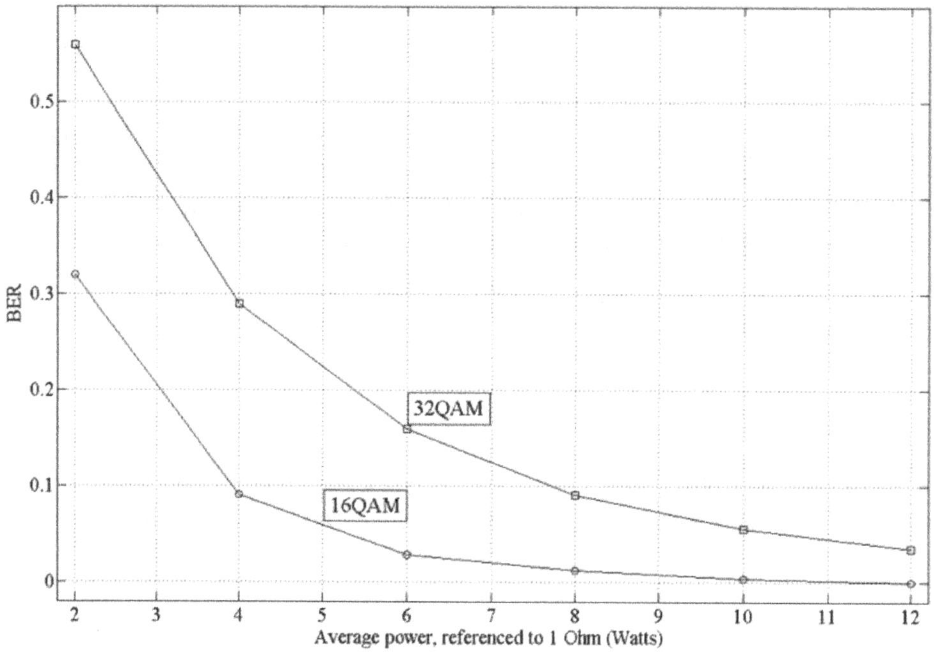

3. Dependencies of a BER on aircraft and ground station antennas diameters. Diameters of the transmitting and receiving antennas were simultaneously changed during the simulation (Figure 4). This caused an increase or decrease in the power of the received signal. Different types of modulation BPSK, 8PSK, and 16QAM were considered. The simulation results indicate that there is a dependence of a BER on the modulation type and antennas diameters. The more antennas diameter, the less is an error probability.
4. Dependencies of a BER on generator sample rate. The sample time is a parameter that indicates when, during simulation, the block produces outputs and updates its internal state. Random Integer Generator in Aircraft Transmitter creates random uniformly distributed integers in the range [0, M-1], where M is the M-ary number. The sample rate is the

Figure 4. Dependencies of a BER for different modulation schemes on aircraft and ground station antennas diameters: satellite and ground receiver noise temperatures are 290 K, phase and frequency offsets are equal to zero, HPA's backoff level is 30 dB, and free space path loss is 120 dB, satellite transponder antennas diameters 1.4 m

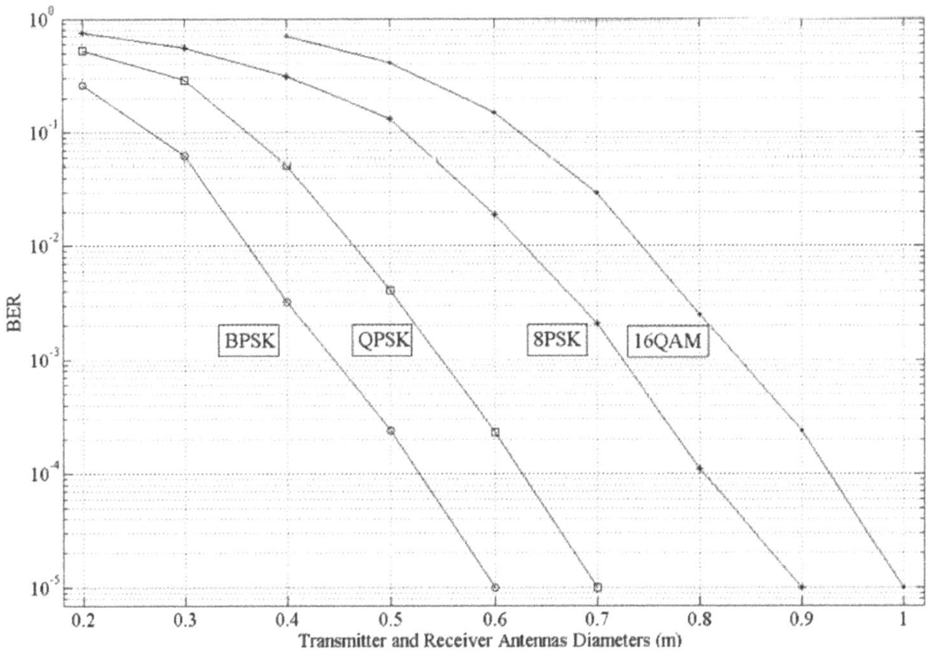

reciprocal of the sample time, and is measured in Hertz. Conditionally, it can be assumed that the sample rate simulates the data transfer rate, measured in bits per second. Dependencies of a BER on generator sample rate for different nonlinearity levels (Figure 5) allow to compare qualitatively the change in a BER as the data transfer rate and the degree of HPA nonlinearity are increased.

5. Spectra of transmitted and received signals. The spectrum of the modulated/filtered signal (blue) and the received signal before demodulation (red) are shown in Figure 6. Comparing these spectra allows to view the effect of the spectral regrowth due to HPA nonlinearities caused by the Memoryless Nonlinearity block.
6. Signal constellations of transmitted and received signals allow studying the influence of different factors on the received signal (Figure 7).

Figure 5. Dependencies of a BER on generator sample rate for different nonlinearity levels: satellite and ground receiver noise temperatures are 290 K, phase and frequency offsets are equal to zero, free space path loss is 120 dB, $d_1 = d_2 = 0.4$ m, satellite transponder antennas diameters 1.4 m

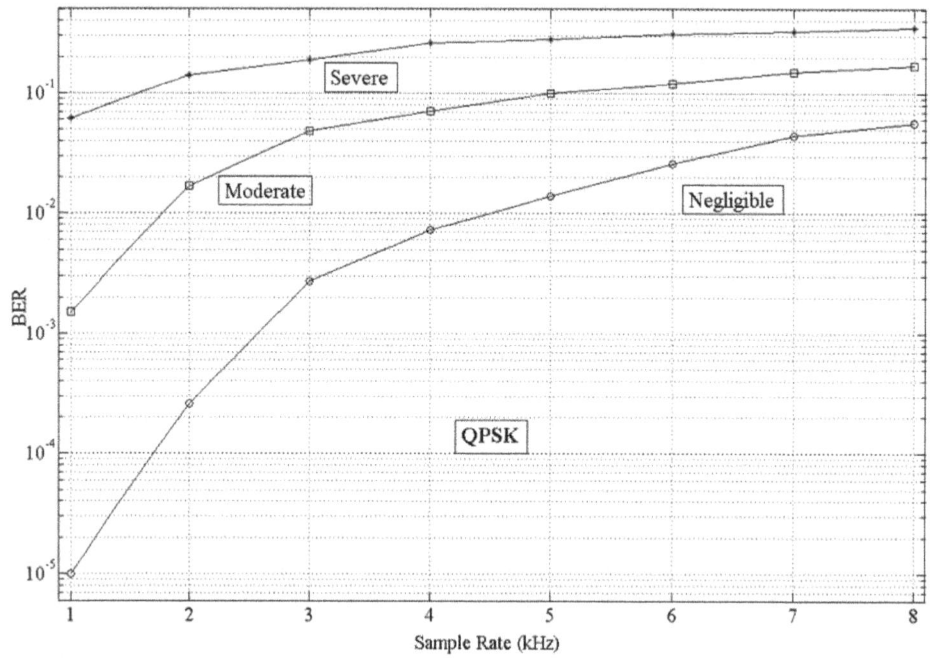

7. The impact of ground receiver noise temperature. Results for a BER dependence on the value of ground receiver noise temperature at different diameters of aircraft and ground station antennas are given in Table 1.
8. The impact of the ground antenna diameter.

Table 2 shows data for aircraft antennas with a diameter from 0.1 m to 0.4 m and corresponding ground station antenna diameter for which a BER becomes zero. For example, when using the BPSK modulation and choosing aircraft antenna diameter to be $d_1 = 0.1$ m, the antenna diameter of the ground station should be not less than $d_2 = 1.8$ m in order to avoid binary errors in the transmitted data. In this case, the condition BER = 0 will be fulfilled for the values of the channel parameters specified in the note to the table. An increase in the diameter of the onboard antenna twice ($d_1 = 0.2$ m), leads to a reduction in the required diameter of the ground antenna by a factor of 2

Figure 6a. Spectra of transmitted (blue) and received (red) signals for different HPA's backoff levels: negligible nonlinearity

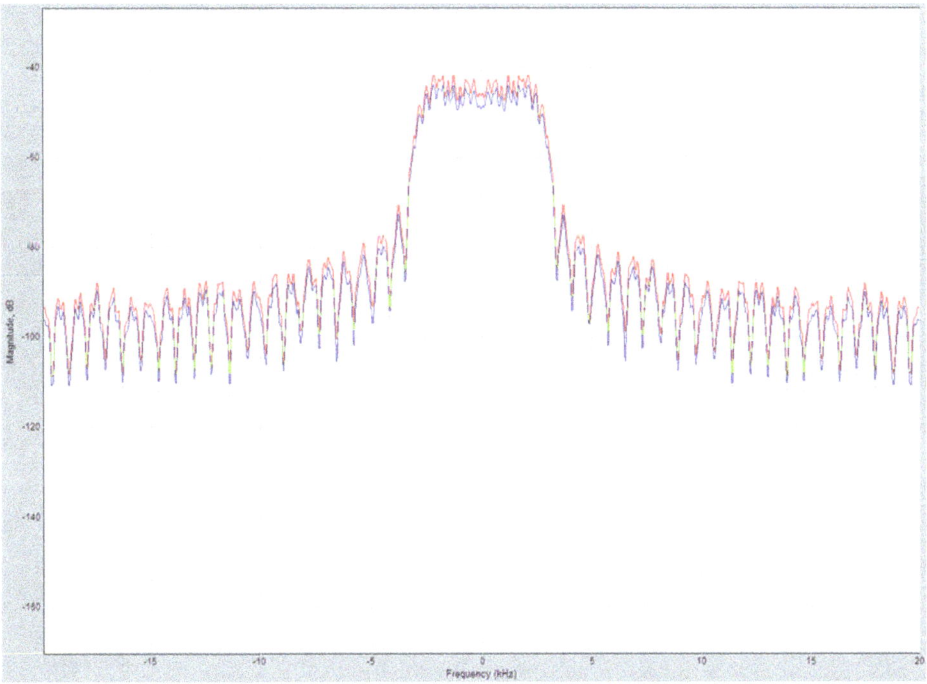

(d_2 = 0.9 m); an increase of three times (d_1 = 0.3 m), leads to a three-fold decrease in the required diameter of the terrestrial antenna (d_2 = 0.6 m); an increase of four times (d_1 = 0.4 m), leads to a decrease in the diameter of the terrestrial antenna by a factor of four (d_2 = 0.4 m). This regularity is traced for all types of modulation considered in Table 2.

9. The impact of a HPA's nonlinearity on a BER. Dependencies of a BER on a HPA's nonlinearity was investigated for different modulations using the aircraft antenna with a diameter of 0.3 m. From Table 2, the values of the terrestrial antenna diameters for which BER = 0 were taken. From Table 3 it follows that with the negligible HPA's nonlinearity a BER is zero for all types of modulation. But even the moderate nonlinearity makes the operation of the channel impossible due to large values of a BER, which increases in the series of modulations BPSK, DQPSK, 8PSK, 16QAM.

Figure 6b. Severe nonlinearity; satellite and ground receiver noise temperatures are 290 K, phase and frequency offsets are equal to zero, free space path loss is 120 dB, $d_1 = d_2 = 0.4$ m, satellite transponder antennas diameters 1.4 m, QPSK modulation

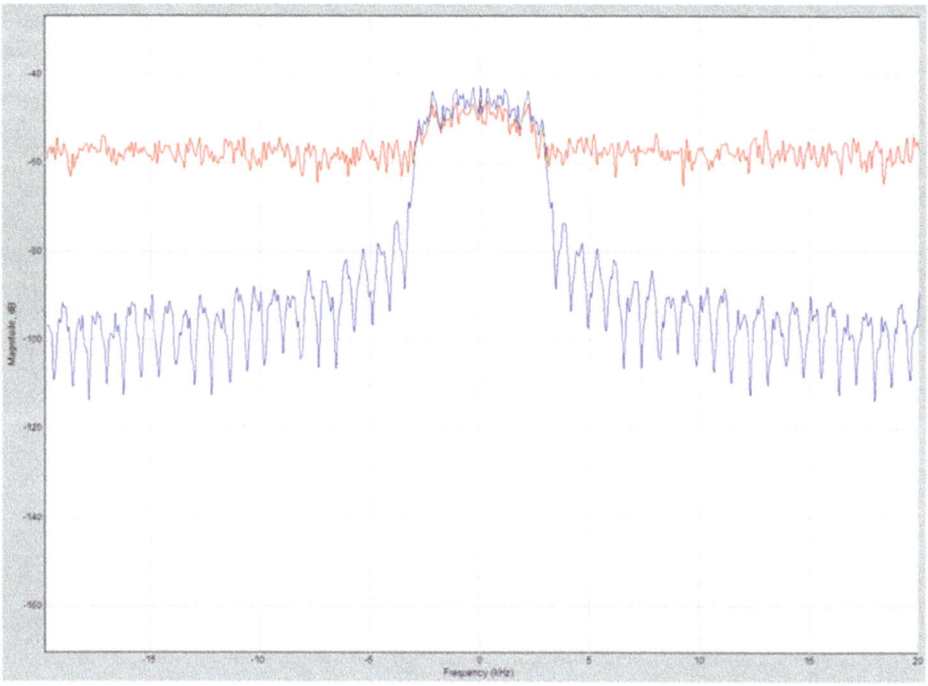

Figure 7. Signal constellations of transmitted (blue) and received (red) signals: satellite and ground receiver noise temperatures are 290 K, HPA's backoff level is 30 dB, free space path loss is 120 dB, $d_1 = d_2 = 0.4$ m, satellite transponder antennas diameters 1.4 m, 16QAM modulation

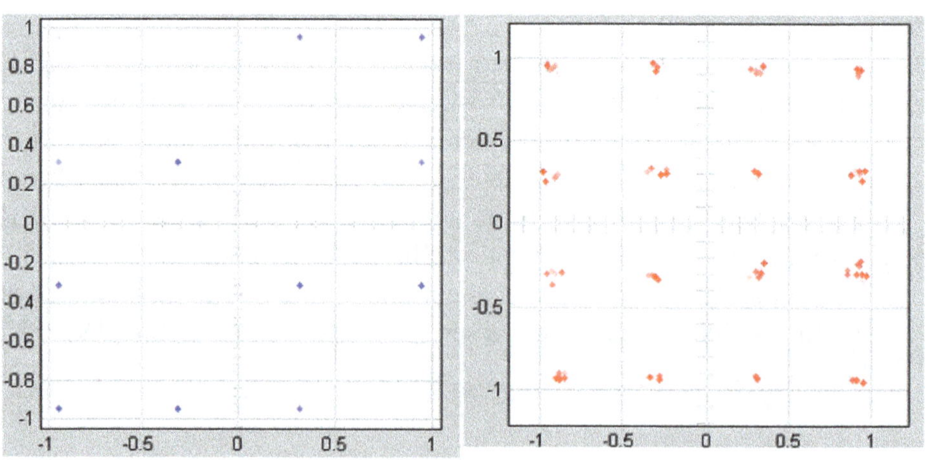

Table 1. Dependencies of a BER on the modulation and ground receiver noise temperature

Modulation	BPSK		QPSK		DQPSK		8PSK		16QAM	
Noise temperature	20 K	290 K	20 K	290 K	20 K	290 K	20 K	290 K	20 K	290 K
Antennas diameters (m)	BER									
[0.1,0.1]	0.21	0.41	0.48	0,69	0,65	0,74	0.71	0.83	0.91	0.93
[0.2,0.2]	8.1e-4	0.20	0.021	0.47	0.079	0.63	0.22	0.70	0.59	0.89
[0.3,0.3]	0.0	0.026	0.0	0.17	0.0	0.31	0.0048	0.45	0.061	0.81
[0.4,0.4]	0.0	2.1e-4	0.0	0.014	0.0	0.074	0.0	0.19	0.0	0.57
[0.5,0.5]	0.0	0.0	0.0	1.1e-4	0.0	4.5e-3	0.0	0.043	0.0	0.22
[0.6,0.6]	0.0	0.0	0.0	0.0	0.0	0.0	0.0	2.6e-3	0.0	0.045
[0.7,0.7]	0.0	0.0	0.0	0.0	0.0	0.0	0.0	0.0	0.0	2.5e-3
[0.8,0.8]	0.0	0.0	0.0	0.0	0.0	0.0	0.0	0.0	0.0	0.0
[0.9,0.9]	0.0	0.0	0.0	0.0	0.0	0.0	0.0	0.0	0.0	0.0

Note. Losses in "uplink" and "downlink" 120 dB; HPA's backoff 30 dB (negligible nonlinearity); the gain of Satellite Transponder 0 dB; in square brackets the first value is aicraft antenna diameter, and the second is the antenna diameter of the ground station; Iridium satellite height - 780 km; the operation frequency is 1616 MHz.

Table 2. The ground antenna diameter, from which BER = 0

BPSK	QPSK	DQPSK	8PSK	16QAM
[0.1,1.8]	[0.1,2.8]	[0.1,3.4]	[0.1,5.1]	[0.1,6.5]
[0.2,0.9]	[0.2,1.4]	[0.2,1.7]	[0.2,2.5]	[0.2,3.3]
[0.3,0.6]	[0.3,0.9]	[0.3,1.2]	[0.3,1.6]	[0.3,2.1]
[0.4,0.5]	[0.4,0.7]	[0.4,0.9]	[0.4,1.2]	[0.4,1.6]

Note. Values of parameters as in Table 1

Calculations similar to those given in Table 3, can be performed for arbitrary losses in free space, receiver noise, antenna diameters, and a satellite transponder gain factor. This means that it is possible to predict the number of errors in the transmission of messages through the communication channel, that is, to assess the reliability of the satellite-based aviation link.

Table 3. Dependencies of a BER on a HPA's nonlinearity

Modulation	BPSK	QPSK	DQPSK	8PSK	16QAM
Antennas diameters (m)	[0.3,0.6]	[0.3,0.9]	[0.3,1.2]	[0.3,1.6]	[0.3,2.1]
Level of nonlinearity	BER				
1 dB (severe nonlinearity)	0.041	0.12	0.044	0.48	0.61
7 dB (moderate nonlinearity)	0.0027	0.011	0.0014	0.18	0.35
30 dB (negligible nonlinearity)	0.0	0.0	0.0	0.0	0.0

Note. Values of parameters as in Table 1

10. Dependencies of a BER on the gain factor G of the satellite transponder. The influence of the gain factor of the satellite transponder G is shown in Table 4. The obvious result is a decrease in the value of a BER by increasing G. However, the nature of this decrease was dependent on the type of signal modulation. The more "complex" is modulation, the higher the gain values are needed to reduce the number of errors to the required level in the transmission of messages. The simulation allowed to determine, that at G = 24 dB the channel is "open" for all types of considered modulations for the chosen parameters of the communication channel "Aircraft-Satellite-Ground Station".

Table 4. Dependencies of a BER on the gain factor G of the Satellite Transponder

G (dB)	BPSK	QPSK	DQPSK	8PSK	16QAM
0	0.20	0.47	0.63	0.70	0.90
2	0.14	0.39	0.56	0.65	0.88
4	0.089	0.3	0.48	0.58	0.85
6	0.039	0.21	0.38	0.5	0.81
8	0.016	0.1	0.28	0.41	0.75
10	0.0031	0.050	0.15	0.3	0.61
12	0.00031	0.013	0.067	0.19	0.55
14	0.0	0.0016	0.021	0.1	0.41
16	0.0	0.0	0.0039	0.038	0.23
18	0.0	0.0	0.00034	0.0099	0.94
20	0.0	0.0	0.0	0.0011	0.022
22	0.0	0.0	0.0	0.0	0.0018
24	0.0	0.0	0.0	0.0	0.0

Note. Diameters of aircraft and ground station antennas - [0.2,0.2]; values of other parameters as in Table 1

Conclusion

1. "Aircraft-Satellite-Ground Station" channel model (Figure 1) allowed to obtain the graphic dependencies of a BER on free space path loss (Figure 2), average signal power (Figure 3), antennas diameters (Figure 4), samples rate (Figure 5), spectra of transmitted and received signals (Figure 6), End-to-End signal constellations (Figure 7), to investigate the impact of modulation types, noise temperatures, HPA's nonlinearities, and satellite transponder gain.
2. The developed model can be taken as the basis for the study of the transfer of ADS-B signals using the low-orbit constellation of Iridium satellites.
3. Modification of this model can be used to study the effectiveness of channel coding to reduce a BER.

CHANNEL CODING IMPACT ON DATA TRANSMISSION VIA IRIDIUM SATELLITES

Satellite communications are very important and there exist international competition in space industry. European R&D in satellite communications has encompassed a large number of activities spanning many programs and organizations. In overview (Castanet, 2011) it was pointed out: "Optimization of satellite systems requires taking into account propagation information early in system design. ...The propagation channel has a strong impact on system performances and relevant channel models should be available to assess by simulation the end-to-end quality of service and the satellite system performances. Propagation impairments may be mitigated by specific techniques. To design and optimize such techniques, it is necessary to simulate or emulate the dynamic behavior of the propagation channel".

To answer the question how it is possible to keep constantly communication channel parameters optimal we need to work out realistic models of satellite aeronautical communication channels and research their behavior. A communication channel can corrupt information sent over it (Huffman & Pless, 2003). An overview of the many practical applications of channel coding theory was presented in (Costello, Hagenauer, Imai, & Wicker, 1998) and included: deep space communication, satellite communication, data transmission, data storage, mobile communication, file transfer, and digital audio/video transmission. In most papers the power of coding was demonstrated on the base of the memoryless Additive White Gaussian Noise (AWGN) channel that formed the basis for Shannon's noisy channel coding theorem (Shannon, 1948).

It is known that BPSK and QPSK modulations in transmission through the AWGN channel have the same bit error probabilities for coherent detection (Sklar, 2001). But the situation in this section is significantly different. First, the models under consideration contain not one AWGN channel, but two (up and down). For two channels in the literature there are no analytical formulas for the probability of bit errors. Secondly, in addition there is a satellite transponder with an amplifier. Thirdly, we do not use coherent detection. Therefore, the bit error probabilities for the BPSK and QPSK modulations differ.

Hamming Encoding of ADS-B Messages for AWGN Link

Error Detection and Correction Analysis

The most significant question about effectiveness of error detection and correction for "Aircraft-Satellite-Ground Station" link was analyzed using classic linear block codes, like Hamming's (7, 4) code. This code is a linear error-correcting code that encodes 4 bits of data into 7 bits by adding 3 parity bits and can correct any single-bit error, or detect all single-bit and two-bit errors. Hamming's (7, 4) code is effective if transmission medium is not extremely noisy and burst errors do not occur.

Hamming's (7, 4) code can be computed in linear algebra terms through matrices - the code generator matrix ***G*** and the parity-check matrix ***H***:

$$\boldsymbol{G} := \begin{pmatrix} 1101 \\ 1011 \\ 1000 \\ 0111 \\ 0100 \\ 0010 \\ 0001 \end{pmatrix} \text{ and } \boldsymbol{H} := \begin{pmatrix} 1010101 \\ 0110011 \\ 0001111 \end{pmatrix}.$$

A generator matrix generates all possible codewords

$$w = c\mathbf{G},$$

where w is a codeword of the linear code C, c is a row vector, and a bijection exists between w and c. A generator matrix can be used to construct the parity check matrix for a code (and vice-versa).

The rows of a parity check matrix are parity checks on the codewords of a code. That is, they show how linear combinations of certain digits of each codeword equal zero. Rows of the parity-check matrix $\boldsymbol{H}$ are used to compute the syndrome vector at the receiving end and if the syndrome vector is the null vector (all zeros) then the received word is error-free; if non-zero then the value indicates which bit has been flipped.

In MATLAB Simulink the Hamming Encoder block creates a Hamming code with message length K and codeword length N. The number N must have the form 2^M-1, where M is an integer greater than or equal to 3, then K equals N-M. The input must contain exactly K elements and if it is frame-based, then it must be a column vector. The output is a vector of length N. The Hamming Decoder block recovers a binary message vector from a binary Hamming codeword vector. For proper decoding, the parameter values in this block should match those in the corresponding Hamming Encoder block.

Model Hamming_AWGN_Sat_AWGN_MPSK for Satellite Link

For satellite communication link modeling a satellite transponder and two AWGN channels for uplink and downlink were included (Kharchenko, Barabanov, Grekhov, Bogunenko, & Rudnyk, 2013a). MATLAB simulation model *Hamming_AWGN_Sat_AWGN_MPSK* is shown in Figure 8 and consists of a source of information (Bernoulli Binary block), Aircraft Transmitter (Hamming Encoder block, Modulator Baseband block), an uplink (AWGN channel), Satellite Transponder (Receiver Dish Antenna Gain, Complex Baseband Amplifier with Noise, Transmitter Dish Antenna Gain), a downlink, Ground Station Receiver and Error Rate Calculation block.

Complex Baseband Amplifier block in satellite transponder generates a complex baseband model of an amplifier with thermal noise. It simulates linear amplifier and allows to specify noise.

During a simulation BPSK, QPSK, 8PSK, and 16QAM modulation schemes were considered. For each of these modulations a BER was calculated without and with coding as function of E_b/N_0 (the ratio of bit energy to noise power spectral density). The value of a ratio E_b/N_0 was changed symmetrically in uplink and downlink channels.

Figure 8. Model Hamming_AWGN_Sat_AWGN_MPSK of aircraft satellite link

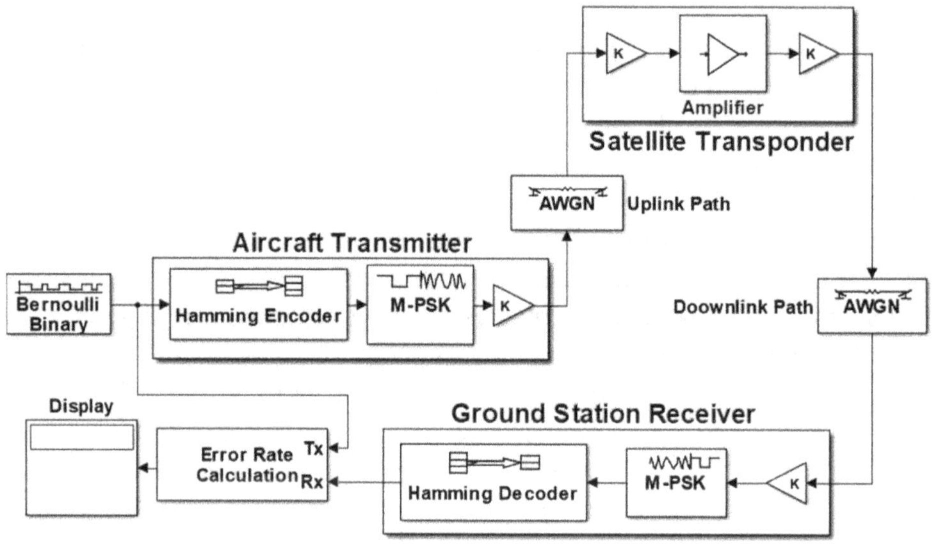

All calculations were done when receiver/transmitter dish antenna gain was equal to unit. Noise temperature in complex baseband amplifier was taken 290 K. When investigating effect of coding for different modulation types we took satellite linear amplifier gain equal to unit. This model also allows investigating a dependence of a BER on satellite transponder amplifier gain without and with coding and such dependencies were obtained and analyzed for BPSK, QPSK, 8PSK, and 16QAM.

Results of Modelling

Results of simulations are shown in Figures 9-15. First, it is interesting to compare a value of a BER for different modulation types without coding. From Figures 9-11 (solid lines) follows that the lowest BER has BPSK modulation and the highest – 16QAM modulation. For example, at $E_b/N_0 = 8$ dB: $BER_{BPSK} = 5.8 \cdot 10^{-3}$, $BER_{QPSK} = 6.0 \cdot 10^{-2}$, $BER_{8PSK} = 3.1 \cdot 10^{-1}$, $BER_{16QAM} = 4.1 \cdot 10^{-1}$.

The higher amount of alternative modulation symbols - the closer plots are arranged. For example, at $E_b/N_0 = 8$ dB: $BER_{QPSK}/BER_{BPSK} \approx 10.3$, $BER_{8PSK}/BER_{QPSK} \approx 5.2$, $BER_{16QAM}/BER_{8PSK} \approx 1.3$.

Secondly, a coding with Hamming code (7, 4), t=1 shows that the effect of BER decreasing is maximal for BPSK. For example, at $E_b/N_0 = 8$ dB:

Figure 9. Effect of channel coding on BPSK and QPSK (dashed line – Hamming code (7, 4), t=1)

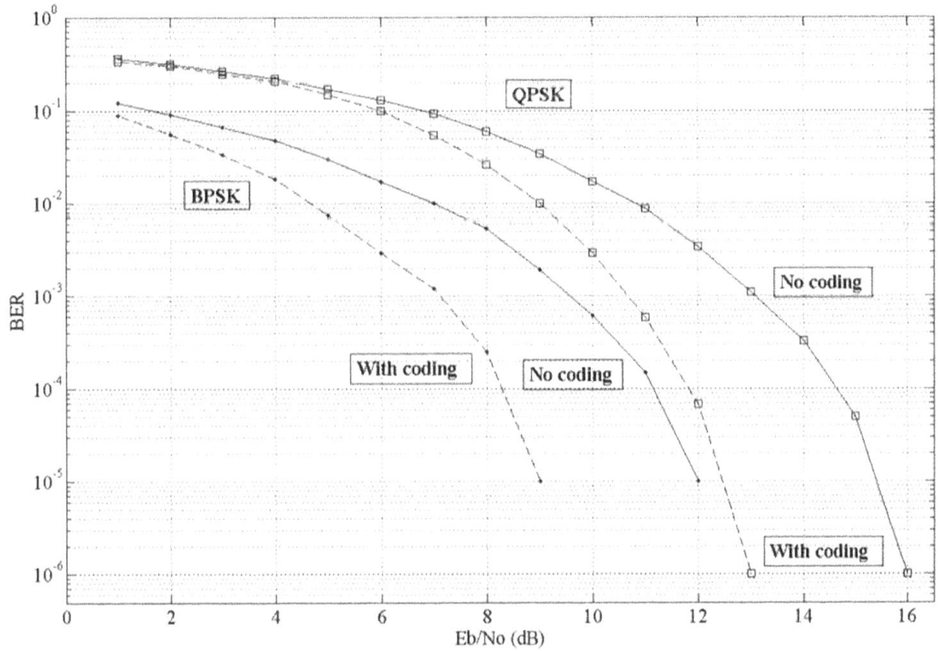

Figure 10. Effect of channel coding on 8PSK (dashed line – Hamming code (7, 4), t=1)

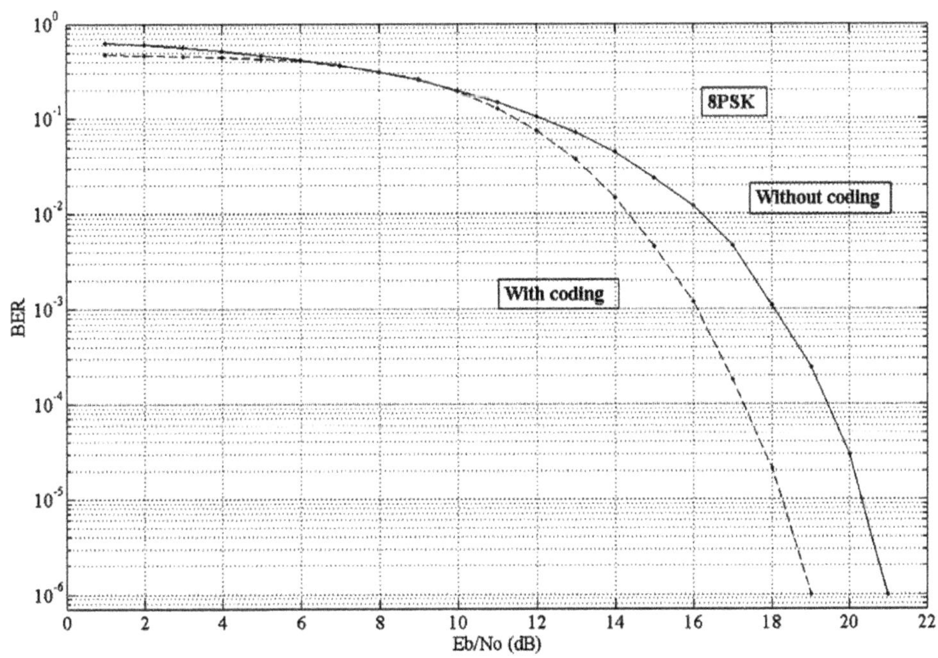

Figure 11. Effect of channel coding on 16QAM (dashed line – Hamming code (7, 4), t=1)

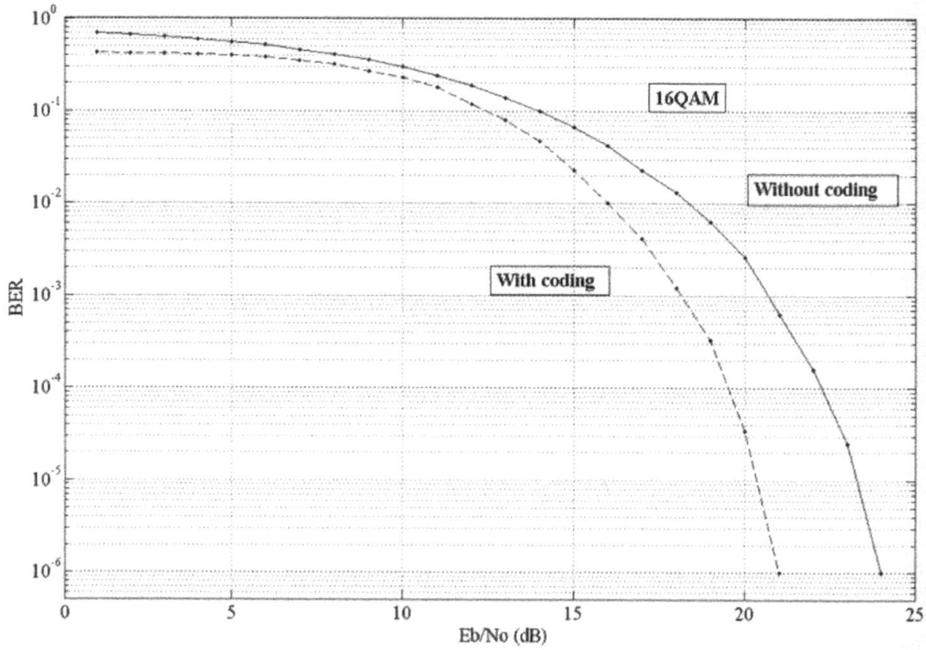

$(BER_{BPSK} - BER_{codedBPSK})/BER_{BPSK} \approx 96\%,$
$(BER_{QPSK} - BER_{codedQPSK})/BER_{QPSK} \approx 57\%,$
$(BER_{8PSK} - BER_{coded8PSK})/BER_{8PSK} \approx 0\%,$
$(BER_{16QAM} - BER_{coded16QAM})/BER_{16QAM} \approx 2\%.$

Notice, that in the range 5 dB $\leq E_b/N_0 \leq$ 10 dB an effectiveness of coding for 8PSK is lower than for 16QAM (compare Figures 10 and 11).

An investigation of BER dependence on satellite transponder amplifier gain without and with coding was provided for two meanings of a ratio E_b/N_0 (1 and 5 dB) in uplink and downlink when receiver/transmitter dish antenna gain was equal to unit, noise temperature in complex baseband amplifier was taken 290 K (Figures 12–15).

The most significant influence satellite transponder amplifier gain has on BPSK modulation (Figure 12). The nature of coding effect on a BER decreasing is similar for BPSK and QPSK modulations but for QPSK the effect is less (compare Figures 12 and 13). For BPSK and QPSK modulations decreasing

Figure 12. Impact of Satellite Transponder Amplifier Gain on BPSK (dots, solid line – E_b/N_0 = 1 dB, without coding; dots, dashed line - E_b/N_0 = 1 dB, Hamming code (7, 4) t=1; squares, solid line – E_b/N_0 = 5 dB, without coding; squares, dashed line - E_b/N_0 = 5 dB, Hamming code (7, 4) t=1)

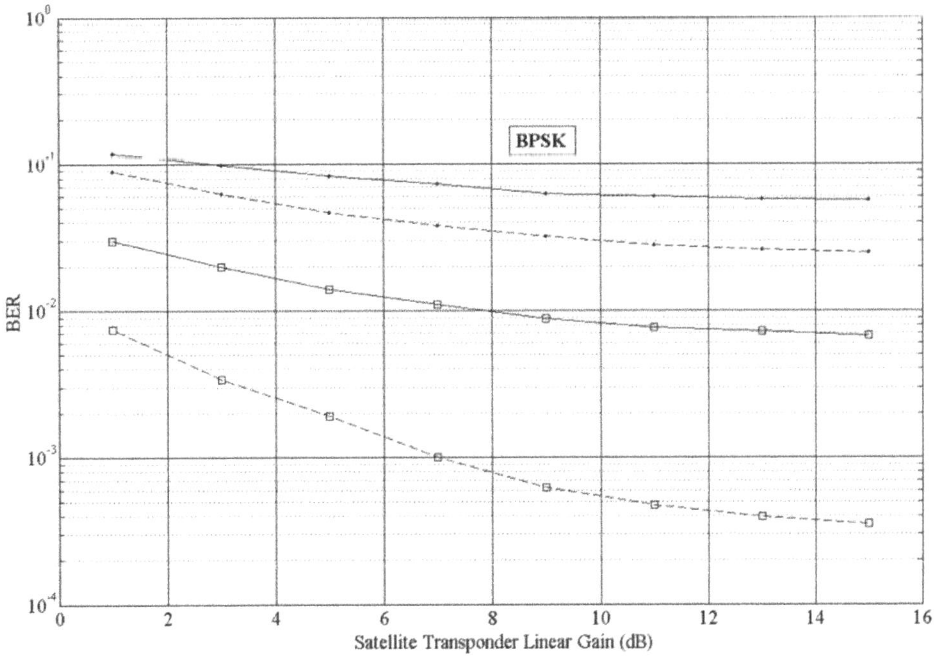

of a BER is bigger for higher ratio E_b/N_0 (compare upper graph for E_b/N_0 = 1 dB and lower graph for E_b/N_0 = 5 dB correspondingly in Figures 12 and 13).

Quite different impact satellite transponder amplifier gain has on 8PSK and 16QAM modulations (Figures 14 and 15). For E_b/N_0 = 1 dB a BER decreasing after coding is bigger than for E_b/N_0 = 5 dB – a converse effect than for BPSK and QPSK modulations. Another interesting feature is that the higher amount of alternative modulation symbols is - the less is impact of transponder amplifier gain increasing on decreasing of a BER (graphs in Figures 14 and 15 when gain is more than 6 dB).

Conclusion

This study deals with the performance analysis of aeronautical satellite baseband link. Original model includes aircraft transmitter, uplink/downlink

Figure 13. Effect of Satellite Transponder Amplifier Gain on QPSK (dots, solid line – E_b/N_0 = 1 dB, without coding; dots, dashed line - E_b/N_0 = 1 dB, Hamming code (7,4) t=1; squares, solid line – E_b/N_0 = 5 dB, without coding; squares, dashed line - E_b/N_0 = 5 dB, Hamming code (7,4) t=1)

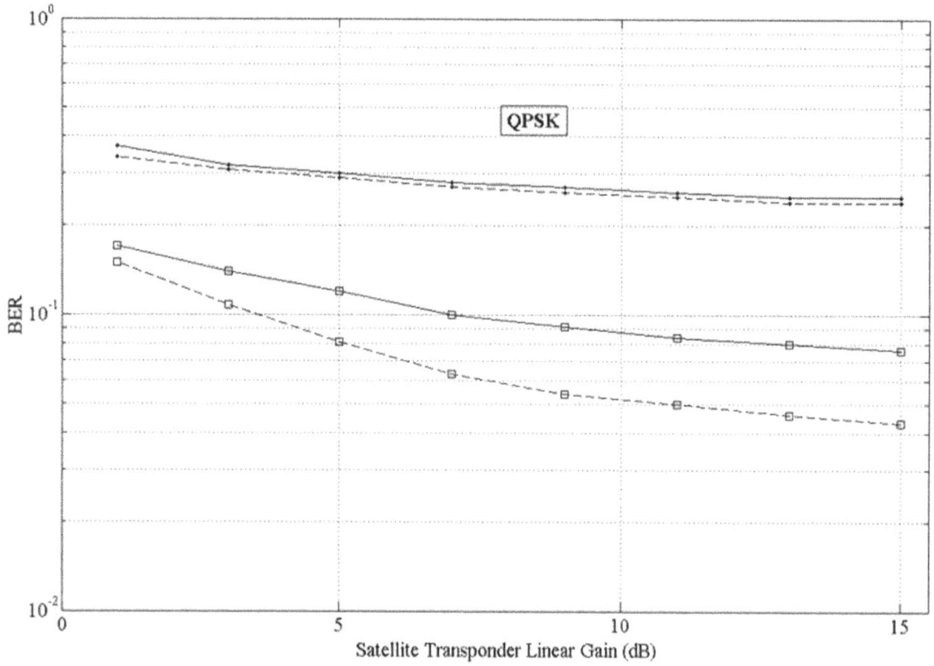

(AWGN channels), satellite transponder, and ground Earth station receiver. Impact of a modulation type (BPSK, QPSK, 8PSK, and 16QAM) and satellite transponder amplifier gain without and with coding on a BER was investigated.

Effectiveness of error detection and correction for "Aircraft-Satellite-Ground Station" link was analyzed using Hamming's (7, 4) code.

The most important results for model *Hamming_AWGN_Sat_AWGN_MPSK* with two AWGN channels and satellite transponder are the following: a) the lowest BER has BPSK modulation and the highest – 16QAM modulation; b) the higher amount of alternative modulation symbols is - the closer plots of a BER as function of E_b/N_0 are arranged in the sequence BPSK, QPSK, 8PSK, and 16QAM; c) a coding with Hamming code (7, 4) shows that the effect of a BER decreasing is maximal for BPSK modulation; d) in the range 5 dB ≤ E_b/N_0 ≤ 10 dB an effectiveness of coding for 8PSK modulation is lower than for 16QAM modulation; e) the most significant influence satellite

Figure 14. Effect of Satellite Transponder Amplifier Gain on 8PSK (dots, solid line – E_b/N_0 = 1 dB, without coding; dots, dashed line - E_b/N_0 = 1 dB, Hamming code (7, 4) t=1; squares, solid line – E_b/N_0 = 5 dB, without coding; squares, dashed line - E_b/N_0 = 5 dB, Hamming code (7, 4) t=1)

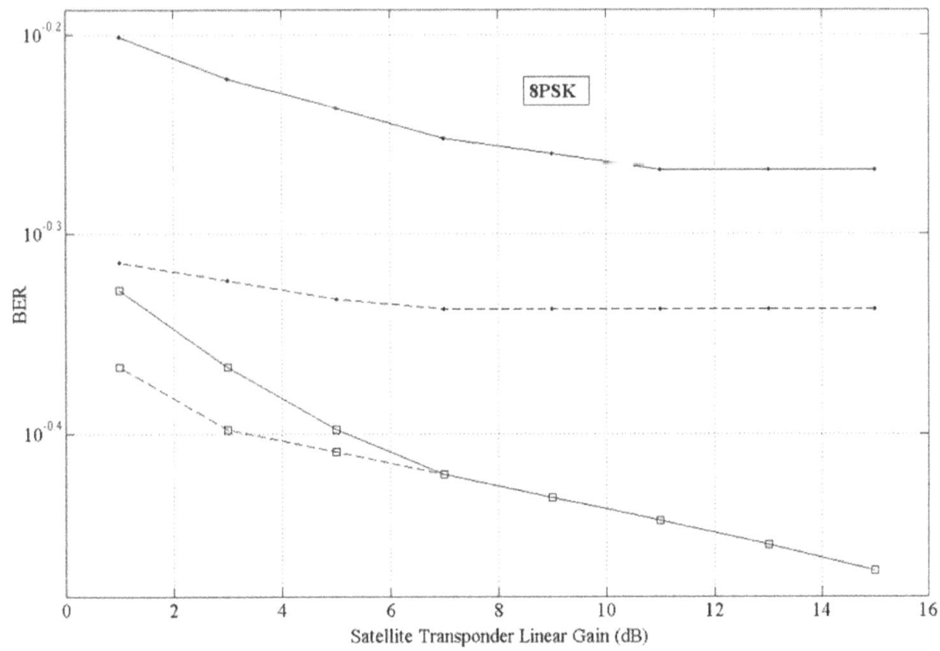

transponder amplifier gain has on BPSK modulation; f) the nature of coding effect on a BER decreasing is similar for BPSK and QPSK modulations but for QPSK modulation the effect is less; g) satellite transponder amplifier gain has quite different impact on 8PSK and 16QAM modulations - the higher is amount of alternative modulation symbols the less is impact of transponder amplifier gain increasing on decreasing of a BER.

CONVOLUTIONAL ENCODING OF ADS-B MESSAGES FOR AWGN LINK

The aim of this study is: 1) to design the model of communication channel "Aircraft-Satellite-Ground Station" with convolutional encoding; 2) on the base of this model investigate a channel integrity and receive dependencies

Figure 15. Effect of Satellite Transponder Amplifier Gain on 16QAM (dots, solid line – $E_b/N_0 = 1$ dB, without coding; dots, dashed line - $E_b/N_0 = 1$ dB, Hamming code (7, 4) t=1; squares, solid line – $E_b/N_0 = 5$ dB, without coding; squares, dashed line - $E_b/N_0 = 5$ dB, Hamming code (7, 4) t=1)

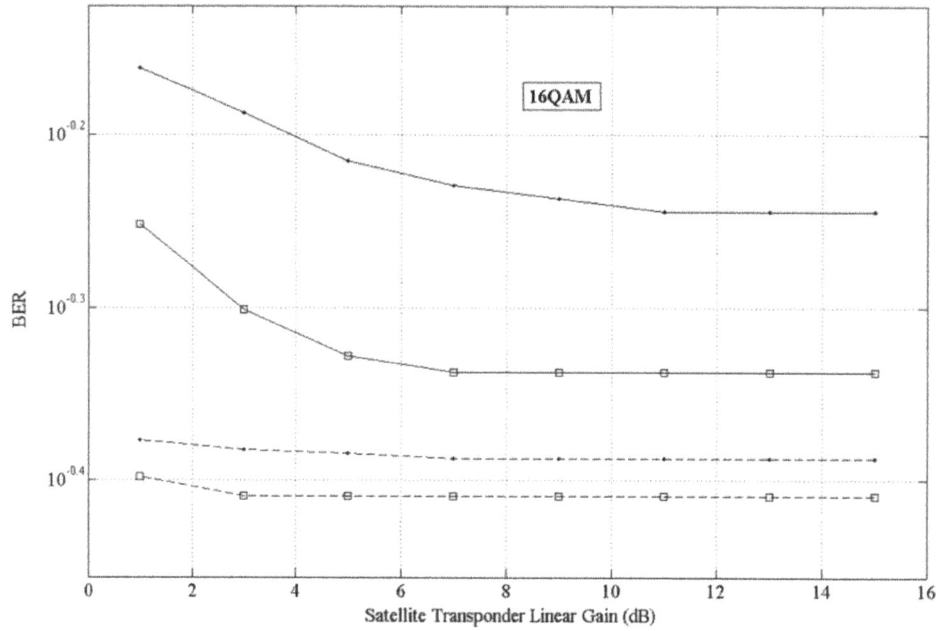

of a BER on a type of signal coding/decoding, ratio E_b/N_0, data rate and satellite transponder gain; 3) to compare convolutional and Hamming coding (Kharchenko, Barabanov, Grekhov, Bogunenko, & Rudnyk, 2013b).

Model Convolutional_AWGN_Sat_ AWGN_MPSK for Satellite Link

The model *Convolutional_AWGN_Sat_AWGN_MPSK* (Figure 16) comprises Source of Data (Bernoulli Binary Generator), Aircraft Transmitter (modulator with or without Convolutional encoder), Uplink Path (AWGN channel), Satellite Transponder (Receiver Dish Antenna Gain, Complex Baseband Amplifier with Noise, Transmitter Dish Antenna Gain), Downlink Path (AWGN channel), Ground Station Receiver (demodulator with or without Viterbi Decoder), Error Rate Calculation block and Display. Complex Baseband

Figure 16. Model Convolutional_AWGN_Sat_AWGN_MPSK of aircraft satellite link

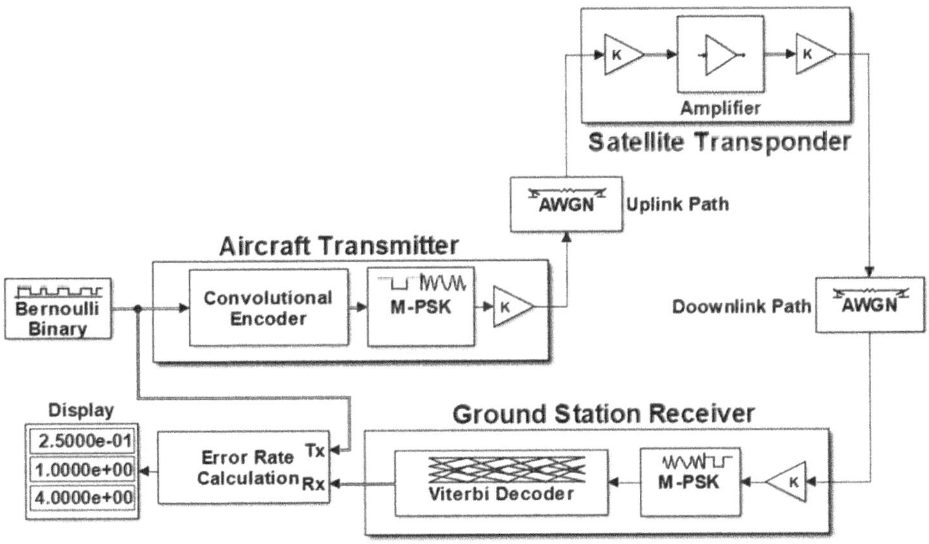

Amplifier block in satellite transponder generates a complex baseband model of an amplifier with thermal noise. It simulates linear amplifier and allows to specify noise.

Modulations BPSK, QPSK, and 8PSK were considered during a simulation. For these modulations a BER was calculated without and with convolutional coding as function of a ratio E_b/N_0. The value of a ratio E_b/N_0 was changed symmetrically in uplink and downlink AWGN channels. All calculations were done when receiver/transmitter dish antenna gain was equal to unit. Noise temperature in complex baseband amplifier was taken as 290 K. When investigating effect of coding for different modulation types we took satellite linear amplifier gain equal to one.

These models also allow investigation of BER dependence on satellite transponder amplifier gain without and with coding and such dependencies were obtained for all modulations.

Aeronautical Satellite Channel Simulation

Iridium system parameters were used for modelling: Forward Error Correction Code in the form of convolutional encoding with Viterbi decoding, a rate 3/4, constraint length 7 convolutional code on both transmission and reception.

The source of information (Bernoulli Binary Generator) generates random binary numbers. It was set to frame-based outputs, with a probability of zero 0.5, initial seed 61, sample time 1 and samples per frame 1.

Convolutional encoder encodes binary data: Viterbi decoder uses the Viterbi algorithm to decode convolutionally encoded input data. Convolutional encoder and Viterbi decoder operated with Trellis structure: *poly2trellis* (7, [171 133]) and with Continuous Operation mode. *Poly2trellis* function is used to create a trellis using the constraint length, code generator (octal) and feedback connection (octal). The *poly2trellis* function accepts a polynomial description of a convolutional encoder and returns the corresponding trellis structure description. The output of *poly2trellis* is a mask parameter for the Viterbi Decoder. Function *poly2trellis* (ConstraintLength, Code Generator) performs the conversion for a rate k/n feed forward encoder. ConstraintLength is a 1-by-k vector (7) that specifies the delay for the encoder's k input bit streams. CodeGenerator is a k-by-n matrix [171 133] of octal numbers that specifies the n output connections for each of the encoder's k input bit streams. In Viterbi Decoder the decision type "Unquantized" was chosen with traceback depth being 96.

Results of Modelling

Results of simulations are shown in Figures 17-20. The effect of coding on BPSK modulation (Figure 17) is manifested in the fact that at $E_b/N_0 > 2$ dB the convolutional coding efficiency is much higher than the Hamming coding. For QPSK modulation (Figure 18), Hamming encoding is more efficient than convolutional coding at 1 dB $< E_b/N_0 < 11$ dB, but at $E_b/N_0 >$ 11 dB convolutional coding is more efficient. For 8PSK modulation (Figure 19) - the situation is similar to QPSK modulation, but the "critical" E_b/N_0 value is 8 dB.

An investigation of BER dependence on satellite transponder amplifier gain without and with coding was provided for two meanings of a ratio E_b/N_0 (5 dB and 8 dB) in uplink and downlink when receiver/transmitter dish antenna gain was equal to unit, noise temperature in complex baseband amplifier was 290 K (Figure 20). For BPSK modulation decreasing of a BER is bigger for higher ratio E_b/N_0 (compare upper graph for $E_b/N_0 = 5$ dB and lower grapf for $E_b/N_0 = 8$ dB correspondingly in Figure 20).

Coding with different coding rates was investigated and the same BER graphs for ½ and ¾ rates were obtained.

Figure 17. Effect of channel coding on BPSK: circles - Hamming (7, 4); asterisks: Convolutional coding (rate 3/4)

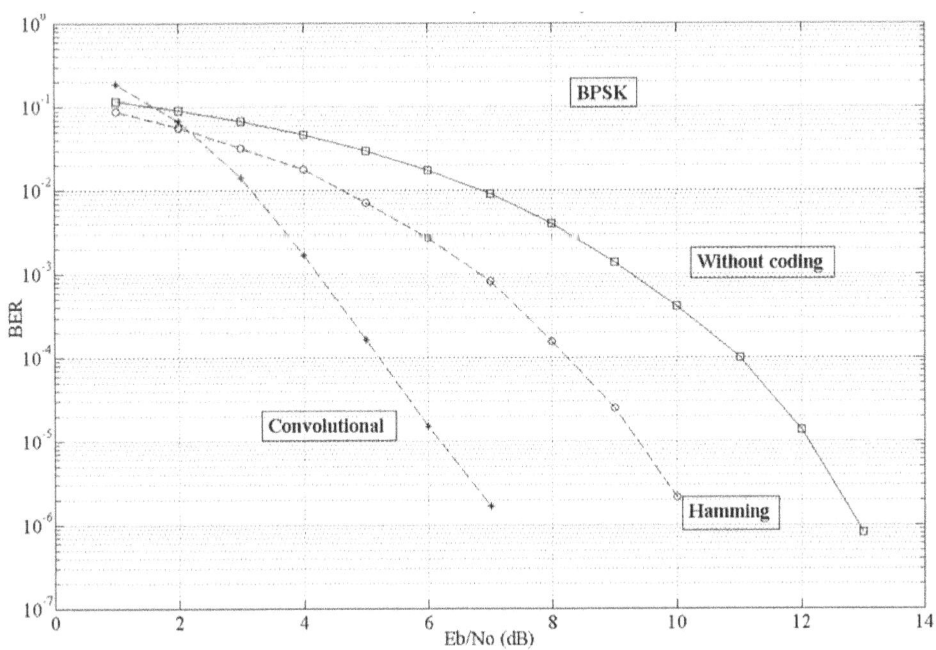

Conclusion

1. For modelling of ADS-B messages transmitting on the base of low-orbit satellite constellation Iridium the model *Convolutional_AWGN_Sat_AWGN_MPSK* of a communication channel "Aircraft-Satellite-Ground Station" was built using MATLAB Simulink.
2. This model allows investigation of BER dependences on signal coding type and satellite transponder gain.
3. Proposed model can be used as basic model for investigation of communication between two airplanes and ground stations using several satellites.
4. Developed model can also be used for finding optimal methods of error-correcting coding and channcl intcgrity.

Figure 18. Effect of channel coding on QPSK: circles - Hamming (7, 4); asterisks: Convolutional coding (rate 3/4)

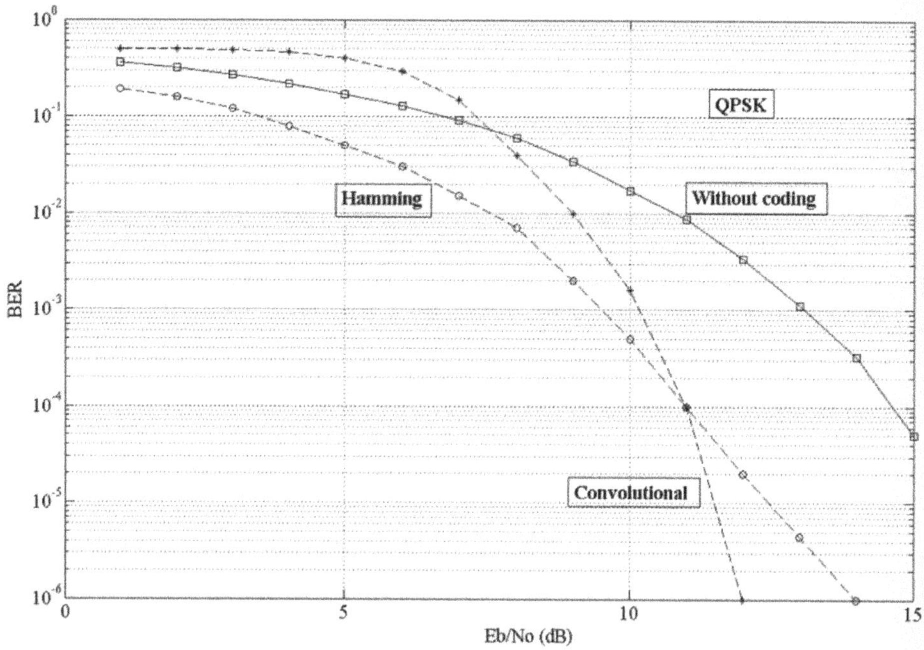

CONVOLUTIONAL ENCODING OF ADS-B MESSAGES FOR LINK WITH FREE SPACE PASS LOSS

Model Convolutional_FSPL_Sat_ FSPL_MPSK for Satellite Link

A model *Convolutional_FSPL_Sat_FSPL_MPSK* for satellite communication channel "Aircraft-Satellite-Ground Station" with error-control coding was built using MATLAB Simulink software. The original model, shown in Figure 21, comprises "Aircraft Transmitter" (Bernoulli Random Binary Generator, Convolutional Encoder, BPSK Baseband Modulator, High Power Amplifier (HPA) with a memoryless nonlinearity, Transmitter Dish Antenna Gain), "Uplink Path" (Free Space Path Loss, Phase/Frequency Offset), "Satellite Transponder" (Receiver Dish Antenna Gain, Complex Baseband Amplifier, Phase/Frequency Offset, Transmitter Dish Antenna Gain), "Downlink Path" (Free Space Path Loss, Phase/Frequency Offset), "Ground Station Receiver"

Figure 19. Effect of channel coding on 8PSK: circles: Hamming (7, 4); asterisks: Convolutional coding (rate 3/4)

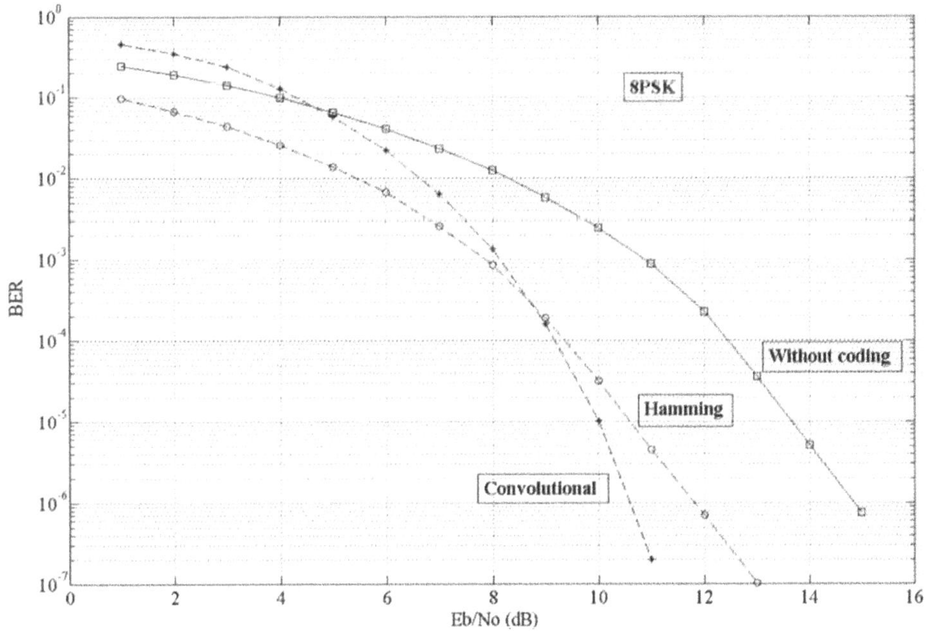

(Receiver Dish Antenna Gain, Phase Noise, Ground Receiver System Temperature, Viterbi Decoder), "Error Rate Calculation block" and "Display".

In the "Aircraft Transmitter" the Bernoulli Binary Generator block generates random binary numbers using a Bernoulli distribution with parameter p, produces "zero" with probability p and "one" with probability 1-p (the value p=0,5 is used). The output signal is a frame-based matrix. The Bernoulli Binary Generator block generates a discrete signal and updates the signal at integer multiples of a fixed time interval, called the sample time. The length of this time interval has the value 1. The output data type is "double".

Iridium system employs a BPSK modulation and forward error correction coding in the form of convolutional encoding with Viterbi decoding (Viterbi, 1971). Iridium uses a rate 3/4, constraint length 7, (r=3/4; K=7) convolutional code on both transmission and reception (Costello, Hagenauer, Imai, & Wicker, 1998). The Convolutional Encoder block is using the *poly2trellis*(7,[171 133],171) function with a constraint length of 7, code generator polynomials of 171 and 133 (in octal numbers), and a feedback connection of 171 (in octal). The puncture vector is [1; 1; 0; 1; 1; 0].

Figure 20. Impact of satellite transponder on BPSK: solid line with asterisks: Eb/N0 = 5 dB, without coding; dashed line with asterisks: Eb/N0 = 5 dB, with Convolutional coding; solid line with circles: Eb/N0 = 8 dB, without coding; dashed line with circles: Eb/N0 = 8 dB, with Convolutional coding

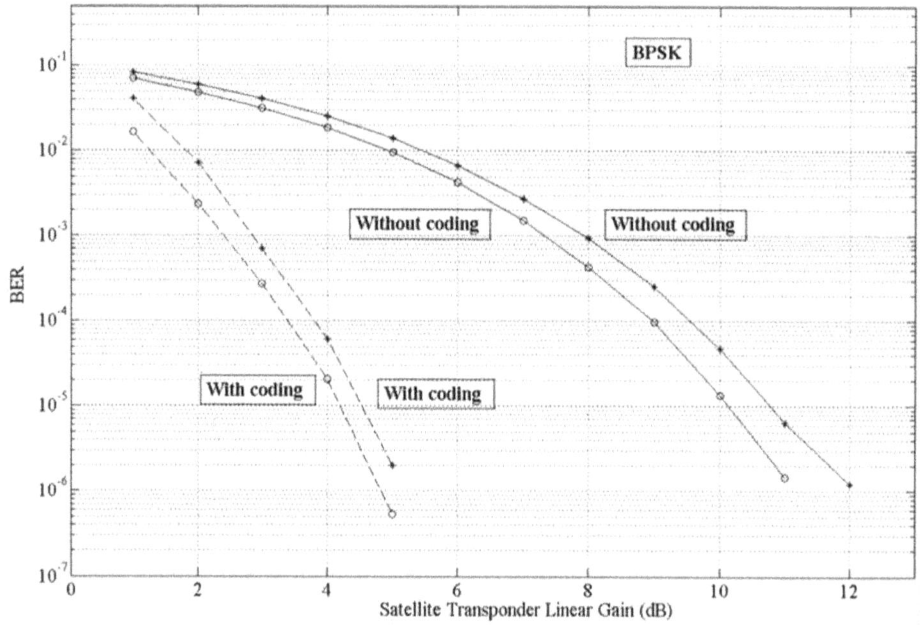

Figure 21. Model Convolutional_FSPL_Sat_FSPL_MPSK of aircraft satellite link

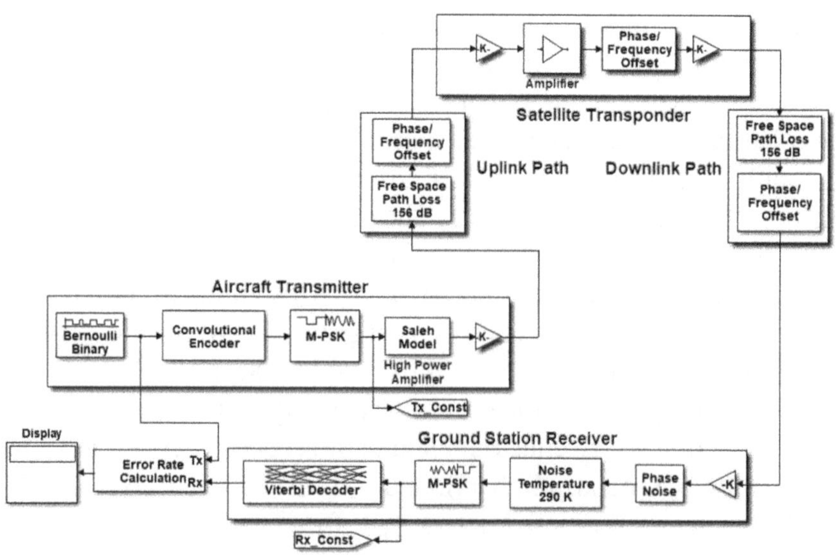

The BPSK Baseband Modulator block modulates a signal using the binary phase shift keying method. The output is a baseband representation of the modulated signal.

The High Power Amplifier block applies memoryless nonlinearity to complex baseband signal and provides five different methods for modeling the nonlinearity. In this paper results only for Saleh model with standard AM/AM and AM/PM parameters are given (Saleh, 1981). A HPA backoff level is used to determine how close the satellite high power amplifier is driven to saturation. The following selected backoff is used to set the input and output gain of the Memoryless Nonlinearity block: 30 dB - the average input power is 30 decibels below the input power that causes amplifier saturation (in this case AM/AM and AM/PM conversion is negligible); 7 dB - moderate nonlinearity; and 1 dB - severe nonlinearity.

The Transmitter (Receiver) Dish Antenna Gain block multiplies the input by a constant value (gain). Dependencies of a BER on transmitting and receiving antennas diameter were obtained using vectors [d_1, d_2] for each pair "transmitter-receiver". The first element in the vector [d_1, d_2] represents the transmitting antenna diameter (in meters) and is used to calculate the gain in the Transmitter Dish Antenna Gain block. The second element represents the receiving antenna diameter and is used to calculate the gain in the Receiver Dish Antenna Gain block. The default setting is [1.0, 1.0] (an antenna gain is 12,4) and diameters of all antennas (transmitting antenna on an aircraft, receiving and transmitting antennas on a satellite, receiving antenna on a ground station) were changed simultaneously.

In the "Uplink (Downlink) Path" the Free Space Path Loss block simulates the loss of signal power due to the distance between the aircraft uplink transmitter and the satellite transponder receiver. The block reduces the amplitude of the input signal by an amount that is determined by the Loss (dB) parameter.

The Phase/Frequency Offset block applies phase and frequency offsets to an incoming signal.

In the "Satellite Transponder" the Satellite Amplifier in addition to a Linear amplifier has five different methods to model the nonlinear amplifier. The amount of noise added to the output signal may be specified in terms of noise temperature. The Phase/Frequency Offset block applies phase and frequency offsets to an incoming signal. Modeling was provided for two values of effective satellite and ground station receiver systems noise temperatures 20 K (very low noise level) and 290 K (typical noise level). These settings were changed simultaneously and the last setting was used to view how typical receivers operate.

The Phase Noise block adds receiver phase noise to a complex baseband signal. The block applies the phase noise as follows: generates additive white Gaussian noise and filters it with a digital filter; adds the resulting noise to the angle component of the input signal. The level of the spectrum is specified by the noise power contained in a one hertz bandwidth offset from a carrier by a certain frequency. Modeling was provided for three levels: negligible (phase noise level: -100dBc/Hz, frequency offset: 100 Hz), low (-55 dBc/Hz, frequency offset: 100 Hz), high (-48 dBc/Hz, frequency offset: 100 Hz).

In the "Ground Station Receiver" the Viterbi Decoder block decodes input symbols to produce binary output symbols. Unquantized decision type parameter was used.

Comparing scatter plots of the signal after BPSK modulation and before demodulation allows viewing the impact of all impairments on the received signal.

Aeronautical Satellite Channel Simulation and Results

For computer modeling a distance between the Iridium satellite and the ground station (satellite altitude) 780 km and an operational frequency 1616 MHz were taken. Changing a carrier frequency of the link updates the Free Space Path Loss block. A dependence of a BER on free space path loss for different noise temperatures without coding and with convolutional coding is shown in Figure 22. Loss values were changed simultaneously in uplink and downlink channels for equal antennas diameter $d_1 = d_2 = d_3 = d_4 = 1{,}0$ m, moderate HPAs nonlinearity (backoff level 7 dB), satellite transponder gain - 0 dB. Convolutional coding considerably decreases errors probability and a BER is vanishing for free space path loss in the range from 0 dB to 161 dB (in case of very low noise temperature 20 K) and from 0 dB to 155 dB (in case of typical noise temperature 290 K).

Changing of all antennas diameter has significant influence on errors probability shown in Figure 23 – the more antennas diameter, the less is an error probability. In this simulation free space path loss was fixed (160 dB), HPA's nonlinearity – moderate (backoff level 7 dB), and two noise temperatures were considered (20 K and 290 K).

Figure 22. Dependencies of a BER on free space path loss in the uplink and the downlink

dots – without coding, circles – with convolutional coding (rate ¾, constraint length K=7); satellite and ground receivers noise temperatures are 20 K (dashed lines) and 290 K (solid lines), HPA's backoff level is 7 dB, phase and frequency offsets are equal to zero, $d_1 = d_2 = d_3 = d_4 = 1{,}0$ m

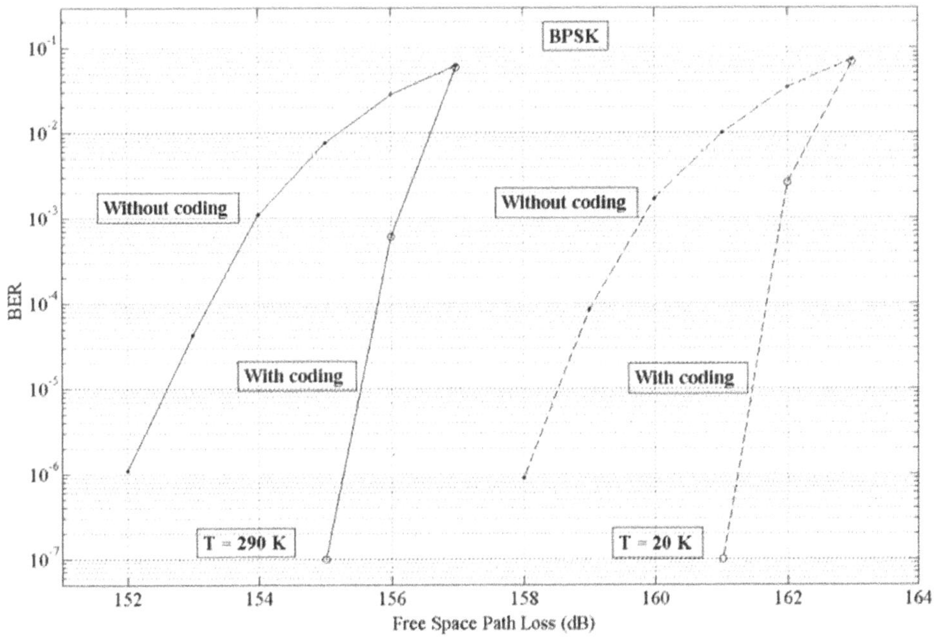

Convolutional coding essentially reduces the error probability that leads to a BER vanishing for antennas with diameter more than 1,2 m. Iridium satellite has three Main Mission Antennas (each of 0,86 m wide and 1,86 m high). The same area has a circle with diameter ≈1,4 m. Apparently, results of our modeling are in the good consent with these data.

In the presence of an arbitrary phase offset introduced by the uplink and the downlink, the demodulator is unable to tell which constellation point is which. A dependence of a BER on phase offset in the uplink and the downlink is shown in Figure 24 for two noise temperatures without convolutional coding and with it. At using convolutional coding a BER is vanishing for phase shifts up to 7° at uplink and downlink losses in free space 157 dB and a noise temperature of a satellite transponder and a ground receiver 290 K.

Our model allows exploring the end-to-end simulation of "Aircraft–Satellite-Ground Station" communications links using original model of a

Figure 23. Dependencies of a BER on satellite and aircraft antennas diameter
dots – without coding, circles – with convolutional coding (rate ¾, constraint length K=7); satellite and ground receivers noise temperatures are 20 K (solid lines) and 290 K (dashed lines), phase and frequency offsets are equal to zero, HPA's backoff level is 7 dB, free space path loss is 160 dB

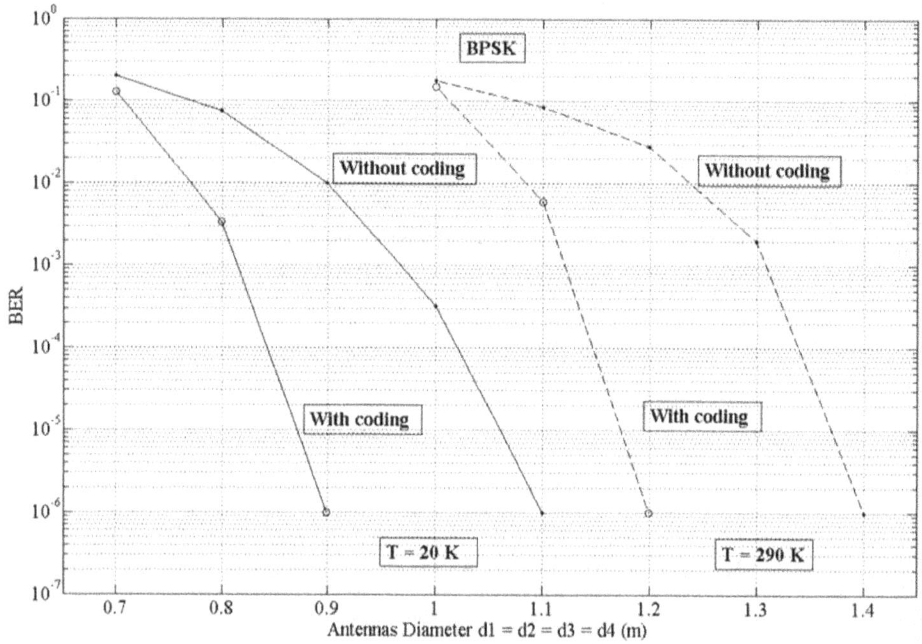

satellite transponder. The Complex Baseband Amplifier block can model a linear amplifier in which the linear method is implemented by a Gain block. A dependence of a BER on satellite transponder gain is shown in Figure 25 for two noise temperatures without coding and with convolutional coding. Convolutional coding essentially reduces the error probability that leads to a BER vanishing for satellite amplifier linear gain more than 8 dB.

Conclusion

For modelling of ADS-B messages transmitting on the base of low-orbit satellite constellation Iridium the original model of a communication channel "Aircraft–Satellite-Ground Station" with error-control coding was built using MATLAB Simulink software.

For studying of a signal transmission through a communication channel without coding and with convolutional coding the following parameters were changed: losses in a free space simultaneously in the uplink and the downlink

Figure 24. Dependencies of a BER on phase offset in the uplink and the downlink
solid line – without coding, dashed line – with convolutional coding (rate ¾, constraint length K=7); free space path loss 157 dB; HPA's backoff level 7 dB; d1 = d2 = d3 = d4 = 1,0 m

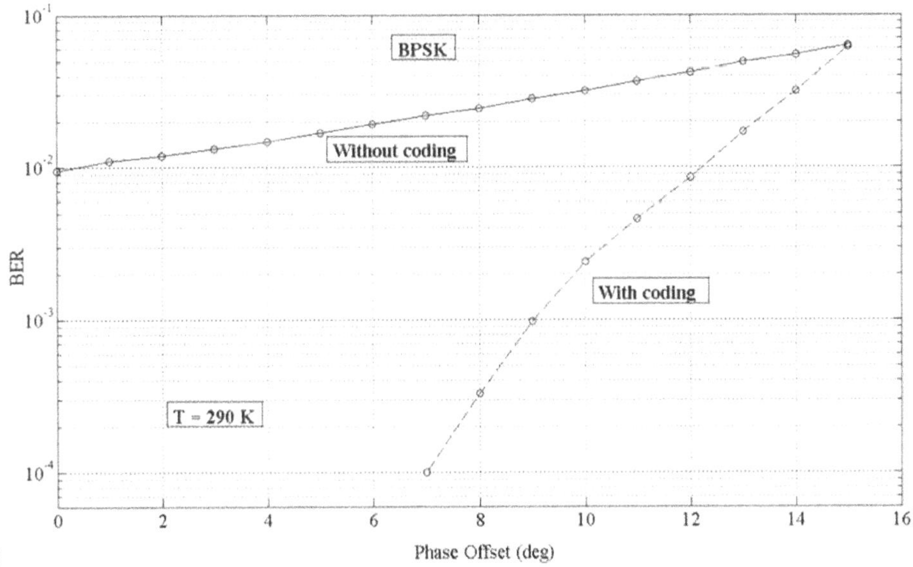

fi

from 0 dB to 250 dB in each channel (Figure 22); noise temperature of a satellite transponder and a ground receiver (20 K, 290 K); symmetrically and asymmetrically diameters of all four antennas that increased or reduced a power of the received signal (Figure 23); phase and frequency (due to Doppler's effect) offsets simultaneously in the uplink and the downlink from 0° to 20° (from 0 Hz to 10 Hz) in each channel (Figure 24); satellite transponder linear gain from 0 dB to 15 dB (Figure 25).

Dependencies shown in Figures 22-25 were obtained for "standard parameters": without coding and with convolutional coding (rate ¾, constraint length K=7); satellite and ground receivers noise temperatures 20 K and 290 K; without phase and frequency offsets; negligible HPA's; $d_1 = d_2 = d_3 = d_4 = 1,0$ m. Values of free space path losses, phase and frequency offsets, antennas diameter were specified in each special case.

For "standard parameters" a BER is vanishing for free space path losses changing from 0 dB to 163 dB at use of convolutional coding and a noise temperature 20K.

Diameters of antennas essentially influence on a BER what is shown in (Figure 23). The probability of errors for "standard parameters" is vanishing for diameters of all antennas more than 1,2 m. This result is in good agreement

Figure 25. Dependencies of a BER on satellite transponder linear gain
solid lines – without coding, dashed lines – with convolutional coding (rate ¾, constraint length K=7); satellite and ground receivers noise temperatures 20 K (dots, free space path loss 160 dB) and 290 K (circles, free space path loss 157 dB), phase and frequency offsets are equal to zero; HPA's nonlinearity is negligible; $d_1 = d_2 = d_3 = d_4 = 1{,}0$ *m*

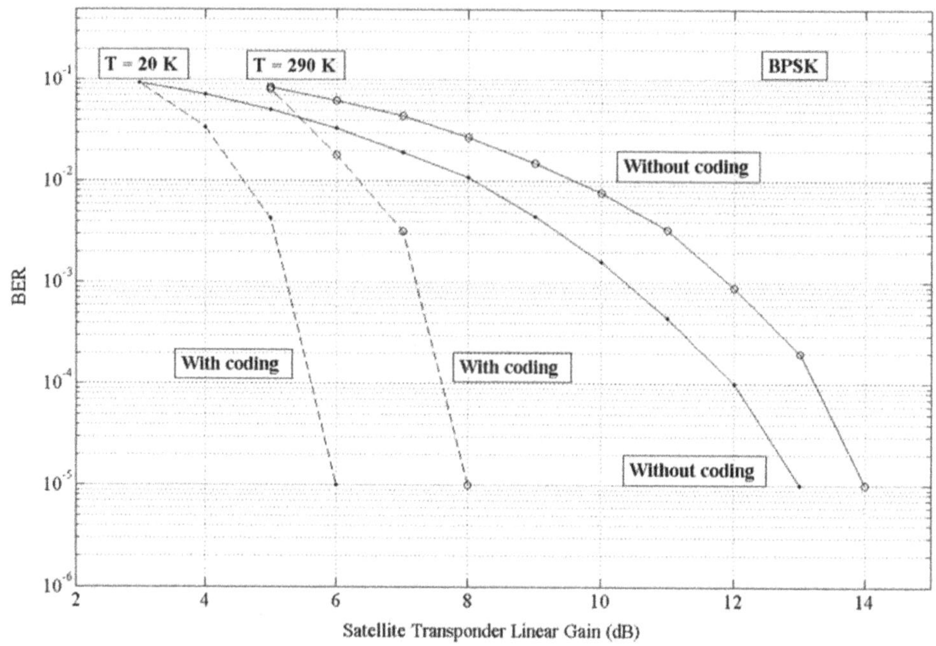

with the size of Iridium satellite antennas. At the same time calculations with a diameter of aircraft transmitting antenna $d_1 = 0{,}3$ m, a diameter of satellite receiving antenna $d_2=1$ m, a diameter of satellite transmitting antenna $d_3 = 1$ m and a diameter of ground station receiving antenna $d_4 = 2$ m (for the noise temperature 0 K) gives BER vanishing for losses in free space up to 220 dB.

Influence of convolutional coding on dependence of a BER on phase offsets in the uplink and the downlink is critical and changes character of this dependence (Figure 24), essentially reducing a level of errors. Under "standard parameters" and noise temperature 20 K a BER is vanishing at phase shifts up to 14°.

Satellite transponder is the central element of a considered communication channel and transponder linear gain along with a choice of a working point (level of TWTA nonlinearity) make strong impact on quantity of errors at data transmission. Under "standard parameters", noise temperature 290 K and free space losses 157 dB in the uplink and 157 dB in the downlink a BER vanishes when transponder linear gain is not less than 8 dB (Figure

25). At a noise temperature 20 K and free space losses 160 dB in the uplink and 160 dB in the downlink a BER vanish when transponder linear gain is not less than 6 dB.

MODELING OF ADS-B MESSAGES TRANSMISSION VIA SATELLITE USING MIMO SYSTEMS

In January 2005 the requirement of ICAO ATC services in accordance with CNS/ATM concept was enhanced using ADS-B function. ADS-B is a surveillance technology for tracking aircraft as part of the Next Generation Air Transportation System (Minimum, 2002). When using ADS-B system both pilots and controllers will see the same radar picture. ADS-B systems based on low earth orbit satellites are of special interest (Osborne & Xie, 2009).

Satellite telecommunication systems are widely used in aviation due to advantages of satellite communication which is connected with possibility of operation with many airplanes at long distances and with independence of communication expenses on distances to airplanes (Doc. 9880-AN/466). During the communication through a wireless channel the transmitted signals are attenuated and faded due to multipath in the channel, making it difficult for a receiver to determine these signals.

Wireless communication using MIMO systems enables increased spectral efficiency for a given total transmit power in aeronautical satellite communication networks. Increased capacity is achieved by introducing additional spatial channels that are exploited by using space-time coding.

MIMO - is a technology used in wireless communication systems, which allows significantly improve the spectral efficiency of the system, the maximum data rate and the capacity of the network. The main way of achieving these benefits is data transmission from source to destination through a few radio connections. MIMO is based on the principle of multiplexing in space for many streaming data in one communication channel. Sensitivity to fading is reduced by the spatial diversity provided by multiple spatial paths (US Patent 5345599, 1994; US Patent 5515378, 1996).

The main advantages of MIMO channels over traditional Single-Input-Single-Output (SISO) channels are the array gain, the diversity gain, and the multiplexing gain. Array gain is the improvement in Signal to Interference-plus-Noise Ratio (SINR) obtained by coherently combining the signals on multiple transmit or multiple receive dimensions and is easily characterized as a shift of BER curve due to the gain in SINR (Alamouti, 1998; Alamouti, Tarokh, & Poon, 1998). Diversity gain is the improvement in link reliability obtained by receiving replicas of the information signal through independently fading links, branches, or dimensions. It is characterized by an steepen slope of the BER curve in the low BER region.

Orthogonal Space-Time Block Codes (OSTBCs) are a technique for MIMO wireless communications. They exploit full spatial diversity order and use symbol-wise maximum likelihood decoding. The combiner for OSTBC at the receiver side provides soft information of the transmitted symbols, which can be utilized for decoding or demodulation of an outer code (Alamouti, 1998; Gong & Letaief, 2002).

MIMO antenna unit is a special case of the Adaptive Antenna Array (AAA). The technology AAA suggests the use of intelligent algorithms implemented as digital signal processors. These algorithms divide the signals in accordance with the vectors of their propagation. Then a pattern of transmit antennas are adaptively adjusted. Widespread use of MIMO devices has substantially changed the world of wireless data transmission, allowed significantly increase the speed and range of channels without increasing the transmitter power. The lower correlation signals at the antenna are, the higher is the efficiency of this technology.

The aim of this investigation is: 1) to design the model of aeronautical satellite MIMO communication channel "Aircraft–Satellite–Ground Station" using MATLAB Simulink software; 2) to analyze MIMO 2×1 and 3×2 fading uplink/downlink channels with antenna diversity; 3) to compare results with AWGN uplink/downlink channels; 4) on the base of these models investigate channels integrity and receive dependences of a BER on the ratio E_b/N_0 (Kharchenko, Wang Bo, Grekhov, & Kostynska, 2014).

Model MIMO_Sat_MIMO_MPSK for Satellite MIMO Link

The model *MIMO_Sat_MIMO_MPSK* (Figure 26) comprises "Aircraft Transmitter" (Source of Data - Bernoulli Binary Generator, M-PSK modulator, High Power Amplifier), "Uplink/Downlink Paths" (MIMO 2×1 (Figure 27), 3×2 (Figure 28) fading uplink/downlink channels with antenna diversity, and AWGN uplink/downlink channels without antenna diversity), "Satellite Transponder"; "Ground Station Receiver" (Noise Temperature block, M-PSK demodulator), Error Rate Calculation block and Display.

Figure 26. Model MIMO_Sat_MIMO_MPSK of aircraft satellite link

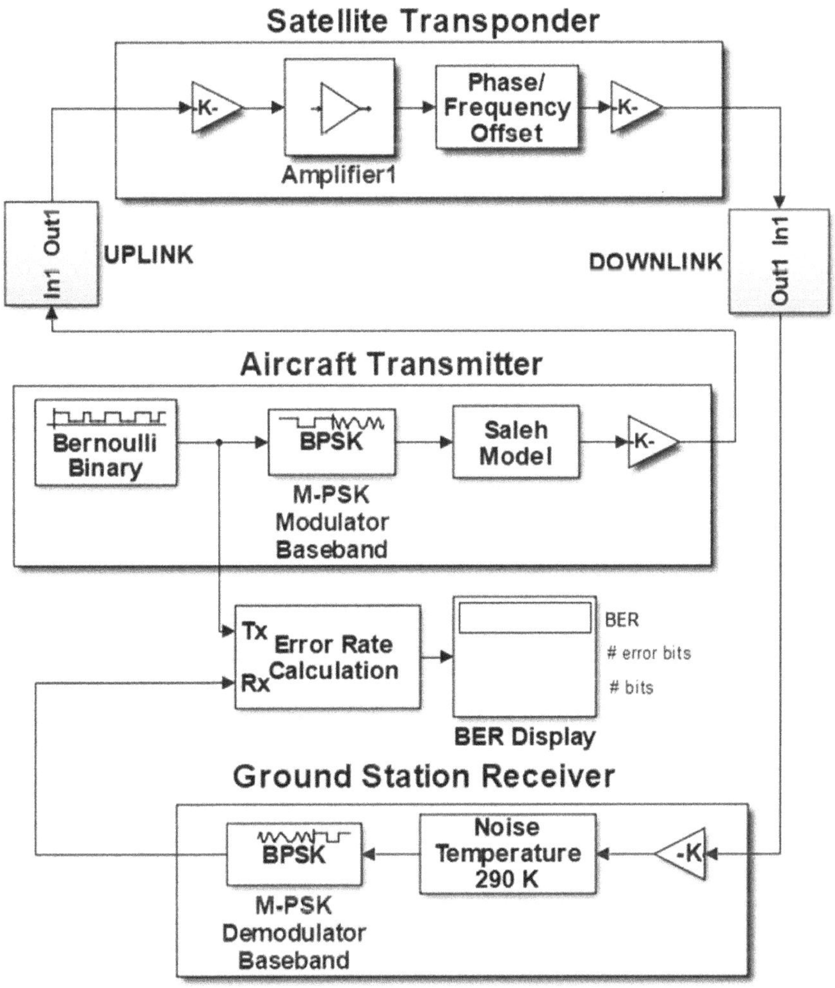

In this study during a simulation BPSK and QPSK modulation schemes were considered only. A BER was calculated as function of a ratio E_b/N_0 (the energy per bit to noise power spectral density ratio). The value of a ratio E_b/N_0 was changed symmetrically in all uplink and downlink channels. Noise temperatures in "Satellite Transponder" and "Ground Station Receiver" were taken as 290 K (typical noise level).

The High Power Amplifier block applies memoryless nonlinearity to complex baseband signal and provides five different methods for modeling the nonlinearity. In this paper results only for Saleh model with standard

Figure 27. MIMO 2×1 channel

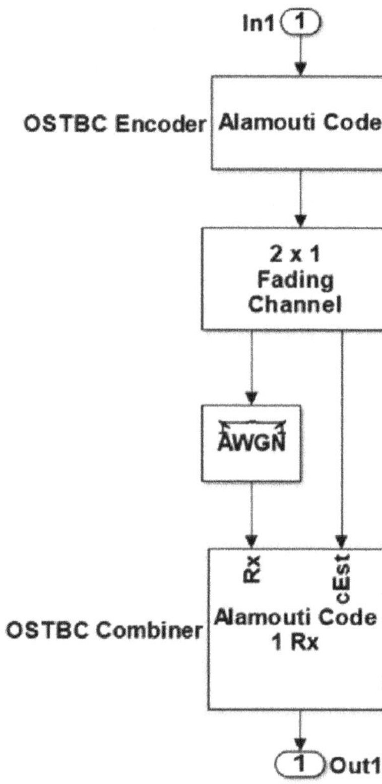

AM/AM and AM/PM parameters are given. HPA's backoff level is used to determine how close the satellite high power amplifier is driven to saturation. The following selected backoff is used to set the input and output gain of the Memoryless Nonlinearity block: 30 dB (negligible nonlinearity) - the average input power is 30 decibels below the input power that causes amplifier saturation (in this case AM/AM and AM/PM conversion is negligible).

Complex Baseband Amplifier block in satellite transponder generates a complex baseband model of an amplifier with thermal noise. It too simulates Saleh model with negligible nonlinearity.

The OSTBC Encoder block encodes an input symbol sequence using OSTBC. The block maps the input symbols block-wise and concatenates the output codeword matrices in the time domain. The MIMO 2x1 uplink/downlink channel uses two transmit antennas and one receive antenna. The MIMO 3×2 uplink/downlink channel uses three transmit antennas and two

Figure 28. MIMO 3×2 channel

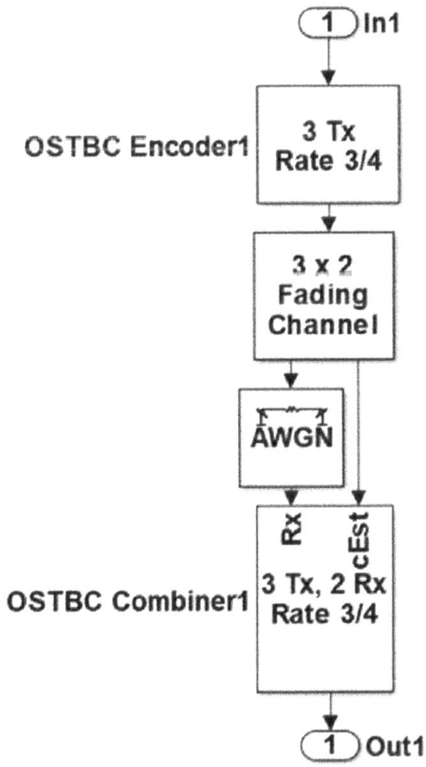

receive antennas. The 2x1 and 3×2 Fading Channels use the Multipath Rayleigh Fading Channel block to simulate the flat Rayleigh fading subchannel from one transmit antenna to the receive antenna.

The Maximum Doppler shift parameters of the Multipath Rayleigh Fading Channel blocks were set to 1 Hz. The reason for using this value is to make the MIMO channel behave like a quasi-static fading channel, i.e., it keeps constant during one frame transmission and varies along multiple frames.

The Initial Seed parameters of the Multipath Rayleigh Fading Channel blocks are set to different values in order to simulate independent fading subchannels. All subchannels have normalized gains.

The AWGN Channel block adds white Gaussian noises at the receiver side. The Mode parameter is set to the ratio Eb/N_0 mode, the number of bits per symbol is 1, the input signal power, referenced to 1 Ohm is set to 1 Watt, and the symbol period is 1 s.

The OSTBC Combiner block combines the input signal (from all of the receive antennas) and the channel estimate signal to extract the soft information of the symbols encoded by an OSTBC.

Aeronautical Satellite MIMO Channels Simulation and Results

During modeling two types of channels were considered: MIMO 2×1 and MIMO 3×2 fading uplink/downlink channels. Obtained dependencies of a BER on a ratio E_b/N_0 for these channels were compared with SISO channels - AWGN uplink/downlink channels (Figures 29-32). These dependencies allow clarifying the effect of each channel on the probability of errors during data transmission.

Figure 29. Dependencies of a BER for BPSK modulation on a ratio Eb/N0 in uplink/downlink channels: squares – MIMO 2×1 uplink/downlink channels, circles – AWGN uplink/downlink channels; satellite and ground receivers noise temperatures are 290 K, HPA's backoff level is 30 dB, phase and frequency offsets are equal to zero

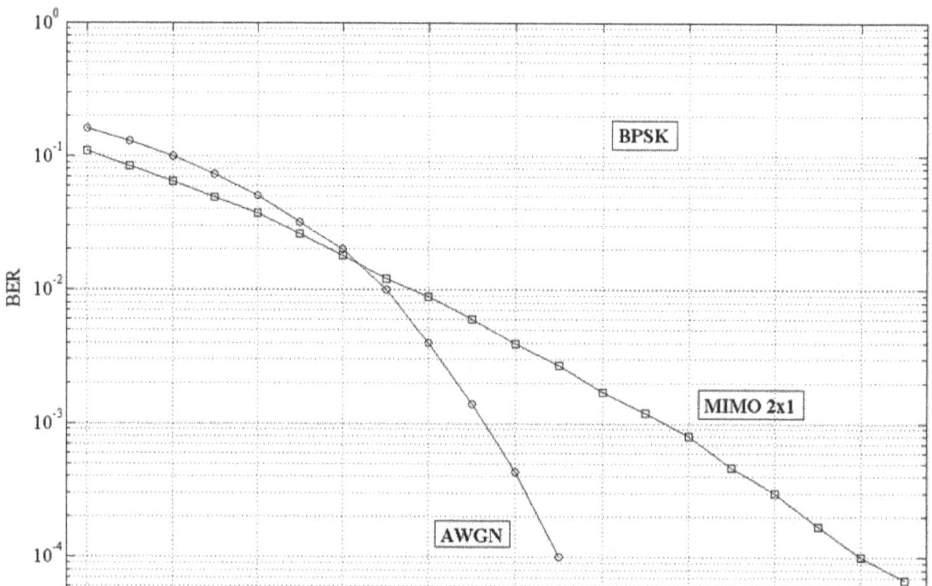

Alamouti scheme provides considerable signals diversity and high-speed transmission by introducing phase orthogonality between the signals simultaneously transmitted and the pairs of signals emitted successively from each antenna. The generalization of this code in case of more transmitter antennas led to the creation of another codes.

In Figure 29 the dependencies of an error probability for BPSK modulation on a ratio E_b/N_0 for MIMO 2×1 and AWGN channels are compared. For small values of a ratio E_b/N_0 MIMO 2×1 uplink/downlink channels give lower values of a BER, but for E_b/N_0 > 16 dB AWGN uplink/downlink channels give lower values of a BER.

In Figure 30 the dependencies of an error probability for QPSK modulation on a ratio E_b/N_0 for MIMO 2×1 and AWGN channels are compared. In this case a situation is similar - for E_b/N_0 > 22 dB AWGN uplink/downlink channels give lower values of a BER.

Figure 30. Dependencies of a BER for QPSK modulation on a ratio Eb/N0 in uplink/downlink channels: squares – MIMO 2×1 uplink/downlink channels, circles – AWGN uplink/downlink channels; satellite and ground receivers noise temperatures are 290 K, HPA's backoff level is 30 dB, phase and frequency offsets are equal to zero

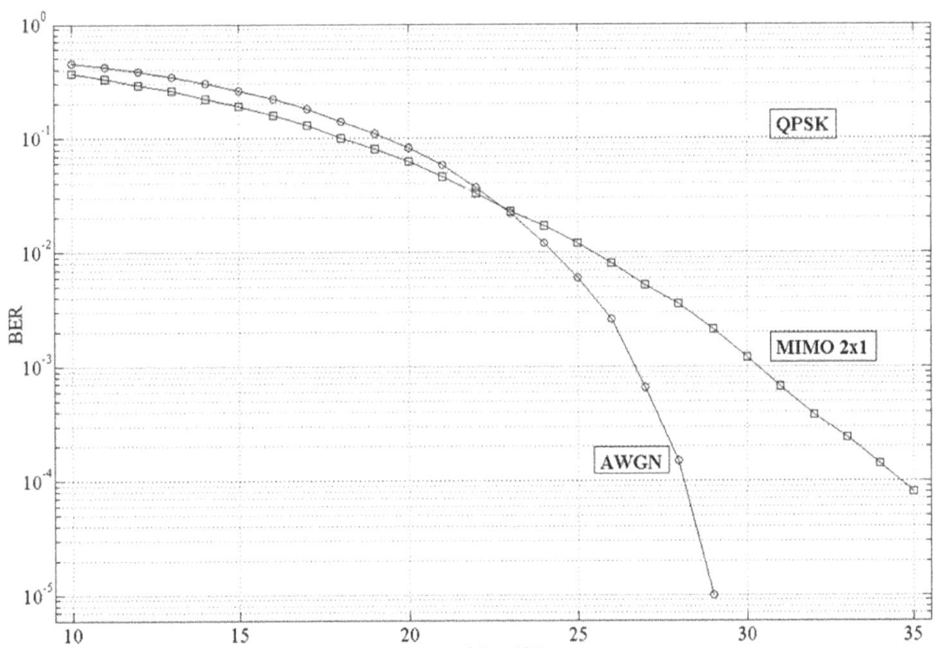

MIMO technology includes broadcasting signal through one channel by a number of transmitters and receivers. The more spaced antennas are, the stronger the improvement of productivity is. In MIMO 3×2 there are 3 transmitting and 2 receiving antennas. Due to such "compression" a channel capacity can be increased by two times or more.

In Figure 31 the dependencies of an error probability for BPSK modulation on a ratio E_b/N_0 for MIMO 3×2 and AWGN channels are compared. In this case a situation is quite different. In the whole range of changes E_b/N_0 the channel MIMO 3×2 provides lower values of a BER.

In Figure 32 the dependencies of an error probability for QPSK modulation on a ratio E_b/N_0 for MIMO 3×2 and AWGN channels are compared. The character of dependencies here are similar to a Figure 31.

Figure 31. Dependencies of a BER for BPSK modulation on a ratio Eb/N0 in uplink/downlink channels: squares – MIMO 3×2 uplink/downlink channels, circles – AWGN uplink/downlink channels; satellite and ground receivers noise temperatures are 290 K, HPA's backoff level is 30 dB, phase and frequency offsets are equal to zero

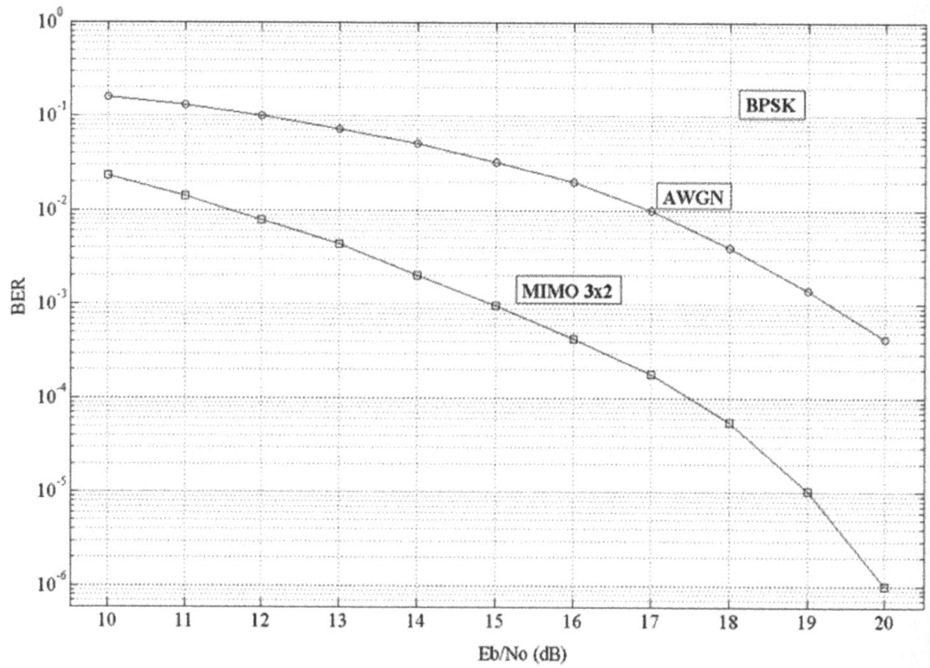

Figure 32. Dependencies of a BER for QPSK modulation on a ratio Eb/N0 in uplink/downlink channels: squares – MIMO 3×2 uplink/downlink channels, circles – AWGN uplink/downlink channels; satellite and ground receivers noise temperatures are 290 K, HPA's backoff level is 30 dB, phase and frequency offsets are equal to zero

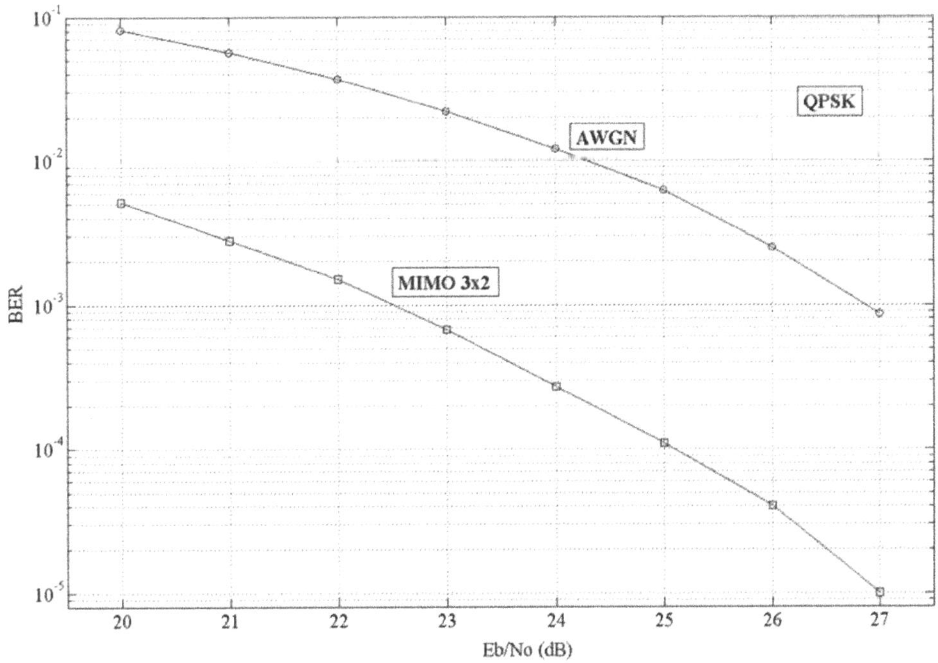

After modeling let's estimate the reduction of a BER for MIMO channels in comparison with SISO AWGN channels. From Figures 29-32 follows that:

ΔBER_{BPSK} (AWGN-MIMO 2×1) = 0,050, for E_b/N_0 = 10 dB;

ΔBER_{QPSK} (AWGN-MIMO 2×1) = 0,080, for E_b/N_0 = 10 dB;

ΔBER_{BPSK} (AWGN-MIMO 3×2) = 0,0043, for E_b/N_0 = 20 dB;

ΔBER_{QPSK} (AWGN-MIMO 3×2) = 0,077, for E_b/N_0 = 20 dB.

Received reduction of a BER is a rather big value. Note that the decrease in a BER for modulation QPSK is more than for BPSK modulation.

Conclusion

Original model *MIMO_Sat_MIMO_MPSK* of aeronautical satellite MIMO communication channel "Aircraft–Satellite–Ground Station" was created.

This model was used for transmission data modeling from aircraft to ground station via satellite using MIMO 2×1 and 3×2 channels with antenna diversity.

On the base of these models channel's integrity was investigated and dependences of a BER on the ratio E_b/N_0 were received.

COMPUTER MODELING OF RADIO FREQUENCY SATELLITE TRANSPONDER

Satellite telecommunications use artificial satellites, which relay analog and digital signals carrying voice, video, and data, in order to provide communication links between various points on Earth and aircraft. Satellite communication systems provide secure and essential communications, navigation, weather, and imaging services around the world. The important aspect of the satellite communications network is that it continues in operation under conditions when other methods of communications are inoperable.

The provision of safe, regular and efficient operation of air transport is the primary task of the ICAO. The ICAO is currently developing a satellite system, which will satisfy future needs of civil aviation in communications, navigation, radar surveillance and air traffic control. Today, the increase airport traffic is constrained by the fact that in order to determine the coordinates of object and display information on the radar screen required from 6 to 12 seconds, during which flying plane has time to change his position. Therefore air traffic controllers to provide safety flight have to increase time intervals between aircraft landing which leads to an incomplete use of airport infrastructure.

As a solution of increasing performance requirements in the developed system there are used the latest satellite and computer technology, data links and on-board avionics. Due to the high cost of the space segment for the construction of a satellite communication system such design principles are applied, which should allow the use a satellite transponder by a large number of ground users. Resource allocation of *satellite transponder* can be accomplished by forming multiple trunks by using multiple satellite transponders operating on different frequency bands (Roddy, 2006).

A transponder is a broadband radiofrequency channel used to amplify one or more carriers on the downlink side of a communication satellite. It is part of the microwave repeater and antenna system that is housed onboard the operating satellite. Even a small degradation of satellite link will affect the system data rate or coverage, both of which are related to operating expenses. It is crucial to have all of the system design parameters optimized before implementation. Furthermore, when things go wrong during construction or initial operation, a simulation model can be used to track down the offending element. The simulation will also be useful for pre-testing any corrective action before attempting it either in space or on the ground.

Modeling of satellite baseband channels was realized in papers (Kharchenko, Barabanov, & Grekhov, 2012a; 2012b; 2013a; 2013b). Issues related to the ADS-B messages transmission via radiofrequency satellite link still are not investigated in detail. The aim of this study is: 1) to design a model for "Aircraft-Satellite-Ground Station" link with radio frequency satellite transponder using MATLAB Simulink software; 2) to investigate dependencies of a BER on free space path losses for different carrier frequencies, noise temperatures and satellite transponder amplifier gain (Kharchenko, Wang Bo, Grekhov, & Stahovskyi, 2014).

Model RF_FSPL_Sat_FSPL_OQPSK for Radio Frequency Satellite Link

Satellite communication link was analyzed using original model *RF_FSPL_Sat_FSPL_OQPSK* designed on the base of MATLAB Simulink demo model *simrf_friis* (Figure 33). Model consists of "Aircraft Transmitter", "Uplink / Downlink", "Satellite Transponder", and "Ground Station Receiver".

"Aircraft Transmitter" comprise: Random Integer Generator block, which generates uniformly distributed random integers in the range [0, M-1], where M is the M-array number defined in the dialog box; OQPSK Modulator Baseband block, which modulates data using the Offset Quadrature Phase Shift Keying method; Raised Cosine Transmit Filter block, which upsamples and filters the input signal; and Transmitter Dish Antenna Gain block, which multiplies the input by a constant value (gain).

"Uplink/Downlink" consists of Normalization block, which multiplies the input by a constant value (gain); Free Space Path Losses block, which

Figure 33. Model RF_FSPL_Sat_FSPL_OQPSK of aircraft satellite link

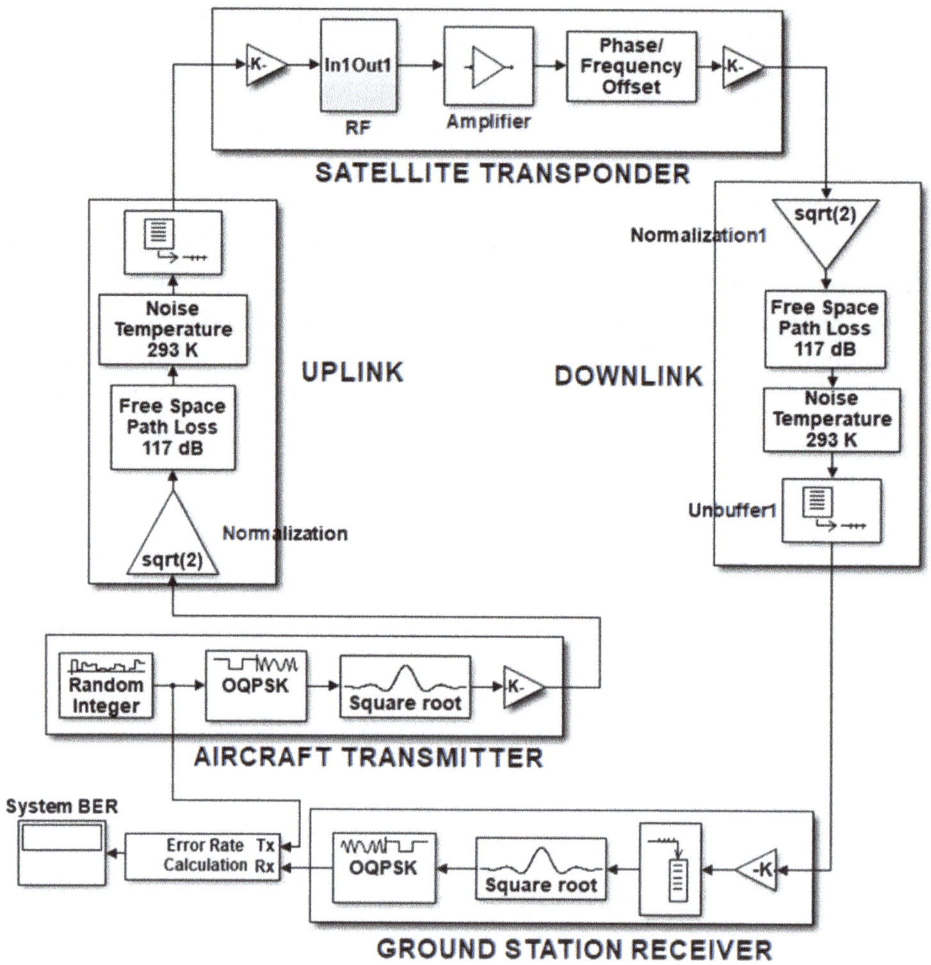

simulates the loss of signal power due to the distance between transmitter and receiver and reduces the amplitude of the input signal by an amount that is determined; Noise Floor block, which apply receiver thermal noise to complex baseband signal; Unbuffer block, which unbuffer input frame into sequence of scalar outputs.

"Satellite Transponder" consists of Receiver Dish Antenna Gain block; Radio Frequency (RF) block; Amplifier; Phase/Frequency Offset block, and Transmitter Dish Antenna Gain block.

Radio Frequency (RF) block (Figure 34) consists of SimRF Inport block, which convert Simulink input signal to SimRF signal; Low-Noise Amplifier (LNA) block; Mixer block, which models a mixer in the SimRF circuit-envelope simulation environment; SimRF Outport block, which converts SimRF signal to Simulink output signal; ContinuousWave block, which models a constant modulation on a carrier in the SimRF circuit-envelope simulation environment; Solver ConFigureuration block, which represents Physical Networks environment and solver conFigureuration; SimRF Parameters block, which specifies system-wide parameters for circuit-envelope analysis.

Amplifier block allows selecting linear and five different methods to model the nonlinear amplifier. In this paper the linear method was chosen. During modeling linear amplifier gain was 1 dB. Phase/Frequency Offset block applies a frequency and phase offset to the input signal. Transmitter Dish Antenna Gain block multiplies the input by a constant value (gain).

"Ground Receiver" comprises Receiver Dish Antenna Gain block and RF System Demodulator block. RF System Demodulator block consists of Receiver Buffer block, which buffer input sequence to smaller or larger frame size; Raised Cosine Receiver Filter block, which filters the input signal; OQPSK Demodulator block, which demodulates a signal that was modulated using the Offset Quadrature Phase Shift Keying method.

"Error Rate Calculation" block displays three-element vector consisting of the error rate, followed by the number of errors detected and the total number of symbols compared. "Error Rate Calculation" block shows the bit error rate as a percentage and should always equal 0 during investigations.

Figure 34. Scheme of RF block

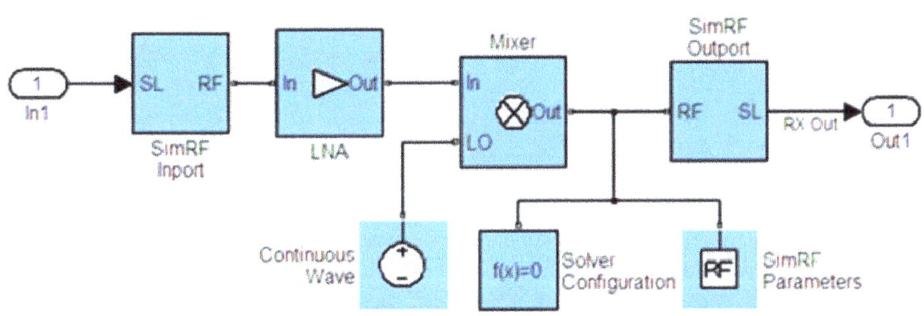

Aeronautical Satellite Channel Simulation and Results

A simulation was provided for carrier RF 1,6 GHz (L-band) and 3,1 GHz (S-band). LNA block has modeled linear amplifier with gain 100 dB, noise coefficient 6 dB and input/output impedance 50 Ohms. Mixer block had power gain -5 dB, noise coefficient 15 dB and input/output impedance 50 Ohms. SimRF Outport block had converted SimRF voltage/current to Simulink signal with carrier frequency 500 MHz. Continuous Wave block had carrier frequencies of Local Oscillator (LO) 1,1 GHz, 1,6 GHz and 2,6 GHz. Satellite transponder amplifier linear gain was 1 dB. All antennas gain was G=1.

In Figure 35 dependencies of a BER on free space path loss for different RF and LO are given. In Figure 36 dependencies of a BER on free space path

Figure 35. Dependencies of a BER on free space path loss for different RF and LO: squares: RF=1,6 GHz, LO=1,1 GHz; diamonds: RF=3,1 GHz, LO=2,6 GHz; noise temperature T=20 K; satellite transponder linear gain 1 dB; antennas gain G=1

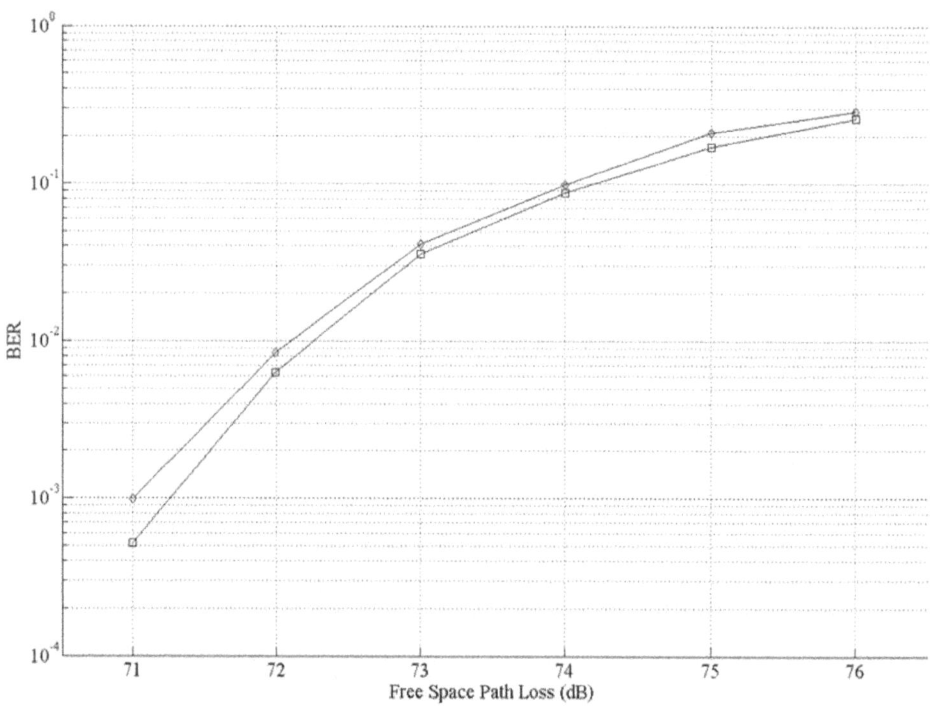

Figure 36. Dependencies of a BER on free space path loss for different noise temperatures: squares: T = 290K, asterisks: T = 100K, diamonds T = 20K ; RF=1,6 GHz, LO=1,1 GHz; satellite transponder linear gain 1 dB; antennas gain G=1

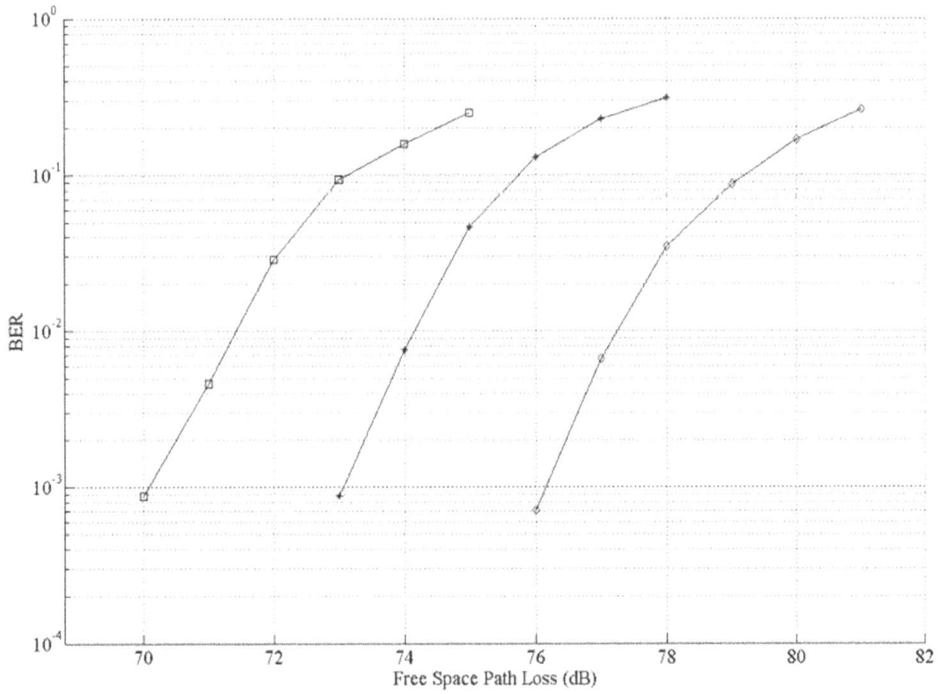

loss for different noise temperatures are shown. In Figure 37 dependencies of a BER on free space path loss for different satellite amplifier linear gain are given.

CONCLUSION

In multiple trunks satellite transponders operate on different frequency bands. The satellite transponder is a central element in the end-to-end communications link, illustrated in Figure 33. That's why it is important to investigate ADS-B messages transmission via radio frequency satellite link. The original model

Figure 37. Dependencies of a BER on free space path loss for different satellite amplifier linear gain: squares: 1 dB, asterisks: 2 dB, diamonds: 3 dB ; RF=1,6 GHz, LO=1,1 GHz; noise temperature T=20K; antennas gain G=1

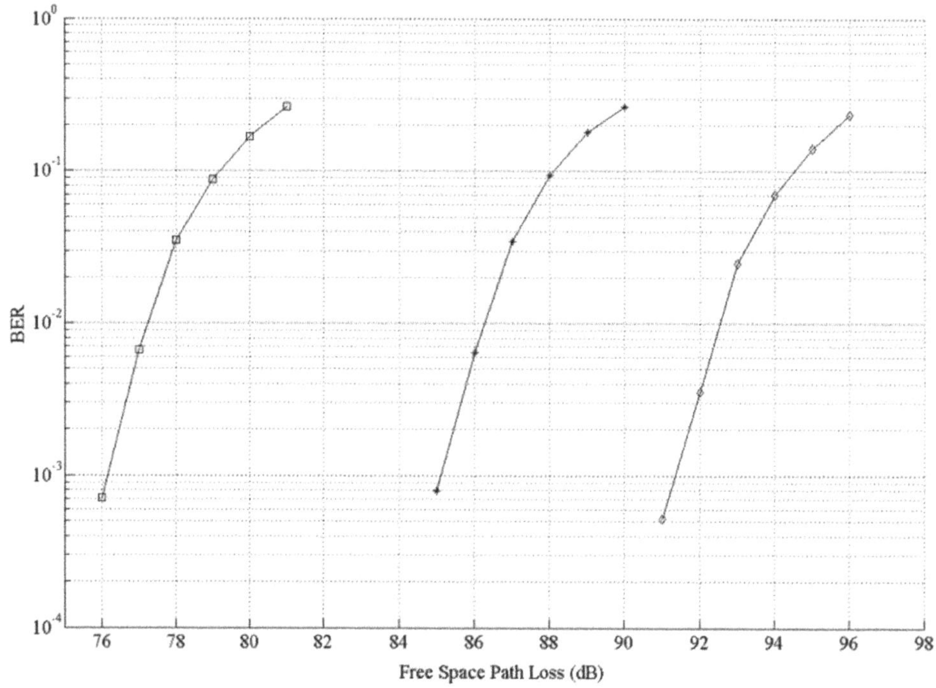

of RF communication channel "Aircraft-Satellite-Ground Station" was created in this study for the first time and fundamental dependencies of a BER on free space path losses, noise temperatures and satellite transponder amplifier gain were received.

REFERENCES

ADS-B. (2014). *Implementation and operations guidance document*. Edition 7.0.

Aircraft-to-Satellite Links Suitable for Transmitting USDA Forest Service Infrared Forest Fire Data. (2003). Retrieved from https://fsapps.nwcg.gov/nirops/docs/upload/USFS_satcom_report.pdf

Alamouti, S. M. (1998). A simple transmit diversity technique for wireless communications. *IEEE Journal on Selected Areas in Communications*, *16*(8), 1451–1458. doi:10.1109/49.730453

Alamouti, S. M., Tarokh, V., & Poon, P. (1998). Trellis-coded modulation and transmit diversity. Design criteria and performance evaluation. *Proceedings of IEEE International Conference on Universal Personal Communications (ICUPC'98)*, 703-707. Available: www.cdg.org/technology/cdma_technology/a_ross/Standards.asp

Castanet, L. (2011). Special issue on 'Channel modelling and propagation impairment simulation activities within the SatNEx project'. *International Journal of Satellite Communications and Networking*, *29*(1), 1–6. doi:10.1002at.949

Costello, D. J., Hagenauer, J., Imai, J., & Wicker, S. B. (1998). Applications of Error-Control Coding. *IEEE Transactions on Information Theory*, *44*(6), 2531–2560. doi:10.1109/18.720548

Doc. 9880-AN/466. (2005). Manual on detailed technical specifications for the aeronautical telecommunication network (ATN).

Doc. 9896. (2011). ICAO manual for the ATN using IPS standards and protocols.

Elbert, B., & Elanix, M. (2009). *Simulating the Performance of Communication Links with Satellite Transponders*. Retrieved from http://docplayer.net/34455242-Simulating-the-performance-0f-communication-links-with-satellite-transponders.html

EUROCONTROL. (2013). *ADS-B & WAM Implementation in Europe*. Retrieved from http://www.eurocontrol.int/publications/ads-b-wam-implementation-europe

Gong, Y., & Letaief, K. B. (2002). Concatenated space-time block coding with trellis coded modulation in fading channels. *IEEE Transactions on Wireless Communications*, *1*(4), 580–590. doi:10.1109/TWC.2002.804180

Gupta, L., Jain, R., & Vaszkun, G. (2015). Survey of important issues in UAV communication networks. *IEEE Communications Surveys and Tutorials*, 1-32.

Hanscom, A. F. B., & Bedford, M. A. (2013). *Unmanned aircraft system (UAS) service demand 2015–2035, literature review & projections of future usage. Res. Innov. Technol. Admin., U.S. Dept. Transp.* Washington, DC: Tech. Rep. DOT-VNTSC-DoD. Retrieved from https://www.icao.int/APAC/Documents/edocs/cns/ADSB_AIGD7.pdf

Huffman, W. C., & Pless, V. (2003). Fundamentals of Error-Correcting Codes. Cambridge.

IridiumN. E. X. T. (2018). Retrieved from https://www.iridiumnext.com

Iridium. (2007). *Manual for ICAO aeronautical mobile satellite (route) service. Part 2*. Retrieved from https://www.icao.int/safety/acp/Inactive%20working%20groups%20library/ACP-WG-M-Iridium-8/IRD-SWG08-IP05%20-%20AMS(R)S%20Manual%20Part%20II%20v4.0.pdf

Jawhara, I., Mohamed, N., Al-Jaroodi, J., Agrawald, D. P., & Zhange, S. (2017). Communication and networking of UAV-based systems: Classification and associated architectures. *Journal of Network and Computer Applications, 84*, 93–108. doi:10.1016/j.jnca.2017.02.008

Kharchenko, V. P., Barabanov, Y. M., & Grekhov, A. M. (2012a). Modeling of aviation telecommunications. *Proceedings of the National Aviation University, 50*(1), 5–13.

Kharchenko, V. P., Barabanov, Y. M., & Grekhov, A. M. (2012b). Modeling of satellite channel for transmission of ADS-B messages. *Proceedings of the National Aviation University, 52*(3), 9–14.

Kharchenko, V. P., Barabanov, Y. M., & Grekhov, A. M. (2013a). Modeling of ADS-B data transmission via satellite. *Aviation, 17*(3), 119–127. doi:10.3846/16487788.2013.840057

Kharchenko, V. P., Barabanov, Y. M., & Grekhov, A. M. (2013b). Modeling of "Satellite-to-Aircraft" link for ADS-B messages transmission. *Transport, 28*(4), 361–367. doi:10.3846/16484142.2013.864699

Kharchenko, V. P., Barabanov, Y. M., Grekhov, A. M., Bogunenko, M. M., & Rudnyk, J. N. (2013a). Performance analysis of "Aircraft-to-Satellite-to-Ground" link using forward error correction. *Proceedings of the National Aviation University, 54*(1), 7–14. doi:10.18372/2306-1472.54.3856

Kharchenko, V. P., Barabanov, Y. M., Grekhov, A. M., Bogunenko, M. M., & Rudnyk, J. N. (2013b). Error-control coding of ADS-B messages for Iridium satellites. *Proceedings of the National Aviation University*, *57*(4), 7–11. doi:10.18372/2306-1472.57.5510

Kharchenko, V. P., Bo, W., Grekhov, A. M., & Kostynska, J. V. (2014). Modeling of ADS-B messages transmission via satellite using MIMO systems. *Proceedings of the National Aviation University*, *60*(3), 15–21. doi:10.18372/2306-1472.60.7598

Kharchenko, V. P., Bo, W., Grekhov, A. M., & Stahovskyi, P. A. (2014). Computer modeling of radio frequency satellite transponder for transmission of ADS-B messages. *Proceedings of the National Aviation University*, *61*(4), 14–20. doi:10.18372/2306-1472.61.7581

Manual for ICAO Aeronautical Mobile Satellite (Route) Service Part 2-Iridium Draft V4.0. (2007).

Minimum Aviation System Performance Standards for Automatic Dependent Surveillance-Broadcast (ADS-B). RTCA. Inc. 2002. DO-242A.

Motlagh, N. H., Taleb, T., & Arouk, O. (2016). Low-altitude unmanned aerial vehicles-based internet of things services: Comprehensive survey and future perspectives. *IEEE Internet of Things Journal*, *3*(6), 899–922. doi:10.1109/JIOT.2016.2612119

Osborne, W. P., & Xie, Y. (2009). Propagation characterization of LEO/MEO satellite systems at 900-2100 MHz. *Conference Proceedings*. Retrieved from http://opensiuc.lib.siu.edu/ece_confs/52

Roddy, D. (2006). *Satellite communications* (4th ed.). The McGraw Hill Companies, Inc.

Saleh, A. A. M. (1981). Frequency-independent and frequency-dependent nonlinear models of TWT Amplifiers. *IEEE Transactions on Communications*, *29*(11), 1715–1720. doi:10.1109/TCOM.1981.1094911

Shannon, C. E. (1948). A mathematical theory of communication. *The Bell System Technical Journal*, *27*(3), 379–423. doi:10.1002/j.1538-7305.1948.tb01338.x

Sklar, B. (2001). *Digital vommunications*. Prentice Hall PTR.

STANAG 4607/AEDP-7. (2010). NATO ground moving target indicator format. (GMTIF).

STANAG 4609/AEDP-8. (2009). NATO digital motion imagery format.

STANAG 7023/AEDP-9. (2009). NATO primary image format.

Szabolcsi, R. (2015). Establishment of the unmanned aerial vehicle system cluster: Mission, motivation, vision, goals. *Review of the Air Force Academy*, *28*(1), 27–30.

Tanil, C., Warty, C., & Obiedat, E. (2013). Collaborative mission planning for UAV cluster to optimize relay distance. *2013 IEEE Aerospace Conference*. 10.1109/AERO.2013.6497407

UAS. (2012). *EUROCAE WG-73: Unmanned aircraft systems*. Retrieved from https://www.icao.int/Meetings/UAS/Documents/05_Kallevig_EUROCAE_WG73_Status_Update_2012-04-18_LIMA.pdf

UAV. (2012). *Unmanned aircraft systems integration in the national airspace system*. Retrieved from https://ntrs.nasa.gov/search.jsp?R=20120015471#

US Department of Transportation Federal Aviation Authority. (2013). *Integration of civil unmanned aircraft systems (UAS) in the national airspace system (NAS) roadmap*. Retrieved from http://www.faa.gov/uas/media/uas_roadmap_2013.pdf

US Patent 5345599. (1994). *Increasing capacity in wireless broadcast systems using Distributed Transmission/Directional Reception (DTDR)*. US Patent Office.

US Patent 5515378. (1996). *Spatial division multiple access wireless communication systems*. US Patent Office.

Vikranth, D. R. (2017). UAV Swarm co-ordination and control using grossberg neural network. *International Journal of Computer Science Trends and Technology*, *5*(4), 1–7.

Viterbi, A. (1971). Convolutional Codes and their Performance in Communications Systems. *IEEE Transactions on Communication Technology*, *5*(5), 751–772. doi:10.1109/TCOM.1971.1090700

Chapter 2
Parameters Estimation of Aircraft and RPAS Satellite Channels Based on IEEE 802.11a Standard

ABSTRACT

This chapter deals with parameters estimation of satellite channels based on IEEE 802.11a standard. Dependencies of a signal to noise ratio on free space path loss for different types of modulation (BPSK, QPSK, 16QAM, 64QAM), noise temperatures, number of OFDM symbols, Doppler frequency offsets, satellite amplifier gain, and aircraft antenna diameter were received using model "OFDM_FSPL_Sat_FSPL_802.11a." A method for parameters estimation of satellite OFDM communication channel was proposed. The spectrums and signals constellations of received signals were compared for different types of the amplifier nonlinearity. The developed model allows predicting spectral regrowth of digitally modulated OFDM signals due to the amplifier nonlinearity. Channel parameters were received for the Rayleigh and Rician fading, different types of Doppler spectrum, the gain of multipath channels, the delay time of message flow using models "OFDM_Multipath_Sat_Multipath_802.11a," "OFDM_AWGN_Sat_Multipath_802.11a," and "OFDM_FSPL_Sat_Rician_802.11a."

DOI: 10.4018/978-1-5225-8214-4.ch002

INTRODUCTION

WiFi Standards Group IEEE 802.11

The IEEE is a developer of standards for WiFi 802.11. IEEE 802.11 is the basic standard for Wi-Fi networks, which defines a set of protocols for the lowest data transfer rates.

IEEE 802.11b. Was ratified in 1999, describes large transmission speeds, and introduces many technological limitations. This standard was originally called Wi-Fi. Frequency channels in the 2.4 GHz spectrum are used. Used radio frequency technology: DSSS. Coding: Barker 11 and CCK (complementary code keying). Modulation: DBPSK and DQPSK. The maximum data transfer rates in the channel are 1, 2, 5.5, 11 Mbps.

IEEE 802.11a. Was ratified in 1999, describes significantly higher transfer speeds than 802.11b. Frequency channels in the frequency spectrum of 5 GHz are used. Protocol is not compatible with 802.11b. Used radio frequency technology: OFDM. Coding: Convoltional Coding. Modulation: BPSK, QPSK, 16QAM, 64QAM. The maximum data rates in the channel are 6, 9, 12, 18, 24, 36, 48, 54 Mbps.

*IEEE 802. 11g.*Was ratified in 2003, describes data transfer rates equivalent to 802.11a. Frequency channels in the 2.4 GHz spectrum are used. The protocol is compatible with 802.11b. Used radio frequency technologies: DSSS and OFDM. Coding: Barker 11 and CCK. Modulation: DBPSK and DQPSK. Maximum data transfer rates in the channel: 1, 2, 5.5, 11 Mbps on DSSS and 6, 9, 12, 18, 24, 36, 48, 54 Mbps on OFDM.

IEEE 802.11n. Was released in 2009 and is the most advanced commercial WiFi standard. Frequency channels are used in the frequency spectra of WiFi 2.4GHz and 5GHz. Compatible with 11b/11a/11g. Supports frequency channels with a width of 20MHz and 40MHz (2x20MHz). Used radio frequency technology: OFDM. The OFDM MIMO technology is used up to the 4x4 level. At the same time, a minimum of 2x Transmitter per Access Point and 1x Transmitter per user device.

In addition to the basic standards of WiFi 802.11a, b, g, n, there are used additional standards for the implementation of various service functions:

IEEE 802.11d. For the adaptation of various WiFi standard devices to the specific conditions of the country. Within the regulatory field of each state, the ranges are often different and may be different even in depending on the geographic location. The IEEE 802.11d WiFi standard allows adjusting bandwidths in devices from different manufacturers with the help of special options entered in the protocols of media access control.

IEEE 802.11e. Describes QoS quality classes for the transfer of various media files and, in general, different media content. Adapting the MAC layer for 802.11e, determines the quality, for example, of simultaneous transmission of sound and image.

IEEE 802.11f. It is aimed at unifying the parameters of Wi-Fi access points of various manufacturers. The standard allows the user to work with different networks when moving between the zones of operation of individual networks.

IEEE 802.11h. Used to prevent problems with meteorological and military radars by dynamically reducing the radiated power of Wi-Fi equipment or dynamically changing to another frequency channel when a trigger signal is detected (in most European countries, ground stations for tracking meteorological satellites and communications satellites, as well as military radar in the bands close to 5 MHz). This standard is an essential requirement of ETSI for equipment that is approved for operation in the territory of the European Union.

IEEE 802.11i. This standard has developed new methods for protecting Wi-Fi networks, implemented at both the physical and software levels. Currently, Wi-Fi Protected Access (WPA) algorithms are recommended for Wi-Fi 802.11 networks. They also provide compatibility between wireless devices of different standards and different modifications. WPA protocols use the advanced RC4 encryption scheme and the mandatory authentication method using EAP. The stability and security of modern Wi-Fi networks is determined by privacy and encryption protocols (RSNA, TKIP, CCMP, AES).

IEEE 802.11k. This standard is aimed at implementing load balancing in the radio subsystem of the Wi-Fi network. In a wireless local area network, the subscriber unit typically connects to the access point that provides the strongest signal. Often this leads to network congestion at one point, when many users join one Access Point at once. To control such situations, the 802.11k standard offers a mechanism that limits the number of subscribers connected to one Access Point, and enables the creation of conditions under which new users will join another Access Point even though weaker signal from it.

IEEE 802.11m. Corrections for the entire group of 802.11 standards are summarized in a separate document with the generic name 802.11m. The first release of 802.11m was in 2007, further in 2011, and so on.

IEEE 802.11p. It determines the interaction of Wi-Fi equipment moving at a speed of up to 200 km/h past fixed WiFi access points, remote at a distance of up to 1 km. Part of the Wireless Access in Vehicular Environment (WAVE) standard. WAVE standards define the architecture and an additional set of service functions and interfaces that provide a secure mechanism for radio communication between moving vehicles. These standards are developed for applications such as traffic management, traffic safety control, automated collection of payments, navigation and routing of vehicles, etc.

IEEE 802.11r. It determines the fast automatic roaming of Wi-Fi devices when moving from the coverage area of one WiFi Access Point to the coverage area of the other. This standard is focused on the implementation of Mobility and it is important for mobile/wearable devices with Wi-Fi, for example, smartphones, tablet computers, Wi-Fi IP phones, etc. Devices with 802.11r support can register in advance with neighboring WiFi Access Points and perform the reconnection process in automatic mode. This significantly reduces the time when the subscriber is not available in Wi-Fi networks.

IEEE 802.11s. The standard for the implementation of full-mesh networks (Wireless Mesh), where any device can serve as a router or an access point. If the nearest access point is overloaded, the data is redirected to the nearest unloaded node. In this case, the packet of data is transmitted from one node to another until it reaches its final destination. In this standard, new protocols are introduced at MAC and PHY levels that support broadcast and multicast delivery over a self-configuring Wi-Fi access point system. To this end, the standard introduces a four-address frame format.

IEEE 802.11t. The standard is designed to institutionalize the testing process of the IEEE 802.11 standard solutions. Methods of testing, methods of measurement and processing of results (treatment), requirements to the test equipment are described.

IEEE 802.11u. Defines procedures for interworking Wi-Fi networks with external networks. The standard should define access protocols, priority protocols and a ban on working with external networks.

IEEE 802.11v. The standard developed amendments aimed improving the network management systems standard IEEE 802.11. Upgrading to MAC and PHY layers should allow centralizing and streamlining the configuration of client devices connected to the network.

IEEE 802.11y. An additional communication standard for the frequency range 3.65-3.70 GHz. Designed for devices of the latest generation, working with external antennas at speeds up to 54 Mbps at a distance of up to 5 km in open space. The standard is not complete.

IEEE 802.11w. Identifies methods and procedures for improving the protection and security of the medium access control layer. The standard introduces the protection of the management frame (MFP: Management Frame Protection), and additional security measures can neutralize external attacks, such as, for example, DoS.

IEEE 802.11ac. A new WiFi standard that only works in the 5GHz frequency band and provides significantly higher speeds for both the individual WiFi client and the WiFi Access Point. 802.11ac is the development of widespread 802.11n technology. Thc main advantages of this standard are the high transmission rates in the radio channel, as well as the more sophisticated mechanisms for monitoring the active and passive state of client devices. All this together leads to significant savings in battery life of the mobile device.

IEEE 802.11ad. This standard in addition to two traditional frequencies has another frequency in the range 60 GHz. While in the frequency bands 2.4 GHz and 5 GHz directional pattern of antenna is close to circular, then in the range of 60 GHz, it is highly directional. Introduction of at 60 GHz is made to provide high-speed wireless direct communication between different devices at a transfer rate of 7 Gbit/s, but ensure the transfer of uncompressed video-awning of high resolution.

OFDM Channel

OFDM is a digital modulation scheme that uses a large number of closely spaced orthogonal subcarriers. Each subcarrier is modulated in a conventional modulation scheme at a low symbol speed, preserving the overall data transfer rate. In practice, OFDM signals are obtained by using Fast Fourier Transform (FFT). Parallel data transmission with frequency separation was invented back in the mid-1960s and was used, like most of today's known technologies, first only in military systems. In those days, the military, using OFDM, already carried out parallel data transmission using 34 subcarriers. In the 1980s, OFDM was used primarily in high-speed modems and digital mobile networks. In the 1990s, OFDM was used in digital broadcasting, in terrestrial broadcasting, in the transmission of video, as well as in the known technologies of the last mile. For a long time OFDM was not very widespread in other communication systems due to the complex technical implementation. Solution of the problem of OFDM signal forming by analog methods is very problematic. The development of computer systems and methods of digital signal processing makes it possible to use OFDM technology today in a wide variety of systems - from radio to wire lines and even fiber-optic ones. Despite the fact that the method literally stands for multiplexing with orthogonal frequency separation, it is primarily referred to as digital modulation methods.

The principle of OFDM transmission: OFDM is a multiplexing technique that divides a channel with a higher relative data rate into several orthogonal subchannels with a lower data rate. At a high data rate, the duration T_s of the symbol is small. Therefore, intersymbol interference due to multipath propagation of the received signal leads to the fact that the duration of the symbol T_s is less than the maximum channel delay. To reduce this effect, a narrowband channel is required, but for high data rates it is necessary to have a broadband channel. To overcome this problem, the common frequency band can be divided into several parallel narrow-band subcarriers. In this case, a block of N consecutive symbols with duration T_s is transformed into a block of N parallel symbols with a duration $T = N \times T_s$. The essence is that the duration of the new symbol of each subcarrier is greater than the maximum channel delay, $T > T_{max}$. Thus, a higher data rate is achieved when there are many subcarriers with a low data rate.

Therefore, for the purpose of creating an OFDM symbol, the conversion of N consecutive symbols into N parallel data symbols is used. Then, each parallel symbol is modulated on different orthogonal frequency subcarriers and combined into an OFDM symbol. This is done using an Inverse Fast Fourier Transform (IFFT). The advantage of using a fast Fourier transform is that the system does not need N oscillators to transmit N subcarriers.

The Cyclic Prefix: Is used in the transmission of OFDM signals as a guard interval and is a copy of the end of the symbol that is inserted at the beginning of each OFDM symbol. The guard interval is used to reduce intersymbol interference due to multipath propagation. These delays create noise and distort the beginning of the next character. One of the ways to overcome this problem is the time shift of the second symbol in relation to the first. But the existence of empty gaps with continuous data transmission is not desirable. In order to solve this problem, a copy of the last part of the symbol is inserted at the beginning of each character. This procedure is called adding a cyclic prefix. The cyclic prefix is added after the inverse fast Fourier transform in the transmitter, and at the receiver the cyclic prefix is removed to obtain the original signal.

OFDM Signal Parameters: IEEE 802.11a standard uses 64 OFDM subcarriers, which include 48 data subcarriers and 4 pilot subcarriers. Four pilot signals are used to track frequency shift and phase noise. Long training symbols (T_1 and T_2), which are located after short training symbols, are used for channel estimation. GI2 is used as a guard interval for a long training training sequence, and GI is used as a guard interval for OFDM symbols. The total duration of the training sequence is 16 μs.

OFDM Signal Formation: To convert a block of transmitted data from N consecutive symbols (each duration T_s) to a block of N parallel symbols (each duration $T = N\ T_s$), a modulation bank is used. The output vector is a set of subcarriers with data in the form of OFDM symbols for each frame. Each OFDM symbol in the IEEE 802.11a standard has four pilot subcarriers. Pilot signals are used to track frequency offset and phase noise. Pilot signals are located on the subcarriers -21, -7, 7 and 21. The pseudo-random noise sequence generator unit creates pilot subcarriers. The sampling time and the number of samples per frame for this generator are determined as follows:

Sampling time = block length/number of OFDM symbols in a frame,

Number of samples per frame = number of OFDM symbols in a frame.

The preamble is used to improve the accuracy of the channel estimation. Four long training OFDM symbols are used instead of two long training symbols in this system. Long training symbols consist of 53 subcarriers. The assembly of OFDM frames is used to insert pilot and training symbols into OFDM symbols. Four pilot sequences are inserted between the subcarriers, and then a training sequence is added to the subcarriers. In the filling block, the input vector expands along the columns. The added values are zero and inserted at the end of the columns to the dimension determined by the number of given points in the inverse fast Fourier transform block. The inverse Fourier transform transforms the frequency domain of the data stream into a corresponding time domain. Then a parallel-to-serial converter is used to send the sample in the time domain for one character. The inverse fast Fourier transform block distributes various orthogonal subcarriers for the transmitted bits, and thus no interference between subcarriers is observed. Subcarriers can be located close enough to each other, which increases the channel capacity.

The cyclic prefix is used as a guard interval to mitigate the effect of intersymbol interference due to multipath propagation. To insert a cyclic prefix for 16 subcarriers, a selective block is used.

The multiplexing unit is the last unit in the transmitter in which the signal is converted from parallel code to serial and the sample is transmitted in the time domain for one character.

Advantages of OFDM: The ability to withstand complex conditions in the radio channel, first to eliminate inter-symbol interference and to deal with narrow-band interference high spectral efficiency. If the number of subcarriers approaches infinity, OFDM systems show an almost doubled spectral efficiency compared to traditional frequency-division multiple-channel systems. OFDM adaptability - the method possibility to use different modulation schemes for different subcarriers, which allows to adapt to the conditions of signal propagation and to various requirements to the quality of the received signal. OFDM has simple implementation by digital processing methods. OFDM ability to withstand interference between subcarriers, which causes good multipath performance.

Disadvantages of OFDM: It requires high-precision synchronization in time and frequency. The OFDM signal has a relatively high value of the peak factor, which leads to excessive energy costs. The use of protective intervals reduces the spectral efficiency of the method. The method is sensitive to the Doppler's effect, which places additional difficulties in its application in mobile networks.

The current use of OFDM: To date, the most widely known application of OFDM modulation in wireless communication systems is Wi-Fi, WiMax, LTE, in terrestrial DVB-T digital television systems, in DVB-C cable television systems, in ADSL technology and this is not all examples.

Nonlinearity

All amplifiers distort the signals they need to amplify. Distortions worsen the transmitted signal, which makes it more difficult for correct transmission. Distortions can create problems for users of one channel, but also for users in adjacent channels. The ideal amplifier produces an ideal copy of the input signal multiplied by a scalar value. In real amplifiers, various nonlinearities appear. They are usually described by the amplitude and phase transfer characteristics of the amplifier. The first is often called AM/AM distortion (Amplitude Modulation/Amplitude Modulation), and the second is AM/PM distortion (Amplitude Modulation/Phase Modulation).

AM/AM Distortions: Are more effective at higher input power levels. In this case, the amplifier starts to enter saturation mode, as a result of which the output current depends little on the level of the input signal. Compression distortions have a significant effect on the noise immunity of QAM signals. To minimize their influence, the linear paths of the transmitter and receiver are designed, as a rule, in such a way that the amount of compression does not exceed 0.1-0.15 dB.

AM/PM Distortions: AM/AM distortions are the cause of another type of nonlinear distortions - AM/PM phase distortions. As a result of compression of the output signal, a constant component appears in its spectrum, which leads to a displacement of the operating point of the amplifier. This leads to the appearance of phase mismatch between the input and output signals.

A nonlinear distortion is the appearance of new frequency components in the output signal spectrum in comparison with the input spectrum. The cause of nonlinear distortion is the nonlinearity of the amplitude characteristic. The magnitude of the nonlinear distortion depends significantly on the amplifier supply voltage and on the voltage of the output signal. To evaluate the nonlinear distortion of the output signal, various nonlinear parameters are used.

Second-Order Intersection Point IP2: Is a measure of linearity that quantifies second-order distortion generated by non-linear systems and devices. The second-order interception point is the point of output power at which the extrapolated lines of the first and second orders intersect on the graph. IP2 is associated with a third-order intercept point.

Third-Order Intercept Point IP3: Is a measure for weakly nonlinear systems and devices. It is based on the idea that the nonlinearity of a device can be modeled using a low order polynomial obtained by Taylor series expansion. The third-order interception point refers to non-linear products caused by a third-order nonlinear term to a linearly amplified signal, in contrast to a second-order intercept point that uses second-order terms. The interception point is a purely mathematical concept and does not correspond to the actually existing physical power level.

The level of third-order non-linear distortions depends on the level of the output signal. Therefore, to evaluate the amplifier by the level of distortions of the third order, enter the parameter of the intersection point of third-order products - IP3. This parameter allows characterizing the amplifier itself, regardless of the level of the useful signal. The application of the IP3 point is possible only in the mode of small signals, when the amplitude characteristic can be approximated by a polynomial of the third degree $y = ax + bx^3$. When approaching the point of inflection of the amplitude characteristic, the level of third-order nonlinearity products increases sharply and calculations using IP3 parameter will be incorrect.

When a nonlinearity is absent, the dependence of the output signal level on the input signal is linear. The dependence of third-order nonlinear distortions on the level of the input signal obeys a cubic law. If these dependencies are expressed on a logarithmic scale (decibels), then both dependences will be straight lines. In this case, the slope of the direct third-order nonlinearity product will be three times larger than the angle of inclination of the output power versus the input power. The higher the quality of the amplifier, the lower the initial level of third-order nonlinearity products and the greater the value of the IP3 point is.

Fading

Fading is the amplitude and phase variation of a signal due to multipath, to the movement of the transmitter, receiver or surrounding objects in the radio communication system and/or signal propagation through an inhomogeneous medium. The fading can be considered as the result of the signal multiplicative interference.

Large-scale fading (shading) is long-term change in the average signal level caused by mobile devices in the shadow of the surrounding objects. Shadow fading is long-term shadow fading due to fluctuations in the power of the radio signal due to collisions with obstacles of the terrain (such as hills, artificial constructions, buildings). The measured signal power substantially varies in different places, although measured on the same radial distance from the transmitter. Many empirical studies show that the obtained mean power fluctuates relatively average power with a logarithmic normal distribution and can be modeled by Gaussian distribution. The shadow complicates the planning of cellular communication.

Small-scale fading (multipath spread) is short-term amplitude signal fluctuation caused by a local multipath signal propagation. Multipath fading is the small-scale fading that describes short-term rapid amplitude oscillations of received signal for a short period. Actual power received on a much smaller distance varies considerably due to interference of several signals that follow in several ways to the receiver. A direct ray actually consists of many rays due to multiple scattering of obstacles along its path. Each of these beams appearing at the receiver will be randomly differ in amplitude and phase due to scattering. Small-scale fading may be further classified into a flat (or frequency-non-selective) fading and frequency selective fading.

Flat Fading: Small-scale fading is determined as flat, if the received multi-beam components of the symbol do not go beyond symbol duration. If the components of the multipath spread in relation to the main component is less than time duration of the symbol, it is considered that the channel undergoes flat fading. In a channel with a flat fading, the intersymbol interference (ISI) is absent. The channel has a constant gain and linear phase response in the passband that is more than bandwidth of the transmitted signal. The channel does not cause any nonlinear distortions due to time dispersion. However, the received signal power usually changes slowly in time due to oscillations caused by multipath signal propagation. In a channel with flat fading the bandwidth transmission of the transmitted signal B_s is much less than the coherence width B_c of the channel. The symbol period of the transmitted signal is much greater than the delay spread. Delay distribution is a variation of multiple distribution delays of scattered rays.

Frequency Selective Fading: Small-scale fading is defined as frequency-selective if received multipath components of the symbol go beyond symbol duration limits. The effect of multipath fading upon signals receiving depends on the bandwidth of the signal. With a relatively large bandwidth, different parts of the spectrum of the transmitted signal are weakened in different ways. This is manifested in the intersymbol interference. If the delay of multipath components in relation to the main component is greater than the symbol duration, it is considered that the channel has frequency-selective fading. The received signal includes several versions of the same symbol, each of which is weakened (damped) and delayed. The received signal is distorted by producing ISI. The channel has a constant gain and linear phase response by band transmission, which is much smaller than the width band of the transmitted signal. Bandwidth of the transmitted signal B_s is much larger than the width of channel coherence B_c. The symbol period of the transmitted signal is much less than delay spread.

For flat fading, it is found that the multipath distribution can be modeled using statistics of Rayleigh/Rice. The Rayleigh fading is considered as a reasonable model for urban environments, where there are many objects in the environment that dissipate the radio signal before it receiving. There is no dominant spread over line of sight between transmitter and receiver. If the medium is such that in addition to scattering in the receiver is observed strongly dominant signal, usually called LOS, then the average value of a random process will no longer be equal to zero, instead there will be a power level of the dominant path. This situation may be better modeled as the fading of Rice.

How quickly fading occurs in a channel will depend on how much fast receiver and/or transmitter move relative to each other. Movement causes a Doppler shift in the received signal components - changing the frequency of the wave for the receiver, moving relative to the transmitter.

Slow or fast fading - depends on the coherence time T_c. The coherence time is a measure of the period during which the fading process is correlated. T_c is related to the delay spread.

Fading is considered slow if the duration symbol T_s is less than the coherence time.

Doppler Frequency Shift

A method of OFDM signal formation involves the sum of harmonic oscillations set, whose frequencies are selected based on the orthogonality condition. In order to fulfill the orthogonality condition, it is necessary that the integer number of harmonic oscillation periods of each frequency subcarriers be laid for the duration of the signal, and that the cross-correlation coefficient between adjacent subcarriers is zero. Such a method of signal formation makes it possible to increase its spectral density by superimposing the spectra of neighboring subcarriers. At the same time, adjacent subcarriers do not interfere. Violation of the orthogonality condition leads to inter-frequency interference. The reason for the violation of orthogonality can be the shift of the signal carrier frequency as a result of the radiating/receiving station motion.

If we consider the Doppler's effect in relation to a narrow-band process, a change in the relative velocity will lead to an increase or decrease in the signal frequency (Doppler frequency shift), due to a change in the period of the signal. The same effect will be manifested for the OFDM signal.

In addition to the Doppler frequency shift, the Doppler spreading of the OFDM signal spectrum takes place, which leads to time scaling of the signal. The spreading of the spectrum occurs as a consequence of the uneven frequency shift of each frequency subchannel included in the OFDM signal. Subchannels frequency with a higher frequency acquire a larger frequency offset, as a result of which the signal band will expand and this will lead to a decrease in the signal duration.

Another effect, called Doppler spectrum scattering, occurs when the propagation channel is characterized by the presence of multiple reflectors. In the real operating conditions the propagation channel parameters change with time because of the transmitter/receiver and the surrounding objects moving. The rate of change in the signal level is described by Doppler scattering, which can be represented as a random phase noise that varies in time. For a narrow-band process, Doppler scattering is defined as the width of the spectrum of the received signal. In the presence of this effect, a signal with a "smeared out" spectrum near the carrier wave is received. This can be applied to the OFDM signal, since on the carrier frequency it can be considered as a narrowband process. The frequency shift, both to the smaller and to the larger sides, is explained by the difference in the mutual radial velocities of the transmitter/receiver and the reflectors. As a result, the direct signal (if any) and all the reflected signals are summed, and the spectrum of the final signal becomes "smeared" at the receiver.

METHOD FOR OFDM LINK PARAMETERS ESTIMATION

IEEE 802.11a standard will be considered in this paragraph. Recall that IEEE 802.11a is one of the first high-speed wireless network standards: used radio frequency technology – OFDM; coding - Convolutional; modulation - BPSK, QPSK, 16-QAM, 64-QAM; the maximum data rates in the channel are 6, 9, 12, 18, 24, 36, 48, 54 Mbps; the working range of the standard is 5 GHz.

OFDM is attracting significant attention for satellite communication systems (Cioni, Corazza, Neri, & Vanelli-Coralli, 2006). The information transmitting by means of OFDM signals became the standard for many modern radio systems in connection with a number of advantages - high spectral efficiency, low level of an intersymbol interference, high quality of transmitting in the conditions of frequency-selective fading.

At the same time OFDM systems are sensitive to the frequency instability of carriers. It is especially important to provide power efficiency for an information transmitting in aviation complexes with rigid restriction of spatially-frequency parametres for onboard radio-electronic equipment.

Nevertheless, issues related to the OFDM satellite channel parameters estimation still are not investigated in detail. OFDM as digital multi-carrier modulation technique has been adopted as physical layer scheme of broadband wireless air interface standards, such as IEEE 802.11/WiFi, IEEE 802.16/ WiMAX. Simultaneously OFDM modulation is attracting more attention for delivering multimedia services over hybrid satellite/terrestrial networks to a variety of small mobile and fixed terminals with compact antennas (Varade & Kulat, 2012). On the other hand, OFDM technique is also being applied in military communications (Baddeley, 2005).

OFDM is a method of encoding digital data on multiple carrier frequencies, which was developed for wideband digital communication (Jeon, Chang, & Cho, 1999). In an OFDM system, the data is divided into multiple parallel sub streams at a reduced data rate, and each is modulated and transmitted on a separate orthogonal subcarrier. This increases symbol duration and improves system robustness.

A large number of closely spaced orthogonal sub-carrier signals are used to carry data on several parallel data streams or channels. Each sub-carrier is modulated with a conventional modulation scheme at a low symbol rate, maintaining total data rates similar to conventional single-carrier modulation schemes in the same bandwidth (Roque & Siclet, 2013).

Channel parameters estimation (Kharchenko, Grekhov, & Ali, 2015a) is a critical component in wireless communications systems. Therefore the aim of this study is: 1) to design model of aeronautical satellite OFDM communication channel "Aircraft/RPAS–Satellite–Ground Station" with adaptive modulation using MATLAB Simulink software; 2) to realize parameters calculations for channels of different types; 3) to develop a method for estimating the parameters of aeronautical satellite communication channel.

Model "OFDM_FSPL_Sat_FSPL_802.11a" for Satellite Channel

Satellite communication channel was analyzed using original model designed on a basis of IEEE 802.11a standard and MATLAB Simulink demo model *commwman80211a*.

The model *OFDM_FSPL_Sat_FSPL_802.11a* (Figure 1) consists of "Uplink/Downlink", "Aircraft/RPAS Transmitter", "Satellite Transponder", "Ground Station" and "Adaptive Modulation Control". Different types of "Uplink/Downlink" were considered: "Multipath", "Rayleigh Fading", "Rician Fading", "Free Path Loss with Phase/Frequency Offset" and AWGN channels. This paragraph is devoted to consideration of "Free Path Loss with Phase/Frequency Offset" type of a link.

Parameter settings for the model are the following: Viterbi traceback depth is 34, hysteresis factor for adaptive modulation (dB) is 3, numbers of OFDM symbols per transmit block are 20 and 1000, number of OFDM symbols in training sequence is 4, Low-SNR thresholds (dB) vector [10 11 14 18 22 26 28] where less than 10 is for BPSK ½, between 10 and 11 - for BPSK ¾, between 11 and 14 - for QPSK ½, between 14 and 18 - for QPSK ¾, between 18 and 22 - for 16-QAM ½, between 22 and 26 - for 16-QAM ¾, between 26 and 28 – for QAM 2/3 and more than 28 - for 64-QAM ¾.

Figure 1. Model OFDM_FSPL_Sat_FSPL_802.11a of satellite link

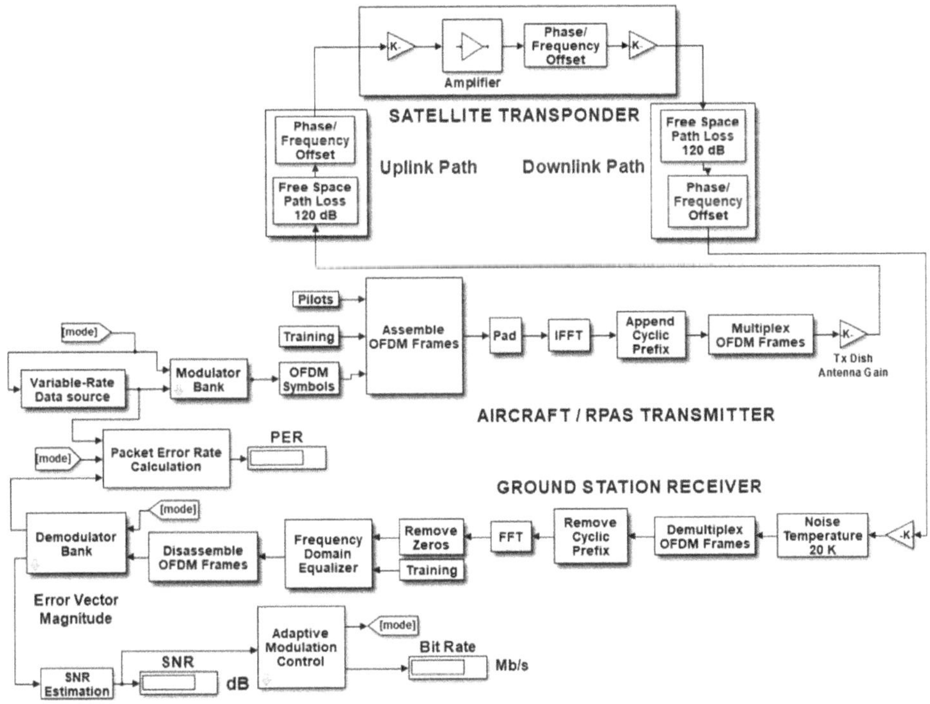

Low-SNR thresholds parameter is a seven-element vector that indicates how the simulation should choose a data rate based on the SNR estimate. The model has eight modes, each associated with a particular modulation scheme and convolutional code. The seven thresholds are the boundaries between eight adjacent regions that correspond to the eight modes. Ideally, the simulation should use the highest-throughput mode that achieves a desired (zero) packet error rate. Determining appropriate thresholds often involves running the simulation multiple times, varying the values of the Low-SNR thresholds parameter.

The communication system in this model performs such tasks as: generation of random data at a bit rate that varies during the simulation; coding, interleaving, and modulation using one of eight schemes specified in the standard; OFDM transmission using 52 subcarriers, 4 pilots, 64-point FFT, and a 16-sample cyclic prefix; physical layer convergence protocol preamble modeled as four long training sequences.

Packet Error Rate (PER) Calculation block shows the packet error rate as a percentage and should always equal zero during investigations.

Aeronautical Satellite Communication Channel Simulation

For modeling of aeronautical satellite link performance, the following parameters in the model (Figure 1) were set up: phase/frequency offsets in uplink/downlink and satellite transponder are equal to zero; the gain of linear amplifier in satellite transponder was taken 10 dB; the number of OFDM symbols per transmit block was taken 20, 200, 400 and 1000. Values of antennas gain were taken $G_{aircraft} = 12.4$, $G_{satellite} = 31.1$, $G_{ground} = 62.1$ (for 4 GHz this corresponds respectively to $d_{aircraft} = 0.4$ m, $d_{satellite} = 1.0$ m, $d_{ground} = 2.0$ m).

Dependencies of a SNR on free path losses for different modulation modes and noise temperatures are given in Figure 2. During modeling the value of a packet error rate was kept at zero by changing the type of modulation (using a SNR estimation and adaptive rate control). In accordance with this a ratio SNR was changed. Free space path loss values were changed simultaneously in uplink and downlink. The channel is "closed" at 134 dB for a noise

Figure 2. Dependencies of a SNR on free space path loss

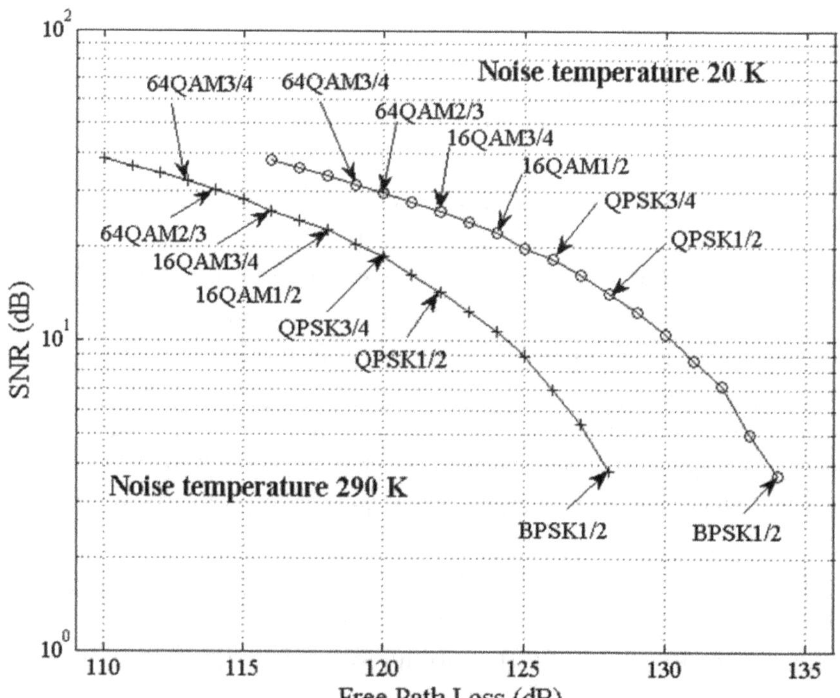

temperature T = 20 K and at 128 dB for T = 290 K. With a decrease in the value of the free path loss up to 128 dB (T = 290 K) a modulation remains BPSK1/2 and bit rate is 6 Mb/s.

At values of free path losses marked by arrows, the modulation type is changed and the data rate increases: BPSK ½ (6 Mb/s), QPSK1/2 (12 Mb/s), QPSK3/4 (18 Mb/s), 16QAM1/2 (24 Mb/s), 16QAM3/4 (36 Mb/s), 64QAM2/3 (48 Mb/s) and 64QAM3/4 (54 Mb/s). Data from Figure 7 show how big a SNR ratio should be and what type of a modulation to be used for data transmission without errors for given free space path losses and a noise temperature. The number of OFDM symbols per transmit block practically does not change the dependencies shown on Figure 7 whereas a noise temperature essentially affects the results.

Data given in Figure 2 confirm the obvious conclusion that 64QAM3/4 modulation has the highest value of a SNR ratio in comparison with other types of a modulation and that the lower a noise temperature the higher the value of a SNR ratio is.

The next investigation was performed for different frequency and phase offsets in the uplink and downlink channels (Figure 3). All parameters in the channel remained the same. A noise temperature and number of OFDM symbols were changed. It was obtained that channel characteristics do not depend on the value of a phase offset. However, a frequency offset influences the dependencies of a SNR on free space path loss.

Dependencies of a SNR on free space path losses for 20, 200 and 400 OFDM symbols was obtained for the frequency offset 15 Hz (Figure 4). Thus, it is possible to conclude that channel is sensitive to number of OFDM symbols when there is a frequency offset.

Dependencies of a SNR on satellite transponder linear gain and aircraft antenna gain are given on Figure 5 and Figure 6 correspondently.

Method of Channel Parameters Estimation

It is possible to summarize the received data and to give a method for estimating the parameters of satellite link:

1. To select the Low-SNR thresholds parameter. (In considered model, a seven-element vector indicates how the simulation should choose a data rate based on a SNR estimate). The model has eight modes, each associated with a particular modulation scheme and convolutional code. The seven thresholds are the boundaries between eight adjacent regions

Figure 3. Dependencies of a SNR on free space path loss for different frequency offsets:
circles – 0 Hz, points – 50 Hz, stars – 100 Hz

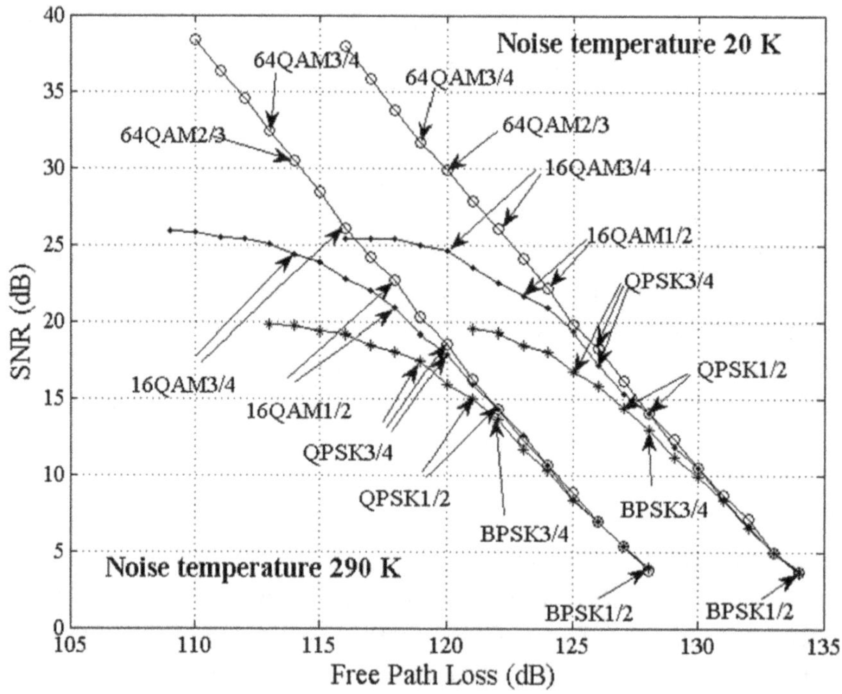

that correspond to the eight modes. The simulation should use the highest-throughput mode that achieves zero packet error rates. Determining appropriate thresholds involves running the simulation multiple times, varying the values of the Low-SNR thresholds parameter).

2. To select the type of the communication channel (Multipath, Rayleigh Fading, Rician Fading, Free Path Loss with Phase/Frequency Offset, AWGN).
3. For the selected communication channel to take the appropriate model from the created model kit.
4. Based on relations received is this paper (under the given conditions: a number of OFDM symbols, noise temperatures, gains of antenna dishes and the satellite transponder amplifier, a phase-frequency shift) to estimate the channel parameters:
    a. The level of free space path loss, for which the satellite communication channel is "open";

Figure 4. Dependencies of a SNR on free space path loss for different number of OFDM symbols per transmit block and a frequency offset 15 Hz: circles – 20, points – 200, stars – 400

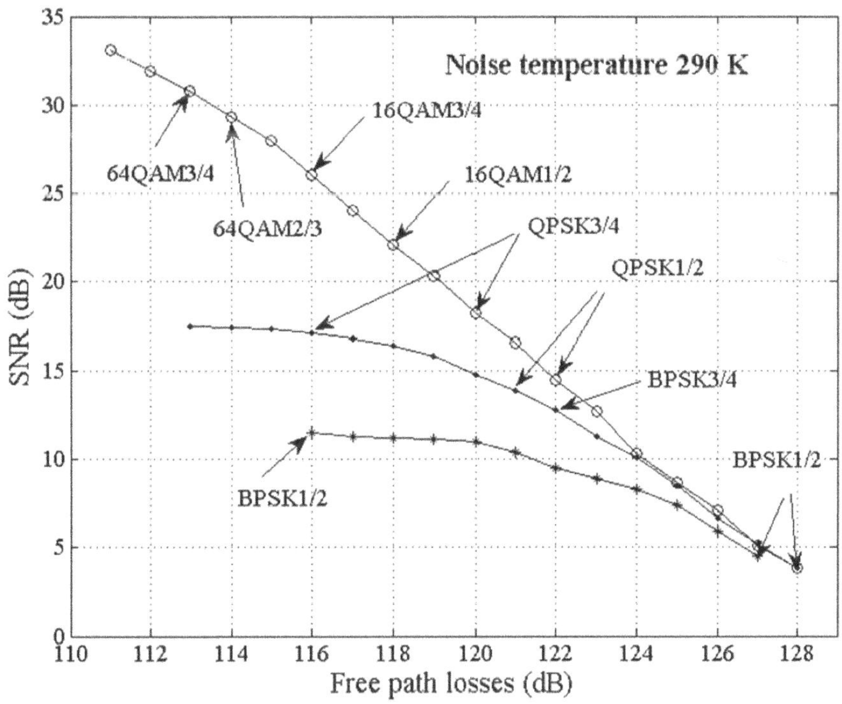

b. The type of a modulation, which is possible under the given conditions;
c. Data transfer rate, which is possible under the given conditions.

5. For the channel with different conditions, it is necessary to make further calculations using the created model kit.

CONCLUSION

On a basis of IEEE 802.11a standard, the realistic model with adaptive modulation for airborne satellite communication channel parameters evaluation is developed for the first time. Different types of uplink/downlink can be taken.

A SNR ratio dependences were received on the type of a modulation (BPSK, QPSK, 16QAM, 64QAM), a bit rate, free space path losses, noise temperatures,

Figure 5. Dependencies of a SNR on satellite transponder linear gain: free space path loss is 120 dB, frequency/phase offsets - zero, number of OFDM symbols – 20

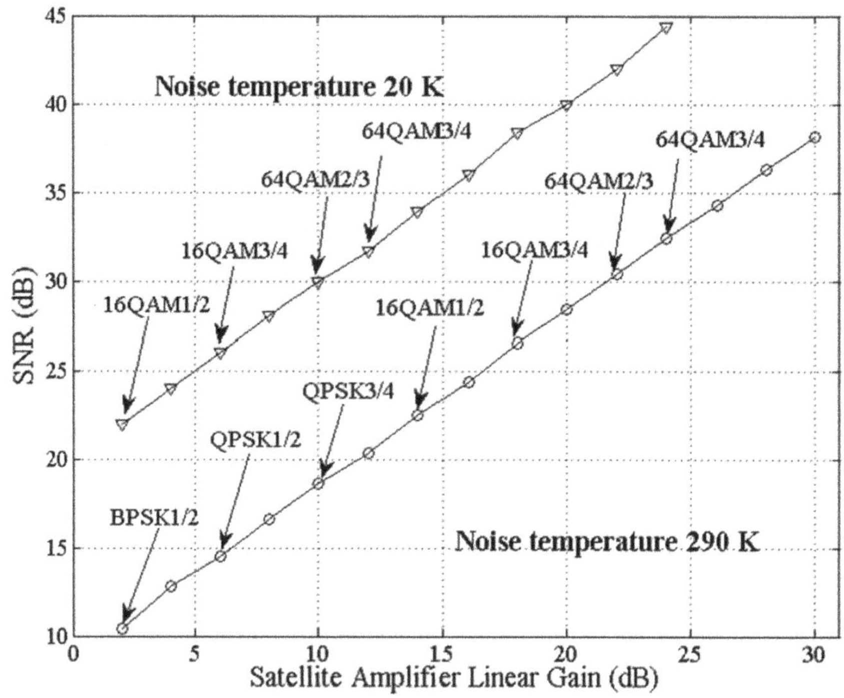

frequency/phase offsets, a number of OFDM symbols in transmitting block, satellite transponder linear gain and aircraft antenna diameter.

A frequency offset essentially influences the dependence of a SNR on free path loss for different noise temperatures and a number of OFDM symbols.

It was shown that channel is sensitive to number of OFDM symbols at presence of a frequency offset.

INFLUENCE OF NONLINEARITY ON OFDM CHANNEL PARAMETERS

Nowadays worldwide communications are available for aviation through communication satellites (Rumanek & Šebesta, 2010). Satellite networks play an important role for the purpose of data delivery over large distances, and are an effective means for reaching remote locations lacking in communication

Figure 6. Dependencies of a SNR on aircraft antenna gain: free space path loss is 120 dB, frequency/phase offsets - zero, number of OFDM symbols – 20

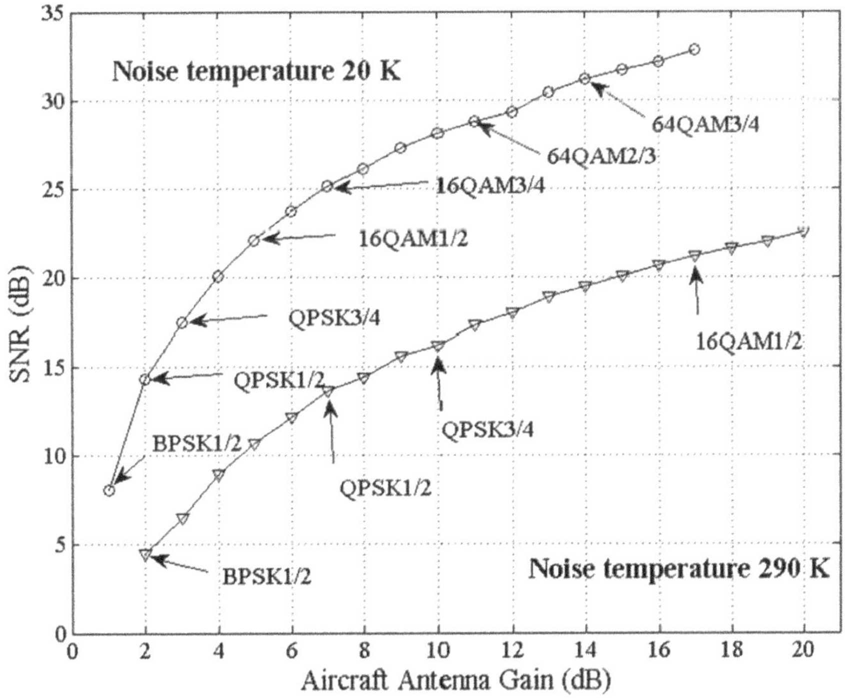

infrastructure (Modiano, 2004). Aviation satellite channels play a critical role in providing a rapidly deployable, reliable, and affordable communications (Global, 1998).

The goal for the aeronautical satellite mobile communications system is to integrate a variety of communication services (voice signals, high-speed data, video and multimedia traffic). The promising approaches is adaptive OFDM. Data traffic for satellite networks must be designed to take into account characteristics of satellite systems: propagation delays, limited energy and power, relatively high channel error rates, and time-varying channel conditions (Sharma & Srivastava, 2013).

The signal transmitted via satellite is affected by many factors: a free path loss, a frequency offset, a phase noise, nonlinearities, a noise temperature, amplifier nonlinearity. Therefore, the main task is successful and reliable data message transmission in spite of these adverse conditions. The information transmitting by means of OFDM signals became the standard for many modern radio systems in connection with a number of advantages - high spectral efficiency, low level of an intersymbol interference, high quality of transmitting in the conditions of frequency-selective fading.

At the same time OFDM, systems are sensitive to phase and frequency instability of carriers. It is especially important to provide power efficiency for an information transmitting in aviation complexes with rigid restriction of spatially-frequency parameters for onboard radio-electronic equipment. For this, simulations are mandatory to infer the performance of mobile satellite communication systems. Nevertheless, issues related to the satellite channel nonlinearities still are not investigated in detail.

Digital multi-carrier modulation technique OFDM has been adopted as physical layer scheme of broadband wireless air interface standards. Nonlinear distortion is a source of major degradation of modulation fidelity in multicarrier systems with OFDM signals. Compared with conventional single carrier communication systems, OFDM signals significantly improve spectrum efficiency and reduce frequency selective fading problems. However their consisting of large numbers of independent QAM subcarriers, means the composite signal's peak to average power ratio (PAPR) can be significant. This makes them sensitive to nonlinear distortion (O'Droma, Mgebrishvili, & Goacher, 2004).

The primary source of this nonlinear distortion is the radio frequency transmitter power amplifier. Nonlinear power amplifiers for wireless communications were modeled (Jantunen, 2004) and nonlinear power amplifier effects in multi-antenna OFDM systems were analyzed (Gregorio, 2007). Modulation schemes effect on radio frequency power amplifier nonlinearity were considered in paper (El-Khatib, MacEachern, & Mahmoud, 2012). A new PAPR reduction technique of OFDM system with nonlinear high power amplifier was proposed (Park & Song, 2007). The use of OFDM radio interface for satellite digital multimedia broadcasting systems (Cioni, Corazza, Neri, & Vanelli-Coralli, 2006), a BER for MIMO-OFDM systems (Varade & Kulat, 2012), performances of weighted cyclic prefix OFDM with equalization (Roque & Siclet, 2013) were studied.

Therefore the aim of this study is: 1) to design model of aeronautical satellite OFDM communication channel "Aircraft–Satellite–Ground Station" with adaptive modulation using MATLAB Simulink software; 2) to calculate parameters of a channel with different types of nonlinearities; 3) to analyze the impact of nonlinearities on parameters of satellite communication channel (Kharchenko, Grekhov, & Ali, 2015b).

Aeronautical Satellite Channel Nonlinearity Simulation

Satellite communication channel was analyzed using the model *OFDM_FSPL_Sat_FSPL_802.11a* (Figure 1). For calculations the following parameters in the model (Figure 1) were set up: phase/frequency offsets in uplink/downlink and a satellite transponder are equal to zero; aircraft antenna gain was taken 12.4 (an antenna diameter ≈ 0.4 m at 4 GHz), satellite antennas gain – 31.1 (an antenna diameter ≈ 1.0 m at 4 GHz), ground station antenna gain – 62.2 (an antenna diameter ≈ 2.0 m at 4 GHz).

The options for the method for modeling amplifier nonlinearity are Linear, Cubic Polynomial, Hyperbolic Tangent, Saleh model (Saleh, 1981), Ghorbani model (Ghorbani & Sheikhan, 1991), and Rapp model (Rapp, 1991).

The linear method is implemented by a Gain block (with a linear gain 10 dB).

All five subsystems for the nonlinear method options apply a memoryless nonlinearity to the complex baseband input signal. Each one multiplies the signal by a gain factor; splits the complex signal into its magnitude and angle components; applies an AM/AM conversion to the magnitude of the signal, according to the selected nonlinearity method, to produce the magnitude of the output signal; applies an AM/PM conversion to the phase of the signal, according to the selected nonlinearity method, and adds the result to the angle of the signal to produce the angle of the output signal; combines the new magnitude and angle components into a complex signal and multiplies the result by a gain factor, which is controlled by the Linear gain parameter.

For Cubic Polynomial Model the Amplifier block models the AM/AM nonlinearity by:

- Using the third-order input intercept point IIP3= 30 dBm parameter to compute the factor f, which scales the input signal before the Amplifier block applies the nonlinearity:

$$f = \sqrt{\frac{3}{IIP3\left(Watts\right)}} = \sqrt{\frac{3}{10^{\left(IIP3\left(dBm\right)-30\right)/10}}}\,.$$

- Computing the scaled input signal by multiplying the amplifier input signal by f;
- Limiting the scaled input signal to a maximum value of 1;

- Applying an AM/AM conversion to the amplifier gain, according to the following cubic polynomial equation:

$$F_{AM/AM}\left(u\right) = u - \frac{u^3}{3},$$

where u is the magnitude of the scaled input signal, which is a unitless normalized input voltage.

The Amplifier block uses the AM/PM conversion (10 degrees per dB) parameter, which specifies the linear phase change, to add the AM/PM nonlinearity within the power limits specified by the Lower input power limit for AM/PM conversion (10 dBm) parameter and the Upper input power limit for AM/PM conversion (infinite dBm) parameter. Outside those limits, the phase change is constant at the values corresponding to the lower and upper input power limits. The Linear gain (10 dB) parameter scales the output signal.

In Hyperbolic Tangent Model, data are processed as in Cubic Polynomial Model with the exception of applying an AM/AM conversion to the amplifier gain, according to the following equation:

$$F_{AM/AM}\left(u\right) = \tanh\left(u\right).$$

For Saleh Model with a negligible nonlinearity the Input scaling (-21.5957 dB) parameter scales the input signal before the nonlinearity is applied. The block multiplies the input signal by the parameter value, converted from decibels to linear units.

The AM/AM parameters [alpha = 2.1587 beta = 1.1517] are used to compute the amplitude gain for an input signal using the following function:

$$F_{AM/AM}\left(u\right) = \frac{\alpha u}{1 + \beta u^2}.$$

The AM/PM parameters [alpha = 4.0033 beta= 9.1040] are used to compute the phase change for an input signal using the following function:

$$F_{AM/PM}\left(u\right) = \frac{\alpha u^2}{1 + \beta u^2},$$

where u is the magnitude of the input signal. The Output scaling (32.9118 dB) parameter scales the output signal.

For Ghorbani Model the Input scaling (-1.5957 dB) parameter scales the input signal before the nonlinearity is applied. The block multiplies the input signal by the parameter value, converted from decibels to linear units.

The AM/AM parameters [x1 = 8.1081, x2 = 1.5413, x3 = 6.5202, x4 = -0.0718] are used to compute the amplitude gain for an input signal using the following function:

$$F_{AM/AM}\left(u\right) = \frac{x_1 u^{x_2}}{1 + x_3 u^{x_2}} + x_4 u.$$

The AM/PM parameters [y1 = 4.6645, y2 = 2.0965, y3 = 10.88, y4 = -0.003] are used to compute the phase change for an input signal using the following function:

$$F_{AM/PM}\left(u\right) = \frac{y_1 u^{y_2}}{1 + y_3 u^{y_2}} + y_4 u,$$

where u is the magnitude of the scaled signal. The Output scaling (32.9118 dB) parameter scales the output signal.

For Rapp Model the amplitude gain for an input signal is computed by the following function:

$$F_{AM/AM}\left(u\right) = \frac{u}{\left(1 + \left(\frac{u}{Q_{sat}}\right)^{2S}\right)^{\frac{1}{2S}}}$$

where u is the magnitude of the scaled signal, S=0.5 is the Smoothness factor and Q_{sat} = 1 is the Output saturation level. The Rapp model does not apply a phase change to the input signal.

Dependencies of a SNR on free space path losses for different models of nonlinearity, modulation modes, noise temperatures and 20 OFDM symbols per transmitting block are given on Figures 7 – 12. During modeling the value of a packet error rate was kept at zero by changing the type of modulation (using a SNR estimation and adaptive rate control). In accordance with this, a SNR was changed. Free space path loss values were changed simultaneously in uplink and downlink.

Data for Figures 7–12 were obtained with model parameters described above and show how big a SNR ratio should be and what type of a modulation to be used for data transmission without errors for given free space path losses and a noise temperature.

At values of free path losses marked by arrows the modulation type is changed and the data rate increases for the lower values of free path losses: BPSK ½ (6 Mb/s), QPSK1/2 (12 Mb/s), QPSK3/4 (18 Mb/s), 16QAM1/2 (24 Mb/s), 16QAM3/4 (36 Mb/s), 64QAM2/3 (48 Mb/s) and 64QAM3/4 (54 Mb/s).

As seen from figures a noise temperature essentially affects the results.

Figure 7. Dependencies of a SNR on free space path loss for Linear Model

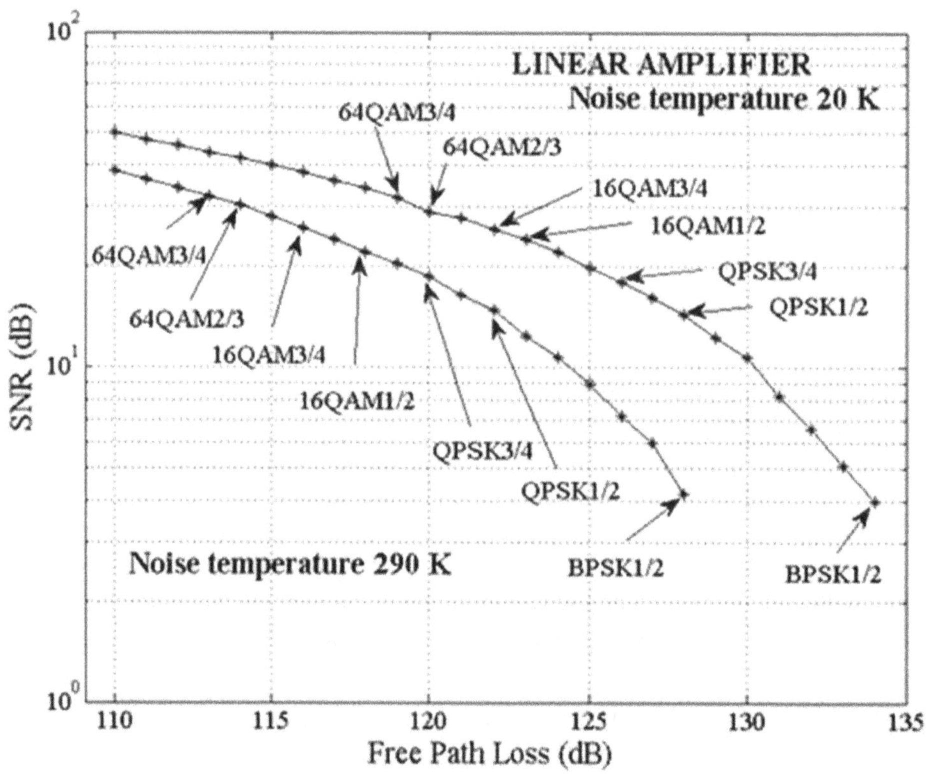

Figure 8. Dependencies of a SNR on free space path loss for Cubic Polynomial Model

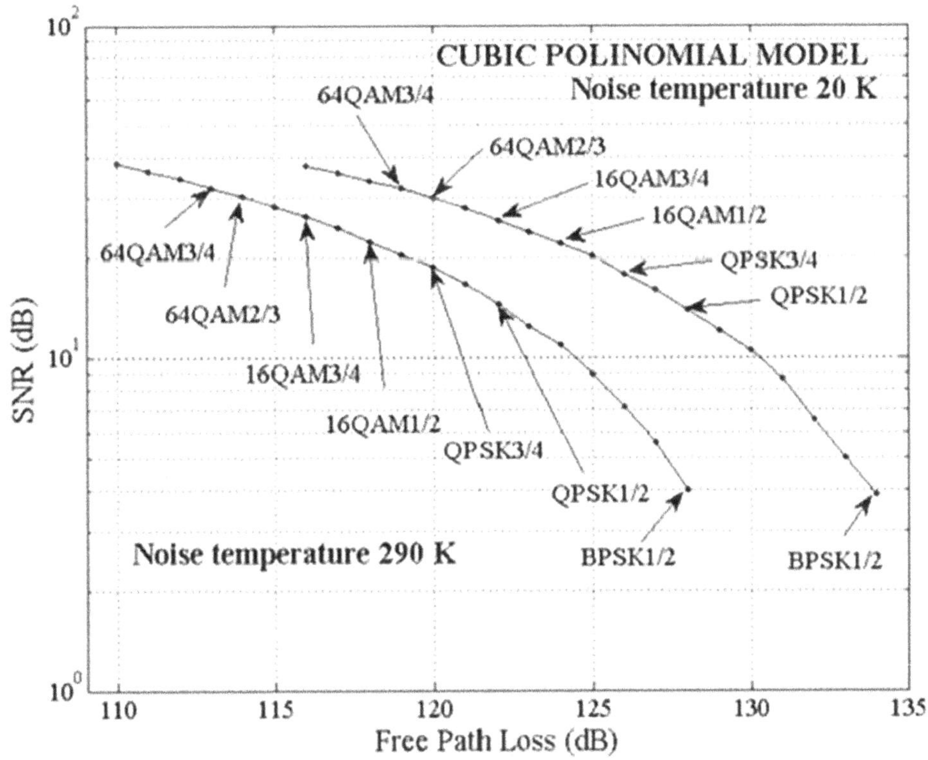

Satellite OFDM communication channels cease to work at different values of free path losses for different nonlinearity models. For example, at noise temperature T = 20 K (290 K) channels are "closed": for the Linear Model (Figure 7) - when free path losses are 134 dB (128 dB); for the Cubic Polynomial Model (Figure 8) – at 134 dB (128 dB); for the Hyperbolic Tangent Model (Figure 9) – at 134 dB (128 dB); for the Saleh Model with a negligible nonlinearity (Figure 10) – at 138 dB (132 dB); for the Ghorbani Model (Figure 11) – at 123 dB (120 dB); and for the Rapp Model (Figure 12) – at 134 dB (128 dB).

According to calculations all the models of non-linear amplifiers (except the Ghorbani model) gave similar curves for the dependence of a SNR on the free path loss in OFDM satellite communication channel. The best from the point of view of the channel "closing" is the Saleh model.

Figure 9. Dependencies of a SNR on free space path loss for Hyperbolic Tangent Model

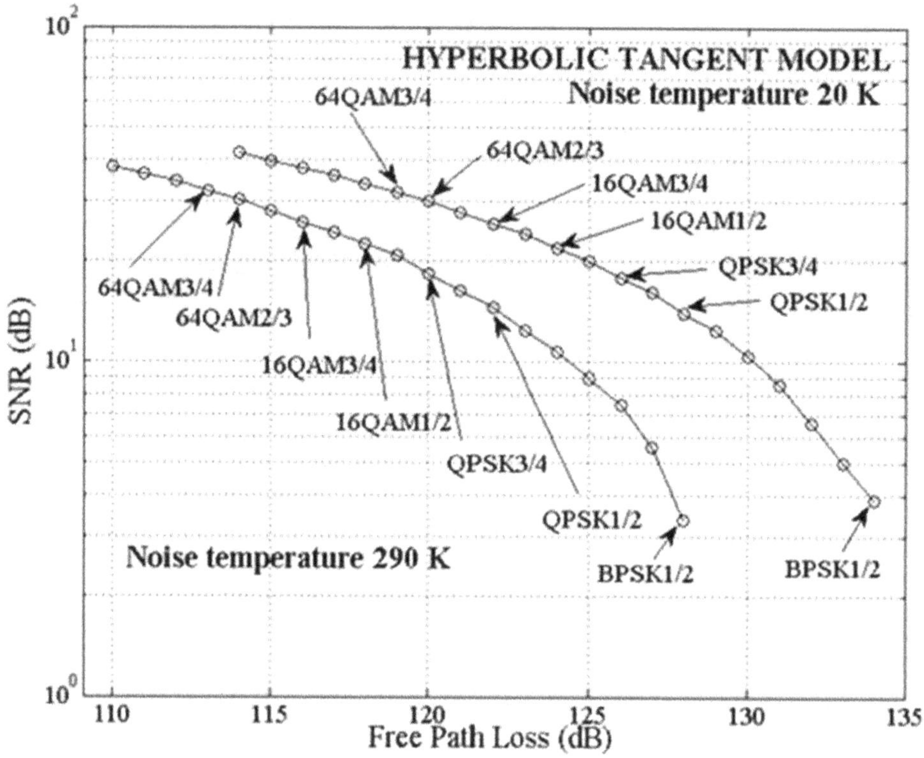

At different values of free space path loss OFDM, satellite communication channels operate at different types of a modulation. For example, at a noise temperature T = 290 K and free path losses 120 dB (115 dB): for the Linear Model – it occurs QPSK3/4 (16QAM3/4); for the Cubic Polynomial Model – QPSK3/4 (16QAM3/4); for the Hyperbolic Tangent Model – QPSK3/4 (16QAM3/4); for the Saleh Model – 16QAM3/4 (64QAM3/4); for the Ghorbani Model – BPSK1/2 (BPSK1/2); and for the Rapp Model – QPSK3/4 (16QAM3/4).

Data for power spectra and signal constellations (Figures 13-20) were received with models parameters described above but the lower value of free path losses. It allows revealing the effect of nonlinear amplifiers. A signal with a bigger power and amplitude shifts the transponder amplifier in a region of nonlinearity (Figures 14, 15, 17-20) and the chosen parameters for AM/AM and AM/PM distortions worsen data transmission conditions.

Figure 10. Dependencies of a SNR on free space path loss (negligible nonlinearity) for Saleh Model

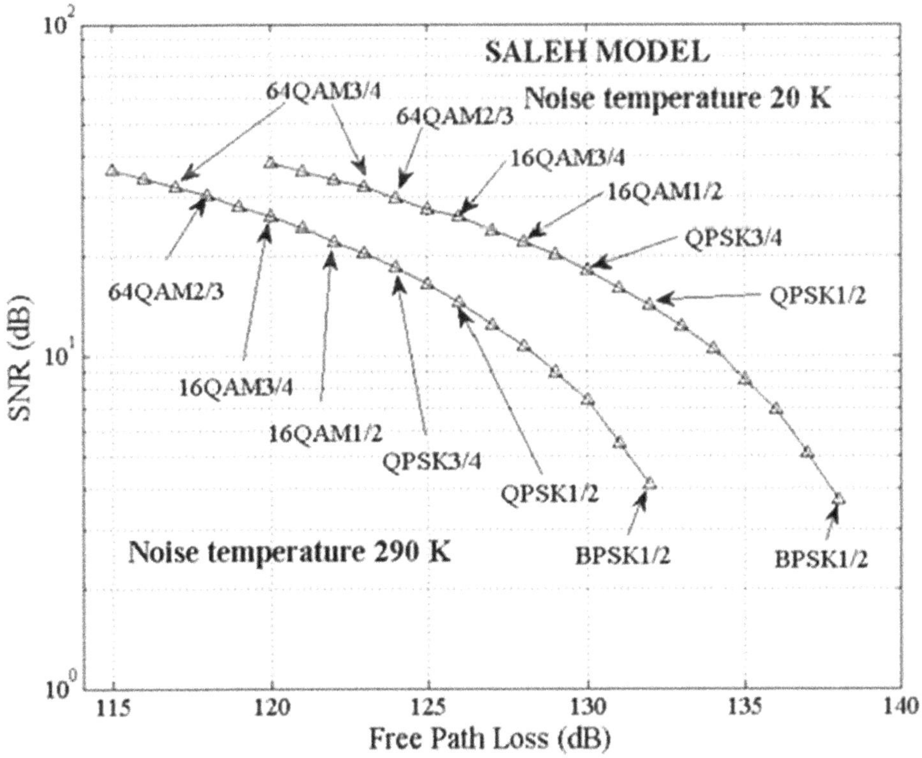

For example, comparing the spectrums of received signals for the linear amplifier model (Figure 13) and the Saleh model (Figure 16) we can see the spectrum power growth in the latter case. This growth is due to the nonlinearity of the amplifier. When digitally modulated signals go through a nonlinear amplifier, spectral regrowth (broadening) appears in the output for the Cubic Polynomial Model (Figure 14), the Hyperbolic Tangent Model (Figure 15), the Saleh Model with moderate (Figure 17) and severe (Figure 18) nonlinearities, the Ghorbani Model (Figure 19) and the Rapp Model (Figure 20).

Comparing the constellations of received signals for the Saleh model (Figures 16-18) with different levels of nonlinearity we can see the signal distortion at the received signal. This distortion is due to the nonlinearity of the amplifier.

Figure 11. Dependencies of a SNR on free space path loss for Ghorbani Model

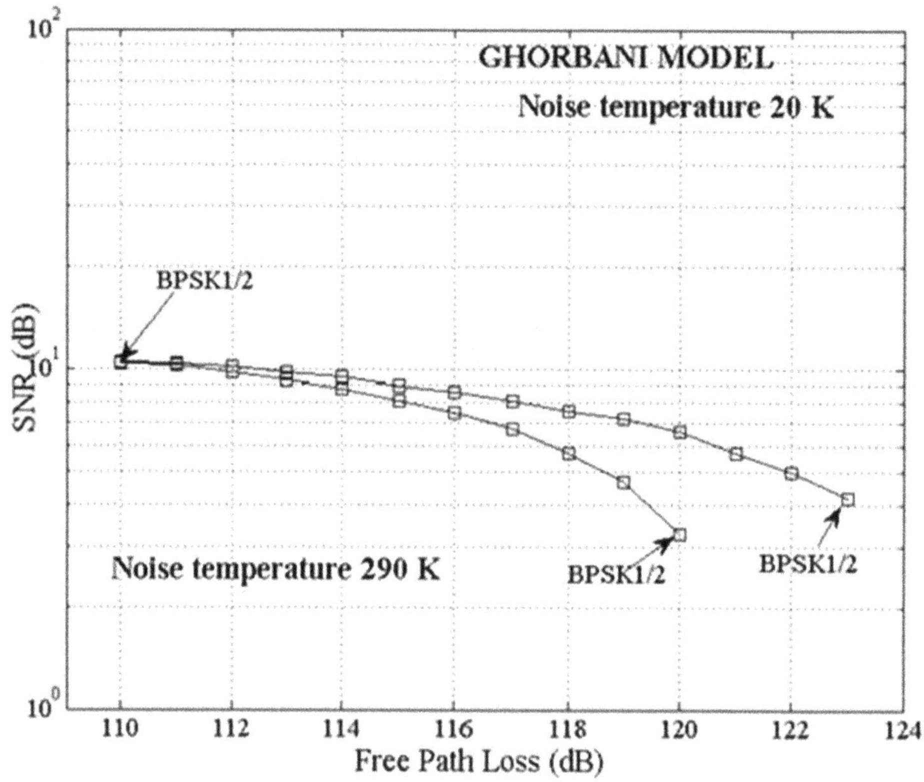

CONCLUSION

To investigate the influence of satellite amplifier nonlinearity the model was developed with adaptive modulation for satellite OFDM communication channel.

Dependencies were received of a SNR ratio on the type of a modulation (BPSK, QPSK, 16QAM, 64QAM), a bit rate, free space path losses, noise temperatures for different models of memoryless nonlinear satellite amplifiers (the Cubic Polynomial Model, the Hyperbolic Tangent Model, the Saleh Model, the Ghorbani Model and the Rapp Model).

On the basis of received is this study data (under the given conditions: a type of nonlinearity, a number of OFDM symbols, noise temperatures, gains of antenna dishes and the satellite transponder amplifier nonlinearity type)

Figure 12. Dependencies of a SNR on free space path loss for Rapp Model

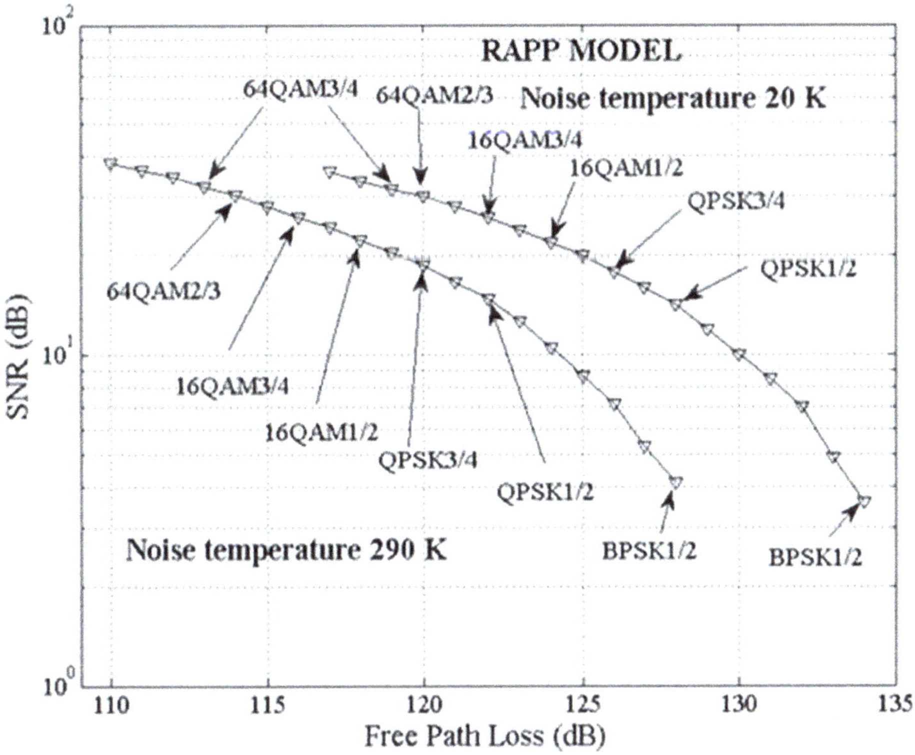

Figure 13. Power spectrum and signal constellation for Linear Amplifier
(SNR = 66.7 dB, bit rate = 54 Mb/s, free path loss in uplink/downlink 70 dB, T=20 K)

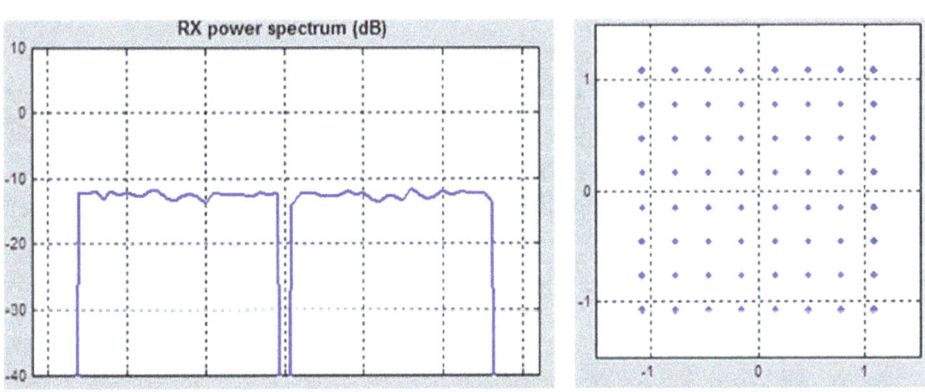

Figure 14. Power spectrum and signal constellation for Cubic Polynomial Model
(SNR = 5.2 dB, bit rate = 6 Mb/s, free path loss in uplink/downlink 70 dB, T=20 K)

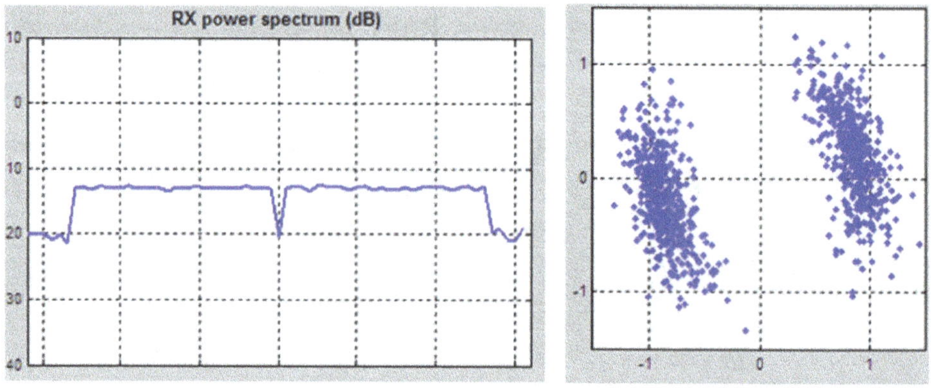

Figure 15. Power spectrum and signal constellation for Hyperbolic Tangent Model
(SNR = 5.7 dB, bit rate = 6 Mb/s, free path loss in uplink/downlink 70 dB, T=20 K)

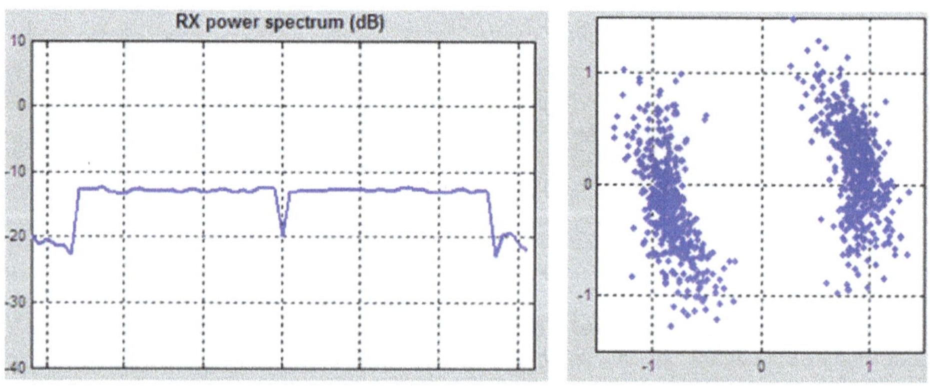

the channel parameters were estimated: the level of free space loss, for which the satellite communication channel is "open"; the type of a modulation and data transfer rate, which are possible under the given conditions.

Developed model allows predicting spectral regrowth of digitally modulated OFDM signals due to amplifier nonlinearity. Prediction of spectral regrowth for a prescribed level of amplifier nonlinearity can be very helpful for designing communication systems.

Figure 16. Power spectrum and signal constellation for Saleh Model with negligible nonlinearity
(SNR = 63.2 dB, bit rate = 54 Mb/s, free path loss in uplink/downlink 70 dB, T=20 K)

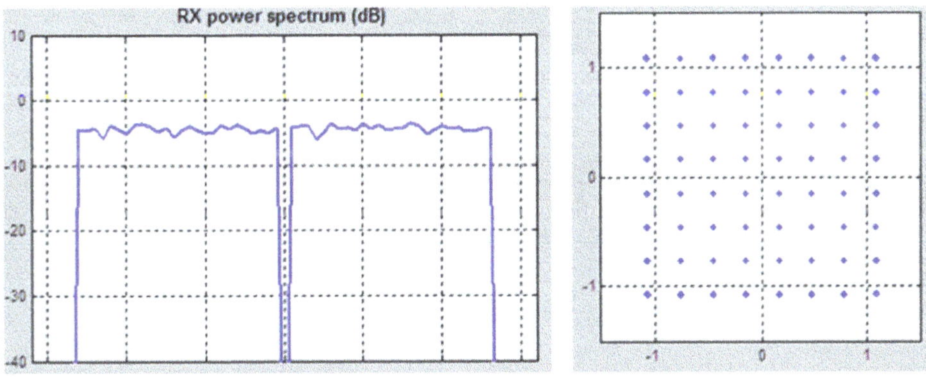

Figure 17. Power spectrum and signal constellation for Saleh Model with moderate nonlinearity
(SNR = 23.1 dB, bit rate = 24 Mb/s, free path loss in uplink/downlink 70 dB, T=20 K)

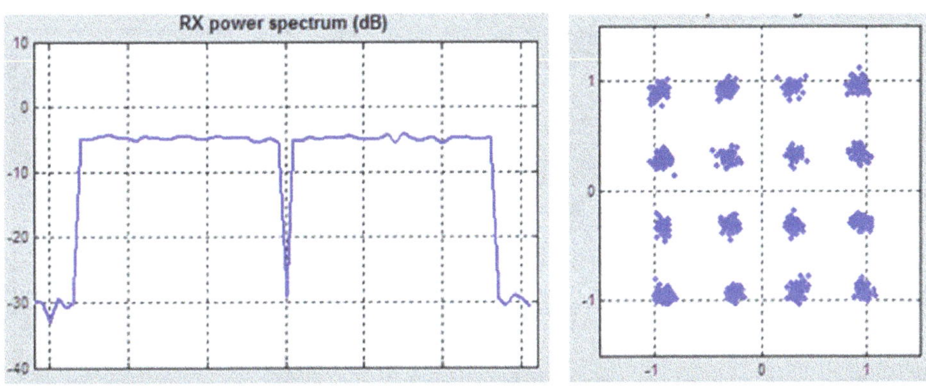

EFFECT OF RAYLEIGH FADING ON OFDM CHANNEL

The technological capabilities of global mobility of aircraft are determined by the ability of satellite communication systems to provide the normative quality of data transmission in multipath channels with Doppler frequency shift. For operation in frequency-selective channels with multipath signal propagation, modern OFDM radio systems are currently used.

Figure 18. Power spectrum and signal constellation for Saleh Model with severe nonlinearity
(SNR = 17.9 dB, bit rate = 18 Mb/s, free path loss in uplink/downlink 70 dB, T=20 K)

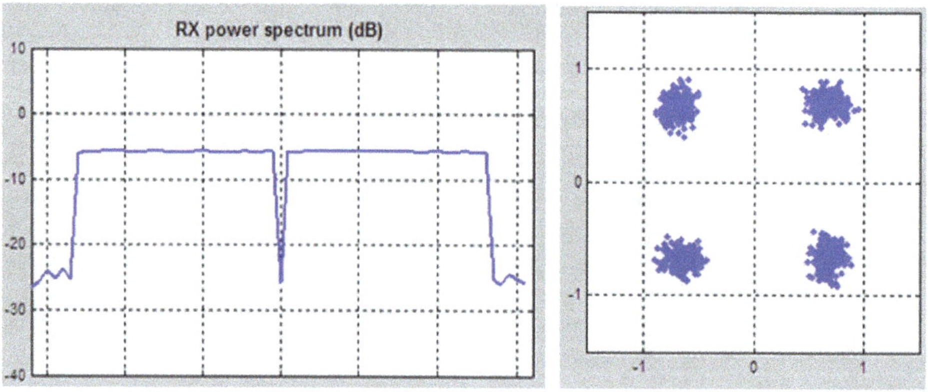

Figure 19. Power spectrum and signal constellation for Ghorbani Model with severe nonlinearity
(SNR = 11.3 dB, bit rate = 6 Mb/s, free path loss in uplink/downlink 70 dB, T=20 K)

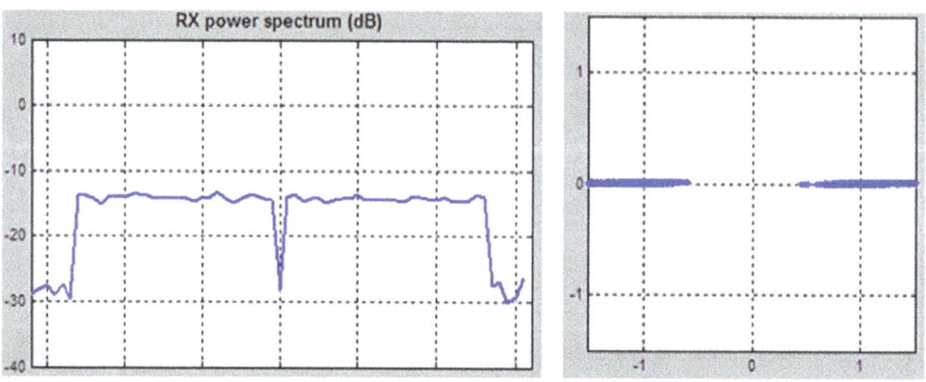

Fading in mobile satellite radio channels with multipath signal propagation was modeled in (Csurgai-Horváth & Bitó, 2007). The operation of channels with additive white Gaussian noise, with frequency-independent and frequency-selective fading, was compared in (Sipon, Mahbubur, Godder, Chandra, & Parvin, 2011). A comparative analysis of data transmission using the physical layer of WiMAX-PHY via the AWGN channel and the fading channel was carried out in article (Islam, Kader, & Julkarnain, 2013).

Figure 20. Power spectrum and signal constellation for Rapp Model
(SNR = 18.0 dB, bit rate = 18 Mb/s, free path loss in uplink/downlink 70 dB, T=20 K)

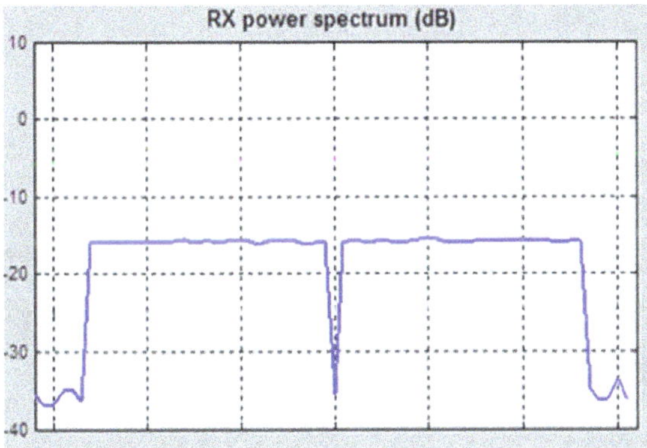

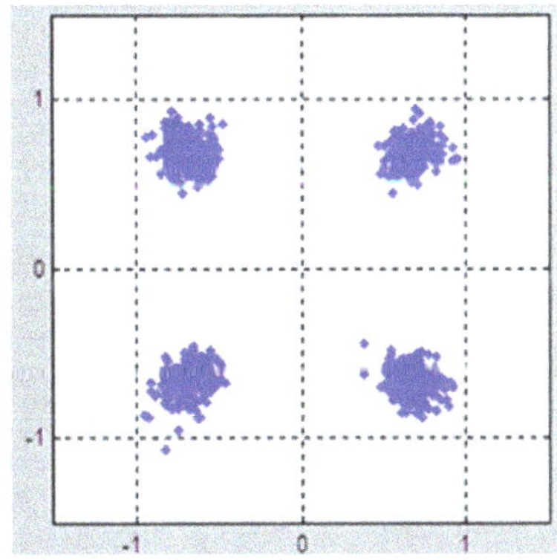

The purpose is to study the effect of fading on messages transmission over an aeronautical satellite OFDM channel with adaptive modulation and the development of a method for estimating the channel parameters (Kharchenko, Grekhov, & Ali, 2015c).

Model "OFDM_Multipath_Sat_ Multipath_802.11a" for Satellite Channel

The original model *OFDM_Multipath_Sat_Multipath_802.11a* (Figure 21) was developed based on the 802.11a standard. The model consists of the channels "Up/Down" (Figure 22), "Aircraft/RPAS Transmitter", "Satellite Transponder", "Ground Station Receiver" and "Adaptive Modulation Control". This study is devoted to the consideration of Rayleigh fading and losses in free space.

The threshold values of the SNR at which a transition to another type of modulation occurs is given in the form of a vector [10 11 14 18 22 26 28]. For SNR values less than 10 dB, BPSK½ modulation is used; for values between 10 dB and 11 dB modulation is BPSK¾; between 11 dB and 14 dB modulation is QPSK½; between 14 dB and 18 dB modulation is QPSK¾; between 18 dB and 22 dB modulation is 16QAM½; between 22 dB and 26 dB modulation is 16-QAM¾; between 26 dB and 28 dB modulation is 64QAM2/3 and more than 28 dB modulation is 64QAM¾. Accordingly, the model has eight modes, each associated with a specific modulation scheme and a convolutional code, which are selected based on estimating the value of the SNR in the channel. Ideally, the simulation should use the highest

Figure 21. Model "OFDM_Multipath_Sat_Multipath_802.11a" for satellite link

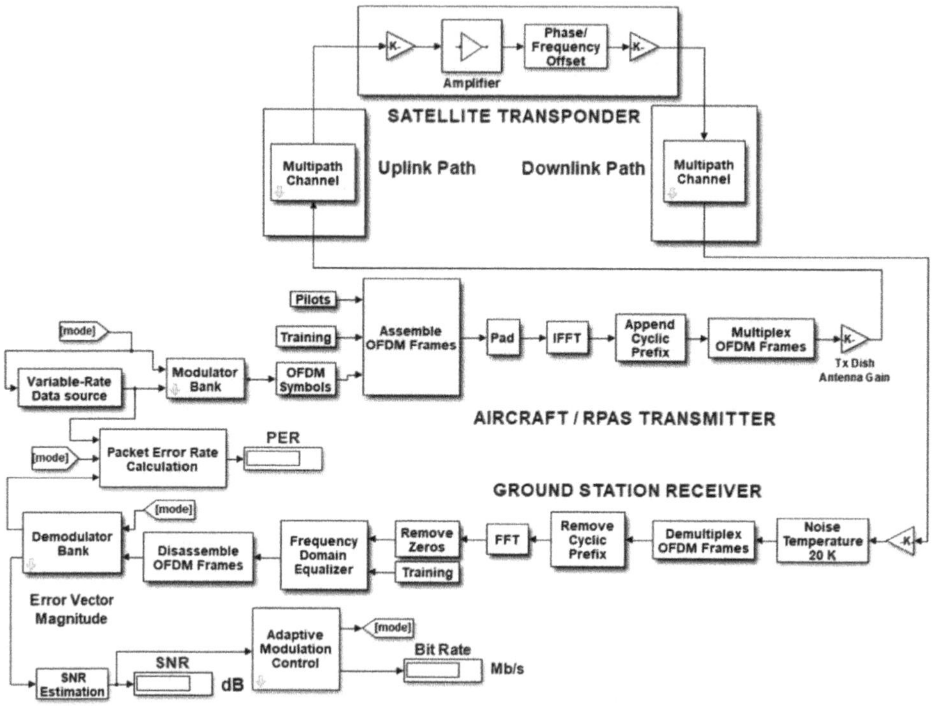

Figure 22. Multipath channel

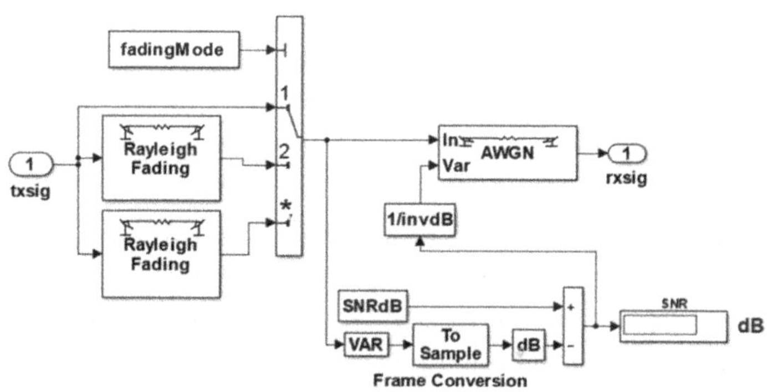

bandwidth mode, which maintains a zero level of errors in the transmission of data packets. The definition of appropriate modes involves repeating the simulation several times with a change in the threshold value.

In "Aircraft/RPAS Transmitter" the following operations are carried out: a generation of data with a certain bit rate, which changes during the simulation; coding, interleaving and modulation using one of the eight modulation schemes; transmitting OFDM signals using 52 subcarriers, 4 pilot sequences, a 64-point fast Fourier transform, and a 16-member cyclic prefix; signal amplification by an antenna amplifier.

In "Satellite Transponder", only the linear amplifier was examined.

In "Ground Station Receiver", the Viterbi decoder unit decodes the input symbols and uses the decision parameter of an unquantized type. The receiver performs reverse operations performed in the transmitter.

For "Uplink/Downlink" signal path, "Multipath Channel" is selected with Rayleigh fading (Figure 22). Frequency-independent and dispersion fading are considered. The Doppler signal frequency shift is described by the MaxDoppler parameter. The type of the Doppler spectrum with Rayleigh fading was of two types: flat, for which the angles of arrival of the signals are uniformly distributed in the azimuth and elevation angle planes, and Gaussian, which is considered a good model for multipath components with long delays for the aviation communication channel.

Due to the multipath of the channel, the signals are reflected in several places and the transmitted signal comes to the receiver along several paths, each of which can have different path lengths and associated different delays in time. In the block parameters dialog box, the discrete signal delay vector on the path determines the delay times (in seconds) for each path. In the block parameters dialog box, a discrete gain vector (in dB) of the signal for each path is also set. Two channels of signal propagation are considered in the study.

In the simulation process, the percentage of errors in the transmission of packets is determined in percent, which should always be zero during the simulation. If errors are not eliminated by modifying the type of modulation, the communication channel is considered to be closed. In this case, the SNR of the channel can have a certain, non-zero value.

Modeling of Channel Operation

The following model parameters were chosen for modeling: the phase-frequency shifts for the signal in the satellite transponder were considered to be zero; the linear amplifier of the satellite transponder had a gain of 10 dB; the Viterbi decoder tracking depth is 34; the hysteresis coefficient for adaptive modulation is 3 dB; number of subcarriers 52; the number of OFDM symbols in the transmitted block - 20; the number of OFDM symbols in the training sequence - 4. The values of the antenna gain were adopted by $G_{aircraft} = 12.4$,

$G_{satellite} = 31.1$, $G_{ground} = 62.1$ (which corresponds to antenna diameters $d_{aircraft} = 0.4$ m, $d_{satellite} = 1,0$ m, $d_{ground} = 2.0$ m at 4 GHz). The noise temperature of the satellite transmitter and the ground receiver is equal to 290 K.

For SNR values marked in the figures with arrows, the modulation type changes and the data rate increases: BPSK½ (6 Mbps), QPSK½ (12 Mbps), QPSK¾ (18 Mbps), 16QAM½ (24 Mbps), 16QAM¾ (36 Mbps), 64QAM2/3 (48 Mbps) and 64QAM¾ (54 Mbps). The data shows how large a SNR ratio should be and what type of modulation should be used for data transmission without errors with the given characteristics of the Uplink/Downlink channels and the noise temperature.

Dependencies of the whole channel SNR on "Downlink" SNR (without fading) for different given SNR values in the "Uplink" (without fading) are shown in Figure 23. The delay time of the first path is 0.0 s and the gain is 0.0 dB, the delay time of the second path is $5 \cdot 10^{-8}$ s and the gain is 4 dB. Since there is no fading, the same dependencies were obtained for Doppler frequency shifts with the parameter MaxDoppler = 10, 500 and 1000 Hz.

Figure 23. Dependencies of channel SNR on "Downlink" SNR (without fading): "Uplink" SNR (without fading): ● 10 dB, ▼ 20 dB, ▲ 30 dB, ■ 40 dB

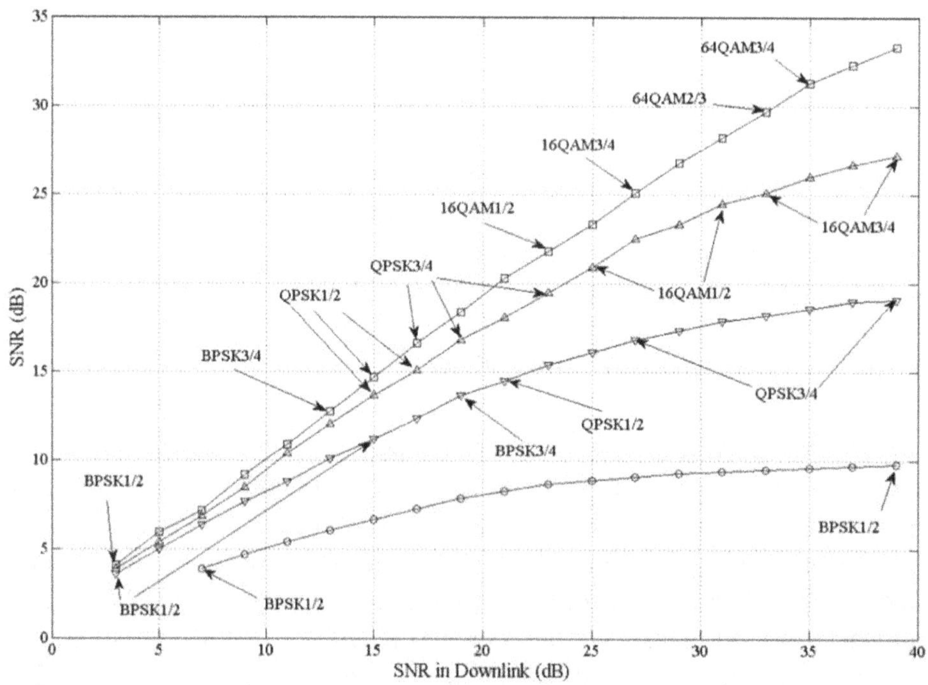

Dependencies of the whole channel SNR on “Downlink” SNR with various fading and Doppler spectrum types are shown in Figure 24. The delay time of the first path is 0.0 s and the gain is 0.0 dB, the delay time of the second path is $5 \cdot 10^{-8}$ s and the gain is 4 dB.

Dependences of the whole channel SNR on the gain of the second “Downlink” path with different fading and Doppler spectrum types are shown in Figure 25. The delay time of the first path is 0.0 s and a gain is 0.0 dB, the delay time of the second path is $5 \cdot 10^{-8}$ s.

Dependences of the whole channel SNR on the delay time τ of the second “Downlink” path with various fading and Doppler spectrum types are shown in Figure 26 ($\tau = 10^{-N}$ seconds). The delay time of the first path is 0.0 s and a gain is 0.0 dB, a gain of the second path is 4 dB.

Dependencies of the whole channel SNR on “Downlink” SNR for different types of “Uplink” are shown in Figure 27. The delay time of the first path

Figure 24. Dependencies of channel SNR on “Downlink” SNR (with fading):
*• - frequency independent fading with a flat Doppler spectrum, o - frequency-independent fading with Gaussian Doppler spectrum, ■ - dispersion fading with a flat Doppler spectrum; * - dispersion fading with a Gaussian Doppler spectrum; “Uplink” SNR (without fading) is 40 dB; MaxDoppler - 1 Hz*

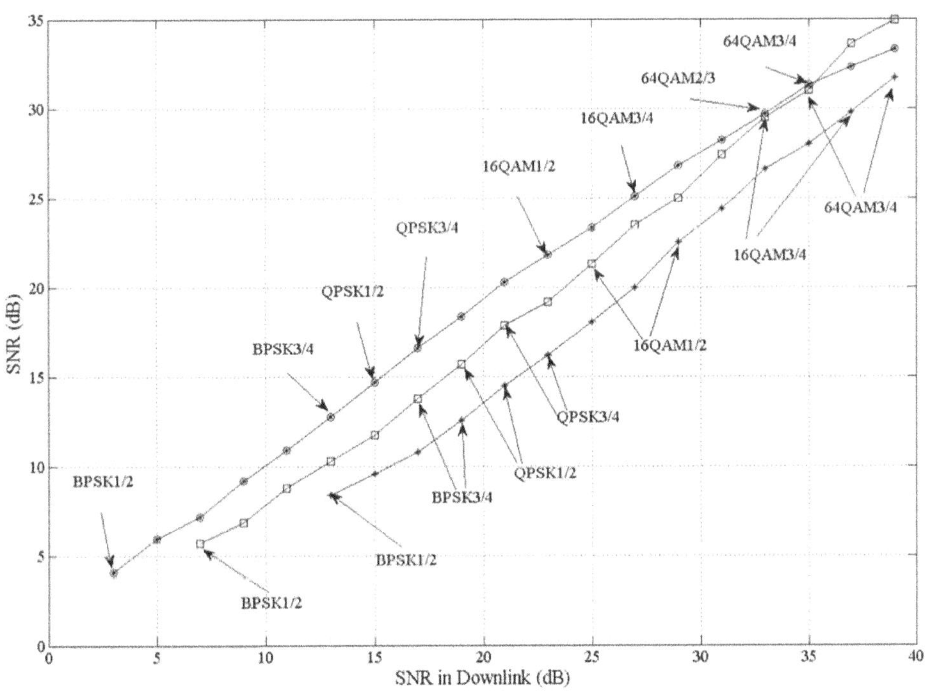

Figure 25. Dependencies of channel SNR on the gain of the second "Downlink" path: o - without fading; • - without fading in "Uplink", with frequency independent fading and with flat Doppler spectrum in "Downlink"; ■ - without fading in "Uplink", with frequency independent fading and with Gaussian Doppler spectrum in "Downlink"; ▼ - without fading in "Uplink", with dispersion fading and with flat Doppler spectrum in "Downlink"; ▲ - without fading in "Uplink", with dispersion fading and with Gaussian Doppler spectrum in "Downlink"; "Uplink" SNR (without fading) is 40 dB; "Downlink" SNR is 20 dB; MaxDoppler - 1 Hz

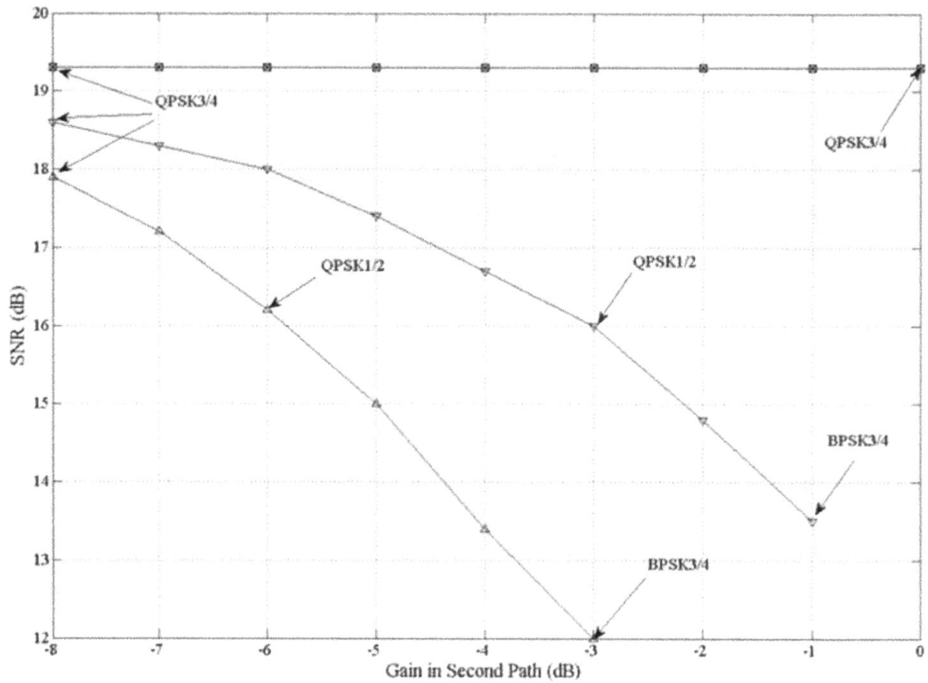

is 0.0 s and a gain is 0.0 dB, the delay time of the second channel is $5 \cdot 10^{-8}$ s and the gain is 4 dB.

SNR_{uplink} = 40 dB; MaxDoppler - 1 Hz.

Figure 28 shows dependencies of a SNR on time for path with dispersion fading and Gaussian Doppler spectrum in "Downlink" (parameter MaxDoppler - 10, 500 and 1000 Hz), with "Uplink" SNR (without fading) is 40 dB and "Downlink" SNR is 20 dB. The delay time of the first path is 0.0 s and the gain is 0.0 dB, the delay time of the second path is $5 \cdot 10^{-8}$ s and the gain is 4 dB.

Figure 26. Dependencies of channel SNR on the delay time of the second "Downlink" path:
o - without fading in "Uplink", with frequency-independent fading and with flat Doppler spectrum in "Downlink"; • - without fading in "Uplink", with frequency-independent fading and with Gaussian Doppler spectrum in "Downlink"; ■ - without fading in "Uplink", with dispersion fading and with flat Doppler spectrum in "Downlink"; ▼ - without fading in "Uplink", with dispersion fading and with Gaussian Doppler spectrum in "Downlink"; "Uplink" SNR (without fading) is 40 dB; "Downlink" SNR is 20 dB; MaxDoppler - 1 Hz; (delay time $\tau = 10^{-N}$ seconds)

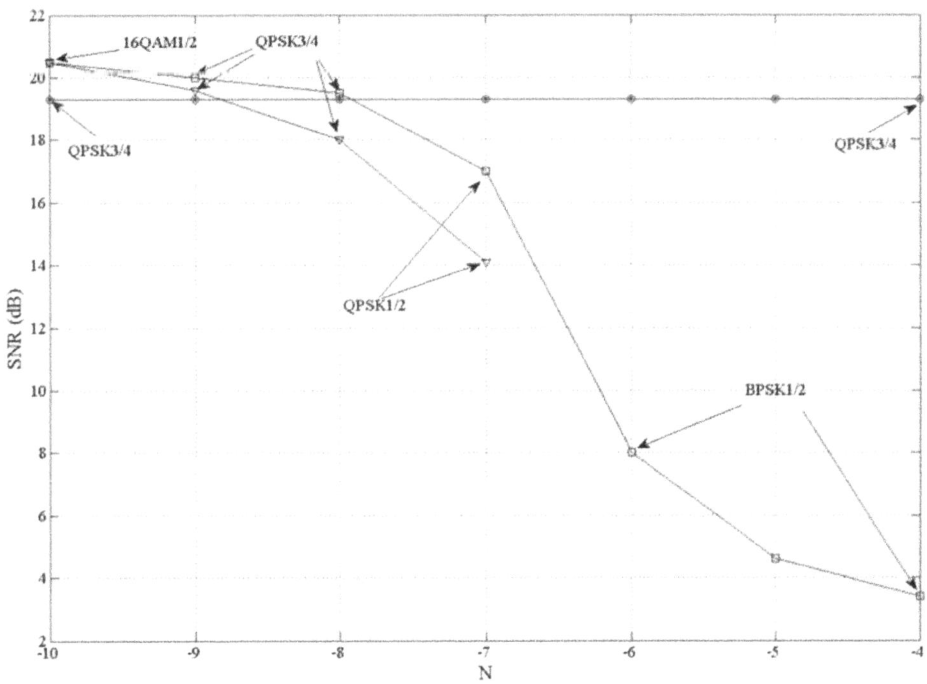

CONCLUSION

The influence of Rayleigh fading on the operation of aeronautical satellite channel with several types of Doppler spectra is considered. The "Uplink" was considered without fading. The channel "Aircraft/RPAS-Satellite-Ground Station" was studied both in "static" (Figures 23-27) with Doppler frequency shift of 1 Hz, and in "dynamics" (Figure 28) with increasing Doppler frequency shift up to 1000 Hz.

From the data in Figure 23 it follows that data transmission using all eight types of modulation in the selected modulation bank is possible, starting from the SNR value in "Aircraft/RPAS-Satellite" uplink of 40 dB.

Figure 27. Dependencies of channel SNR on "Downlink" SNR:
o – "Uplink" is "Free Space Path Loss" channel with SNR_{uplink} = 120 dB, "Downlink" is "Multipath" channel without fading; • - "Uplink" is AWGN channel SNR_{uplink} = 30 dB, "Downlink" is "Multipath" channel without fading; ■ – "Uplink/Downlink" – "Multipath" channels without fading with

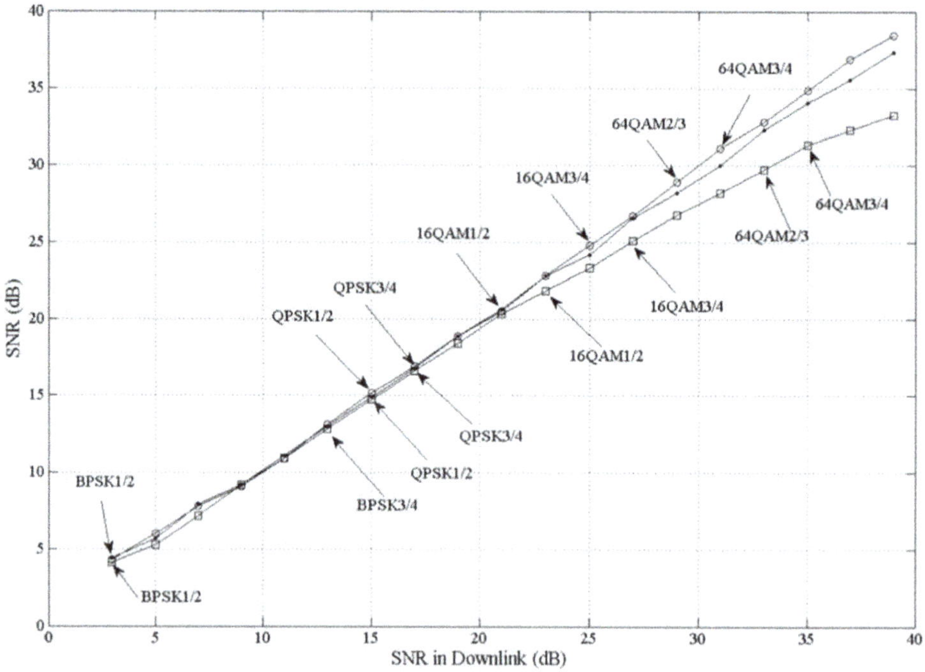

Figure 28. Dependencies of SNR on time for path with dispersion fading and Gaussian Doppler spectrum in "Downlink". MaxDopper parameter: a) 10 Hz, b) 500 Hz, c) 1000 Hz.
"Uplink" SNR (without fading) is 40 dB and "Downlink" SNR is 20 dB

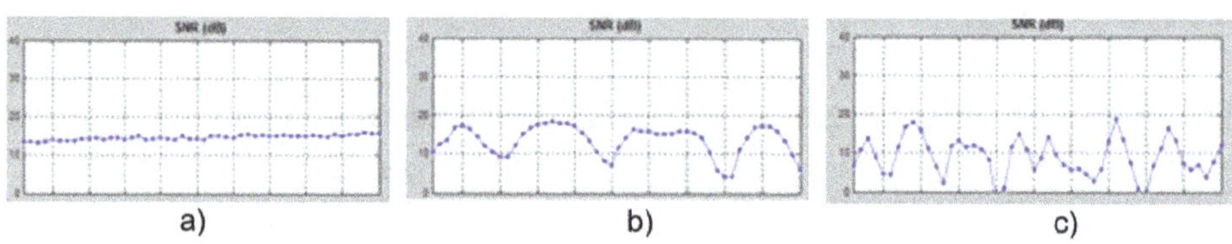

The type of the Doppler spectrum does not affect the frequency independent fading (Figure 24) and, on the contrary, has a significant effect on dispersion fading. At the same time, the lowest values of a SNR are observed for dispersion fading with a Gaussian Doppler spectrum.

The gain alteration of one "Downlink" path with fading does not change a SNR for frequency-independent fading (Figure 25) with different Doppler spectra. At the same time, a SNR for dispersion fading changes significantly with the increase in the gain of one path, leading to the closure of the channel and is very different for different Doppler spectra.

The alteration in the delay time for signal transmission via the second "Downlink" path with fading does not change a SNR for frequency-independent fading (Figure 26) with different Doppler spectra. However, a SNR ratio for dispersion fading significantly varies with the delay time of the second "Downlink" fading path, leading to the closure of the channel and is very different for different Doppler spectra.

From Figures 24-26 follows that a SNR ratio changes most strongly in the channel with dispersion fading and the Gaussian Doppler spectrum.

The effect of "Uplink" type on a transmission is shown in Figure 27. Obtained data indicate that for different "Uplinks" results are similar.

Figure 28 shows the change in the SNR of the entire channel with time as the frequency Doppler shift changes: the larger the parameter MaxDoppler, the stronger fading is. The data are presented only for dispersion fading with a Gaussian Doppler spectrum, but similar changes in the SNR with increasing frequency are observed for all types of fading considered above. Strong changes in the SNR of the entire channel over time are observed when the MaxDoppler parameter is bigger than 100 Hz.

Summarizing the obtained data, we can formulate a method for estimating the effect of fading on the parameters of the satellite communication channel:

1. It is necessary to select SNR threshold values at which a transition to a higher type of modulation will occur (in this model, such a set of threshold values a vector of seven elements, which corresponds to eight modes, each mode associated with a specific modulation scheme and convolutional coding).
2. Based on the results obtained in this study (with a given number of OFDM symbols, noise temperatures, antennas amplification and satellite transponder, phase and frequency shifts), the parameters of the communication channel can be estimated.
3. For a channel with other data transfer conditions, additional calculations must be realized using the models created.

EFFECT OF RICIAN FADING ON OFDM CHANNEL

It is expected that the number of satellite broadband subscribers worldwide will grow to about 6 million in 2020. Fading due to multipath propagation may distort and attenuate received signals on line-of-sight paths and thereby impair the performance of aeronautical radio systems (Recommendation ITU-R F.1093-1, 1997; Recommendation ITU-R P.530-16, 2015). Multipath fading is the dominant propagation factor for digital radio-relay systems operating at frequencies below about 10 GHz.

Satellite communication channels are random and time-variant. The wireless multi-path channel causes in the received signal arbitrary time dispersion, attenuation, and phase shift, known as fading. Fading is caused by interference between two or more versions of the transmitted signal that arrive at the receiver at slightly different times.

In a mobile satellite link (mobile airborne station to a satellite and a satellite to a ground station) signal attenuation is mainly due to the free space loss, shadowing and multipath propagation. A channel model that outputs signal attenuation due to clouds and precipitation as a function of time was presented in a paper (Fiebig, 1999).

A modeling method of the fade duration caused by multipath propagation on a land mobile satellite channel was proposed (Csurgai-Horváth & Bitó, 2007). The model is based on the measurement of a satellite channel and applied to calculate the model parameters. The dependency of the model parameters on the attenuation threshold was obtained and the fade duration distribution for any threshold was calculated.

Modern mobile satellite telecommunication technologies combine the spatial diversity and OFDM (Dahlman, Parkvall, & Skold, 2011; Hanzo, Akhtman, & Wang, 2011).

The aim of this study is to investigate the influence of Rician fading on messages transmission via the aeronautical satellite OFDM channel with adaptive modulation (Kharchenko, Grekhov, Ali, & Udod, 2016).

Model "OFDM_Free_Sat_Rician_802.11a" for Satellite Channel

For aeronautical satellite communication link the original model was built using the IEEE 802.11a standard and the software package MATLAB Simulink. The model ***OFDM_Free_Sat_Rician_802.11a*** (Figure 29) consists of the

Figure 29. Model OFDM_Free_Sat_Rician_802.11a for satellite link

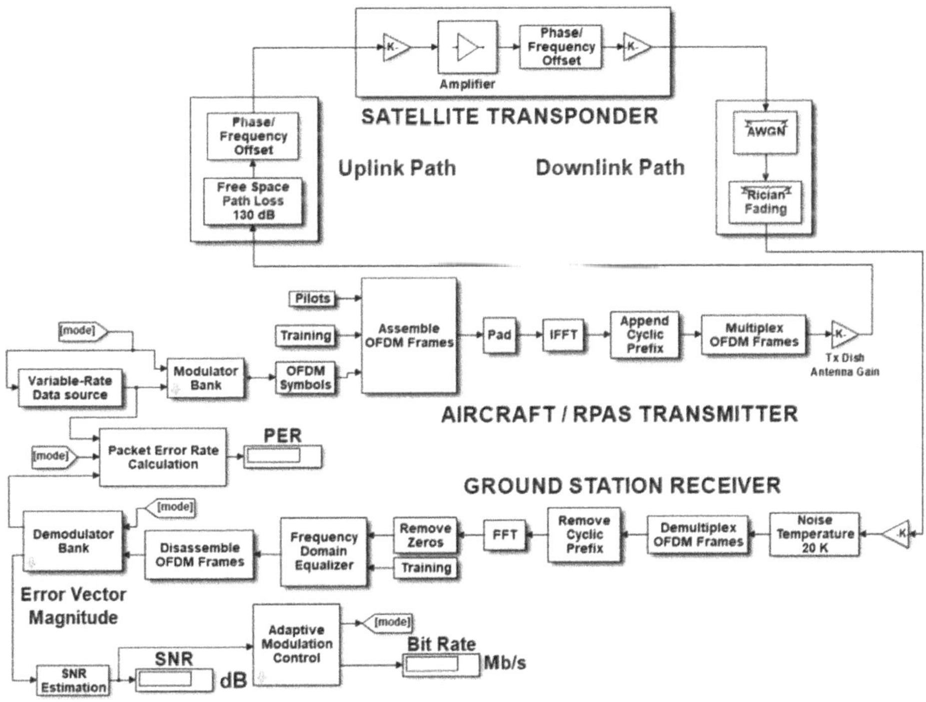

"Aircraft/RPAS Transmitter" (Variable-Rate Data Source, Modulator Bank, OFDM Transmitter, Transmitter Dish Antenna Gain), the "Uplink Path" (Free Space Path Loss, Phase/Frequency Offset), the "Satellite Transponder" (Receiver Dish Antenna Gain, Satellite Receiver System Temperature, Complex Baseband Amplifier, Phase/Frequency Offset, Transmitter Dish Antenna Gain), the "Downlink Path" (Additive White Gaussian Noise and multipath Rician fading), the "Ground Station Receiver" (Receiver Dish Antenna Gain, OFDM Receiver, Demodulator Bank), the "Packet Error Rate Calculation block", the "SNR Estimation", and the "Adaptive Modulation Control".

In the "Aircraft/RPAS Transmitter" are carried out: the generation of data with a certain bit rate that varies during the simulation; coding, interleaving, and modulation using one of the mentioned above eight modulation schemes used in the standard; OFDM signals transmission using 52 subcarriers, 4 pilot sequences, 64-point fast Fourier transform and the 16-membered cyclic prefix; using four long training sequence in the physical layer; a signal amplifying by antenna gain amplifier.

In the "Satellite Transponder", only a linear amplifier was considered in this investigation.

In the "Ground Station Receiver" the Viterbi decoder decodes the input symbols and uses the unquantized type of decision-making. The receiver performs the reverse operations performed in the transmitter.

Packet Error Rate Calculation block shows the packet error rate as a percentage and should always equal zero during investigations.

SNR Estimation block estimates the SNR based on the error vector magnitude.

Adaptive Modulation Control takes into accounts Low-SNR thresholds, Hysteresis factor, and Bit rates. Adaptive modulation systems improve the rate of transmission. The implementation of adaptive modulation is according to the channel information that is present at the transmitter. The method of making adaptive modulation in this model is according to the estimated SNR, a bit rate will be specified and then data source generates binary data according to the specified data rate in adaptive modulation control.

Error rate calculation block calculates the bit error rate, by comparing the received data with transmitted data.

The SNR ratio thresholds (in dB) at which a transition to another type of modulation takes place are given as a vector [10 11 14 18 22 26 28]. For values of the SNR ratio less than 10 dB the modulation is used BPSK½, for values between 10 dB and 11 dB - BPSK¾, between 11 dB and 14 dB - QPSK½, between 14 dB and 18 dB - QPSK¾, between 18 dB and 22 dB - 16QAM½, between 22 dB and 26 dB - 16QAM¾, between 26 dB and 28 dB - 64QAM2/3 and for values of the SNR ration greater than 28 dB - 64QAM¾. In accordance with this, the model has eight modes, each of which is associated with a specific modulation scheme and a convolutional code. The latter are determined by estimated values of the SNR ratio in the channel. Ideally, in the simulation a mode with the highest throughput must be used, which is maintained at a zero level of errors during the transmission of data packets. Determining appropriate mode involves repeating the simulation for several times with the change of a threshold.

Created model supports data rates 6, 9, 12, 18, 24, 36, 48, and 54 Mb/s, uses adaptive modulation and coding over a satellite communication channel with free space path losses and Rician fading, whereby the simulation varies the data rate dynamically.

In the "Uplink Path" Free Space Path Loss block simulates the loss of signal power due to the distance between the aircraft uplink transmitter and the satellite transponder receiver. The block reduces the amplitude of the input signal by an amount that is determined by the Loss (dB) parameter. Phase/Frequency Offset block applies a frequency and phase offset to the input signal.

In the "Downlink Path" Multipath Rician Fading Channel block implements a baseband simulation of mobile wireless communication when the transmitted signal can travel to the receiver along a dominant line-of-sight or direct path and is described by the Rician distribution:

$$p(r) = \begin{cases} \frac{r}{\sigma^2} exp \frac{-(r^2 + A^2)}{2\sigma^2} I_0\left(\frac{A_r}{\sigma^2}\right) & (A \geq 0, r \geq 0) \\ 0 & (r < 0) \end{cases}$$

where A is the amplitude of the dominant component, σ^2 is the average power or dispersion of signal fluctuations and $I_0(.)$ is the modified Bessel function of the first kind and zero-order. The parameter K in Rician distribution is the ratio between the power of the LOS component and the disperse component. Because a multipath channel reflects signals at multiple places, a transmitted signal travels to the receiver along several paths, each of which may have differing time delays and path gains. If K-factor is a vector of the same size as Discrete path delay vector, then each discrete path is a Rician fading process with a K-factor given by the corresponding element of the vector.

Relative motion between the "Satellite Transponder" and "Ground Station Receiver" causes Doppler shifts in the signal frequency. The following types of Doppler spectrum were considered.

The Flat Doppler spectrum is observed in isotropic scattering environment, where the angles of arrival are uniformly distributed in the azimuth and elevation planes. The normalized flat Doppler power spectrum is given analytically by:

$$S(f) = \frac{1}{2f_D}, |f| \leq f_D,$$

where f_D is the maximum Doppler frequency.

The Gaussian Doppler spectrum is a good model for multipath components with long delays in UHF communications and for the aeronautical channel. The normalized Gaussian Doppler power spectrum is given analytically by:

$$S_G(f) = \frac{1}{\sqrt{2\pi\sigma_G^2}} exp\left(-\frac{f^2}{2\sigma_G^2}\right).$$

where $\sigma_G = f_D / \sqrt{2}$.

The Jakes Doppler power spectrum applies to a mobile receiver. The normalized Jakes Doppler power spectrum is given analytically by:

$$S(f) = \frac{1}{\pi f_D \sqrt{1-\left(\frac{f}{f_D}\right)^2}}, |f| \le f_D,$$

where f_D is the maximum Doppler frequency.

The Rounded Doppler spectrum was proposed as an approximation to the measured Doppler spectrum of the scatter component of fixed wireless channels at 2.5 GHz. The normalized rounded Doppler spectrum is given analytically by a polynomial:

$$S(f) = C_r \left[a_0 + a_2\left(\frac{f}{f_D}\right)^2 + a_4\left(\frac{f}{f_D}\right)^4\right], |f| \le f_D$$

where $C_r = \frac{1}{2 f_D \left[a_0 + \frac{a_2}{3} + \frac{a_4}{5}\right]}$ a_0, a_2, a_4 are real finite coefficients.

The Restricted Jakes Doppler spectrum is used when the angles of arrival are not uniformly distributed and the Jakes power spectrum does not cover the full Doppler bandwidth. This exception also applies to the case where the antenna pattern is directional. The normalized Restricted Jakes Doppler power spectrum is given analytically by:

$$S(f) = \frac{A_r}{\pi f_D \sqrt{1-\left(\frac{f}{f_D}\right)^2}}, 0 \le f_{min} \le |f| \le f_{max} \le f_D,$$

where

$$A_r = \frac{1}{\frac{2}{\pi}\left[arcsin\left(\frac{f_{max}}{f_D}\right) - arcsin\left(\frac{f_{min}}{f_D}\right)\right]},$$

f_{min} and f_{max} denote the minimum and maximum frequencies where the spectrum is nonzero.

The Asymmetrical Jakes Doppler power spectrum is given analytically by:

$$S(f) = \frac{A_r}{\pi f_D \sqrt{1 - \left(\frac{f}{f_D}\right)^2}}, 0 \le f_{min} \le |f| \le f_{max} \le f_D,$$

where

$$A_r = \frac{1}{\frac{1}{\pi}\left[arcsin\left(\frac{f_{max}}{f_D}\right) - arcsin\left(\frac{f_{min}}{f_D}\right)\right]},$$

f_{min} and f_{max} denote the minimum and maximum frequencies where the spectrum is nonzero.

The Bi-Gaussian Doppler spectrum consists of two frequency-shifted Gaussian spectra and is used in modeling long echoes for urban and hilly terrain profiles. The normalized Bi-Gaussian Doppler spectrum is given analytically by:

$$S_G(f) = A_G\left[\frac{C_{G1}}{\sqrt{2\pi\sigma_{G1}^2}} exp\left(-\frac{(f - f_{G1})^2}{2\sigma_{G1}^2}\right) + \frac{C_{G2}}{\sqrt{2\pi\sigma_{G2}^2}} exp\left(-\frac{(f - f_{G2})^2}{2\sigma_{G2}^2}\right)\right],$$

where σ_{G1} and σ_{G2} are standard deviations, f_{G1} and f_{G2} are center frequencies, C_{G1} and C_{G2} are power gains, and $A_G = \frac{1}{C_{G1} + C_{G2}}$ is a normalization coefficient.

The Bell Doppler spectrum was proposed for the indoor MIMO 802.11n channel modeling. The normalized bell Doppler spectrum is given analytically by:

$$S(f) = \frac{C_b}{1 + A\left(\frac{f}{f_D}\right)^2},$$

where $|f| \le f_D$ and $C_b \frac{\sqrt{A}}{\pi f_D}$.

Aeronautical Satellite Channel Simulation

To simulate a channel operation the following parameters were chosen for the model: the value of signal phase-frequency shifts in the uplink and satellite transponder is equal to zero; the gain of linear amplifier in satellite transponder was taken 10 dB; Viterbi traceback depth is 34, hysteresis factor for adaptive modulation is 3 dB, the number of OFDM symbols in the training sequence 4, the number of OFDM symbols per transmit block are 20. The values of antenna gains were taken $G_{aircraft} = 12{,}4$; $G_{satellite} = 31{,}1$; $G_{ground} = 62{,}1$ (at the signal frequency of 4 GHz that corresponds to antenna diameters $d_{aircraft} = 0{,}4$ m; $d_{satellite} = 1{,}0$ m, $d_{ground} = 2{,}0$ m). A noise temperature of the satellite transmitter amplifier and ground station receiver is 290 K.

For SNR values, marked by the arrow, the modulation type is changed and the data rate increases: BPSK½ (6 Mb/s), QPSK½ (12Mb/s), QPSK3/4 (18 Mb/s), 16QAM½ (24 Mb/s), 16QAM3/4 (36 Mb/s), 64QAM2/3 (48 Mb/s) and 64QAM3/4 (54 Mb/s). The data show how big should be the SNR and what type of modulation to be used for data transfer without errors for given uplink/downlink characteristics and a noise temperature.

Fading causes the signal to become diffuse. The *K*-factor parameter, which is part of the statistical description of the Rician distribution, represents the ratio between the power in the LOS component and the power in the diffuse component. The ratio is expressed linearly, not in decibels. The *K*-factor parameters control the gain's partition into LOS and diffuse components.

If the *K*-factor parameter is a vector then each discrete path is a Rician fading process with a *K*-factor given by the corresponding element of the vector. It is possible to attribute the LOS component a Doppler shift, through the Doppler shifts of LOS components parameter, and an initial phase, through the Initial phases of LOS components. The Doppler shifts and initial phases of LOS components parameters must be of the same size as the *K*-factor parameter.

Relative motion of the transmitter and receiver causes Doppler shifts in the signal frequency. Data in Figures 30-32 are given for maximum Doppler frequency 1 Hz (that means no relative motion) and in Figure 33 – for 1000 Hz.

In Figure 30 dependencies of Estimated SNR in channel on SNR_{AWGN} in a downlink are shown for different proportions between a power in the LOS component and a power in the diffusive component: K = [1 0], [0.75 0.25] and [0.5 0.5]. From data on Figure 30 (for a free space path loss in Uplink 90 dB) follows that these dependencies coincide. For the lower free space path loss in Uplink (80 dB) this dependence shifts upward, and for the higher one (100 dB) shifts downward. At the same time, a data transmission with a certain modulation for a bigger free space path loss in Uplink will be possible for bigger SNR_{AWGN} in a downlink. For example, for a free space path loss in Uplink 80 dB - QPSK1/2 modulation is observed for SNR_{AWGN} = 3 dB, for a free space path loss in Uplink 90 dB - for SNR_{AWGN} = 11 dB, and for a free space path loss in Uplink 100 dB - for SNR_{AWGN} = 21 dB. Data on Figure 30 are given for the Flat and Gaussian Doppler spectrum types. For other types of Doppler spectra the mentioned dependencies are similar.

In Figure 31 a dependence of Estimated SNR in channel on a Gain in the diffusive component is given for a certain value of a free space path loss in Uplink (90 dB) and K = [0.5 0.5]. Estimated SNR does not depend on a Gain in the diffusive component and varies with SNR_{AWGN}: for SNR_{AWGN} = 10 dB a modulation QPSK1/2 is observed, for SNR_{AWGN} = 20 dB - a modulation 16QAM1/2, and for SNR_{AWGN} = 30 dB - a modulation 64QAM3/4. The Doppler spectrum type has practically no effect on the considered dependence.

Because a multipath channel reflects signals at multiple places, a transmitted signal travels to the receiver along several paths, each of which may have differing lengths and associated time delays. The Discrete path delay vector specifies the time delay for each path.

In Figure 32 a dependence of Estimated SNR in channel on the Delay in Disperse Component is given for a certain value of a free space path loss in Uplink (90 dB), K = [0.5 0.5], SNR_{AWGN} = 20 dB and different types of

Figure 30. Dependence of estimated SNR on SNR in AWGN downlink for different K and free space path loss for Flat and Gaussian Doppler spectrum types, MaxDoppler 1 Hz

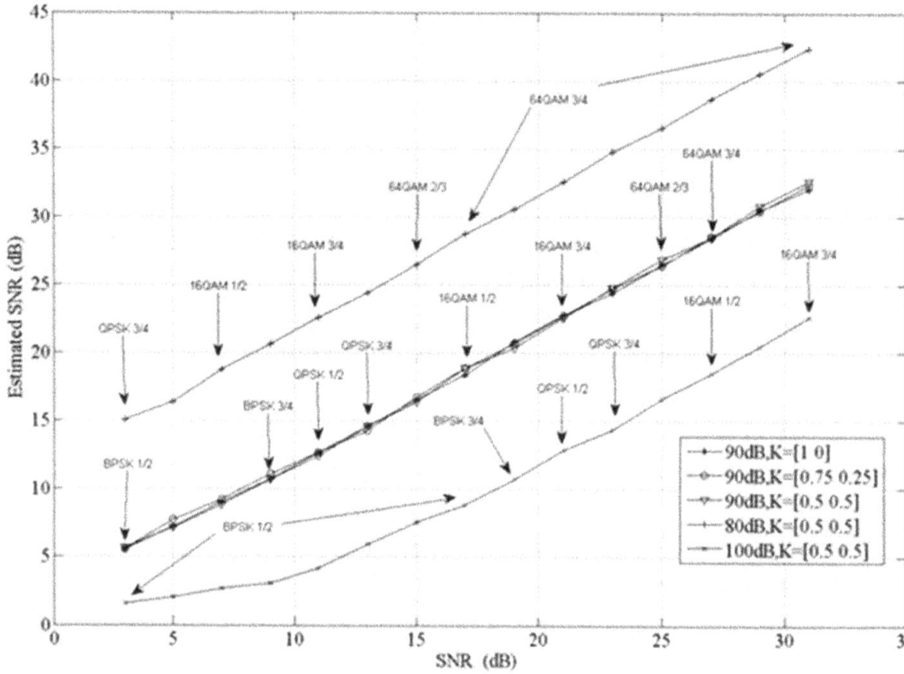

Doppler spectrum. For small Delay times ($\tau = 10^{-10}$ - 10^{-7} s) received data practically coincide for all types of Doppler spectrum being described by almost the same values of Estimated SNR and a modulation type – 16QAM1/2. Increasing of the Delay value in Disperse Component ($\tau = 10^{-7}$ - 10^{-5} s) leads to a dramatic reduction of Estimated SNR (from ≈ 20-21 dB to ≈ 2-5 dB), the emergence of differences between the spectra and the transition to a modulation BPSK1/2 already at the Delay time 10^{-6} s.

The strongest decrease of Estimated SNR for $\tau = 10^{-5}$ s experiences Rounded Doppler spectrum and then in order of Estimated SNR increasing are arranged: Bell, Bi-Gauss, Jakes, Gaussian, Asymmetrical Jakes and Restricted Jakes Doppler spectra.

Figure 33 shows the nature of Estimated SNR and Bit rate changes over time in the presence of the Rician Fading in channel for different types of Doppler spectra.

Figure 31. Dependence of estimated SNR on gain in diffusive component
(free space path loss in Uplink 90 dB, K = [0.5 0.5], SNR in AWGN Downlink 10 dB (QPSK½), 20 dB (16QAM½), 30 dB (64QAM3/4), MaxDoppler 1 Hz)

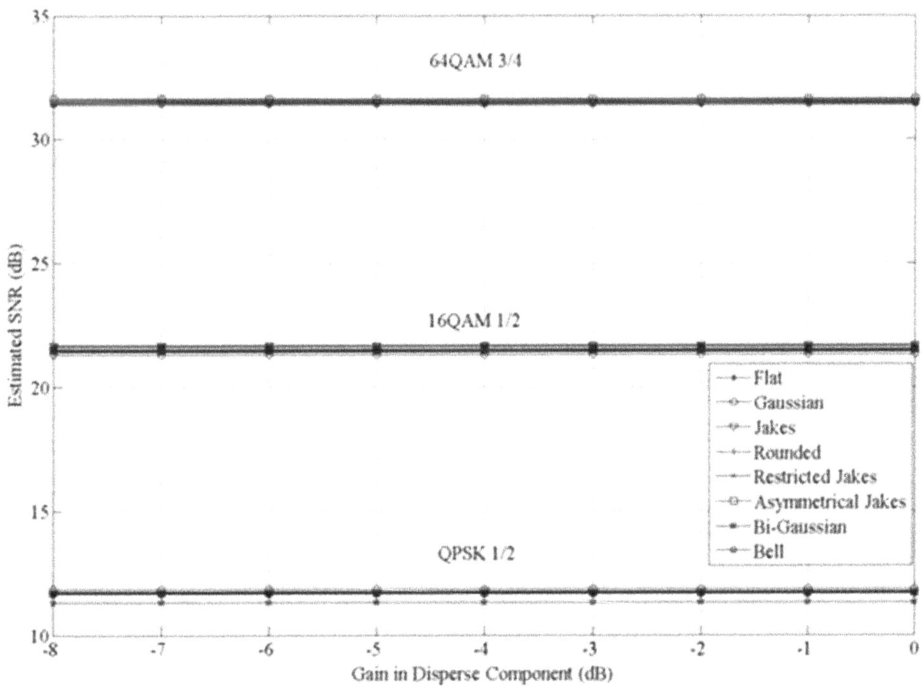

CONCLUSION

The model of aeronautical satellite OFDM link with Rician fading is developed for the first time on a basis of IEEE 802.11a standard and used for channel parameters evaluation. On the basis of data received under given conditions (a number of OFDM symbols, noise temperature, gains of antenna dishes and the satellite transponder amplifier type) the channel parameters were estimated: the level of free space loss, proportions between a power in the LOS component and in the diffusive component for which the satellite communication channel is "open"; the type of a modulation and data transfer rate, which are possible under the given conditions. The impact of a gain and a delay time in diffusive component on SNR and Bit rate was investigated for different Doppler spectrum types. Dependence of Estimated SNR and Bit rate on time for different Doppler spectrum types and Doppler frequency shifts was obtained.

Figure 32. Dependence of estimated SNR on delay in diffusive component
delay time $\tau = 10^{-N}$ s, (free space path loss in Uplink 90 dB, K = [0.5 0.5], SNR in AWGN Downlink 20 dB, MaxDoppler 1 Hz)

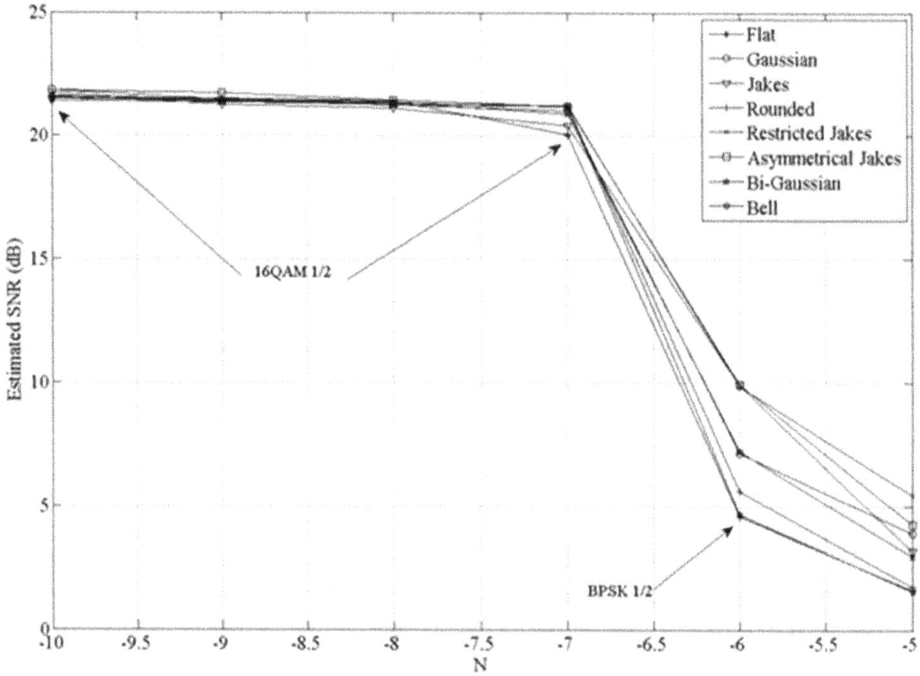

The developed model allows predicting the operation of the communications channel with Rician fading and can be helpful for designing communication systems.

IMPACT OF NONLINEARITY ON OFDM CHANNEL WITH RAYLEIGH FADING

RPAS are a new way of using flying machines. RPAS are distinguished from manned aircraft by the data link connecting the remote pilot station with the remotely piloted aircraft and used for command and control (C2) of RPAS and as a relay for communications between ATC operator and the remote pilot. The combination of these two functions is termed C3 – command, control and ATC communications.

Figure 33. Dependence of estimated SNR and Bit rate on time for different Doppler spectrum types (free space path loss in Uplink 90 dB, K = [0.5 0.5], SNR_{AWGN} in AWGN Downlink 20 dB, average path gain vector (dB): [0 -4] discrete path delay vector (s) [0.0 5.0e-8], modulation QPSK1/2, MaxDoppler 1000 Hz)

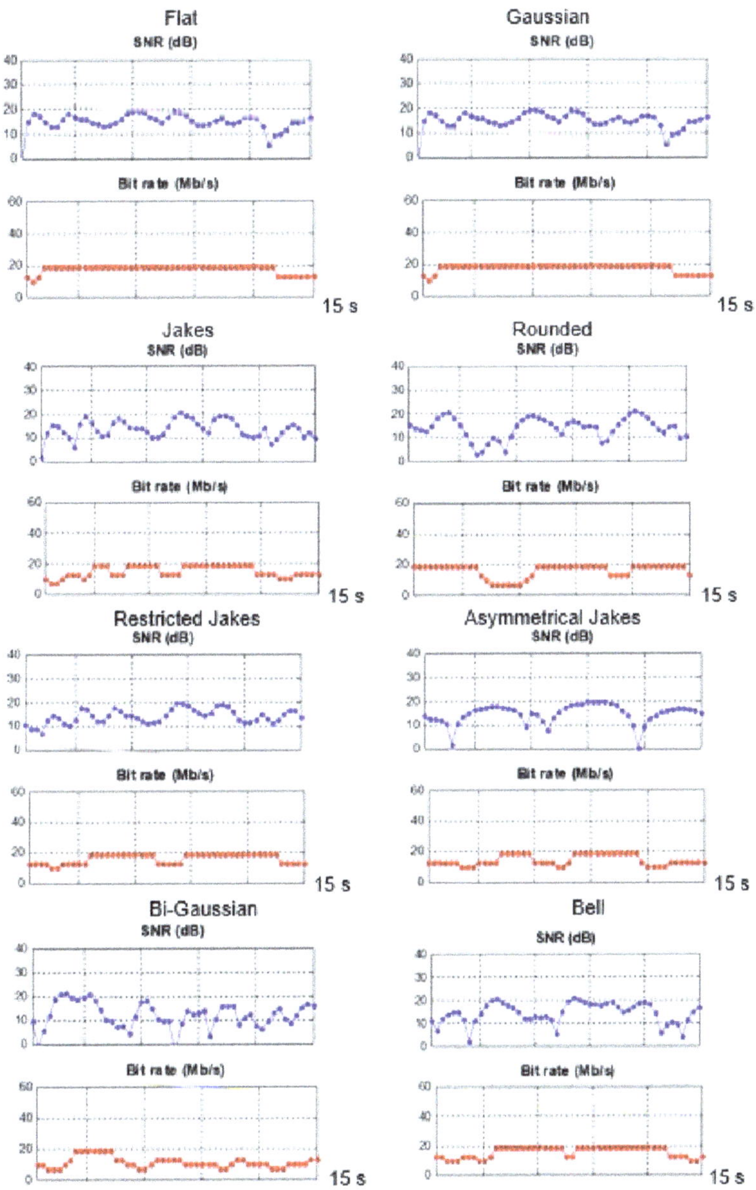

The use of RPAS is characterized by a wide range of applications (military, emergency services, surveying applications, agriculture, etc.) as well as a huge increasing in the complexity of flight tasks in each individual area. These circumstances call for RPAS flight control systems capable of performing a variety of tasks, including complex radio navigation and communication conditions. Basic requirements for RPAS data rate are stated in the NATO standards (STANAG 4609/AEDP-8, 2009; STANAG 7023/AEDP-9, 2009; STANAG 4607/AEDP-7, 2010). Standards define two classes of data: sensor and support. Sensory data comes from sensors forming synthetic aperture radar, infrared and television cameras, etc.

RPAS satellite mobile communications system has to integrate high-speed radar and trajectory control data, video and multimedia traffic. The promising approach for this is adaptive OFDM.

ADS-B signals transmitted from RPAS via satellite are affected by a free path loss, a frequency offset, a phase noise, a fading, a noise temperature, and amplifier nonlinearities. Data traffic for RPAS networks must take into account propagation delays, limited energy and power, relatively high channel error rates, and time-varying channel conditions. Digital multi-carrier modulation technique OFDM has been adopted as physical layer scheme of broadband wireless air interface standards.

Nonlinear distortion is a source of major degradation of modulation fidelity in multicarrier systems with OFDM signals. The source of nonlinear distortions is the radio frequency transmitter power amplifier. RPAS satellite communication channels are random and time-variant. Multipath fading is the dominant propagation factor for RPAS digital communication systems operating at frequencies below 10 GHz. Fading due to multipath propagation may distort and attenuate received signals and impair the performance of RPAS communication systems. Aircraft/RPAS satellite channel fading and nonlinearity are critical for wireless communications systems. Therefore the aim of this study is: 1) to design a model of satellite OFDM communication channel "Aircraft/RPAS–Satellite–Ground Station" with adaptive modulation; 2) to calculate parameters of a channel with Rayleigh fading and different types of nonlinearities; 3) to analyze the impact of fading and the nonlinearities type on parameters of satellite communication channel (Grekhov & Kondratiuk, 2017).

Model "OFDM_AWGN_Sat_Multipath_802.11a" for Satellite Channel

The model *OFDM_AWGN_Sat_Multipath_802.11a* (Figure 34) consists of "Aircraft/RPAS Transmitter", "Ground Station", "Uplink", "Downlink", and "Satellite Transponder".

Parameters settings for the model are the following: a number of OFDM symbols per transmit block is 20, a number of OFDM symbols in training sequence is 4, OFDM transmission uses 52 subcarriers, 4 pilots, 64-point Fast Fourier Transform, and 16-sample cyclic prefix. Low-SNR thresholds (dB) vector is [10 11 14 18 22 26 28] (where the SNR less than 10 dB is for BPSK ½, between 10 dB and 11 dB - for BPSK ¾, between 11 dB and 14 dB - for QPSK ½, between 14 dB and 18 dB - for QPSK ¾, between 18 dB and 22 dB - for 16-QAM ½, between 22 dB and 26 dB - for 16-QAM ¾, between 26 dB and 28 dB – for QAM 2/3 and more than 28 dB - for 64-QAM ¾).

Figure 34. Model of OFDM_AWGN_Sat_Multipath_802.11a for satellite link

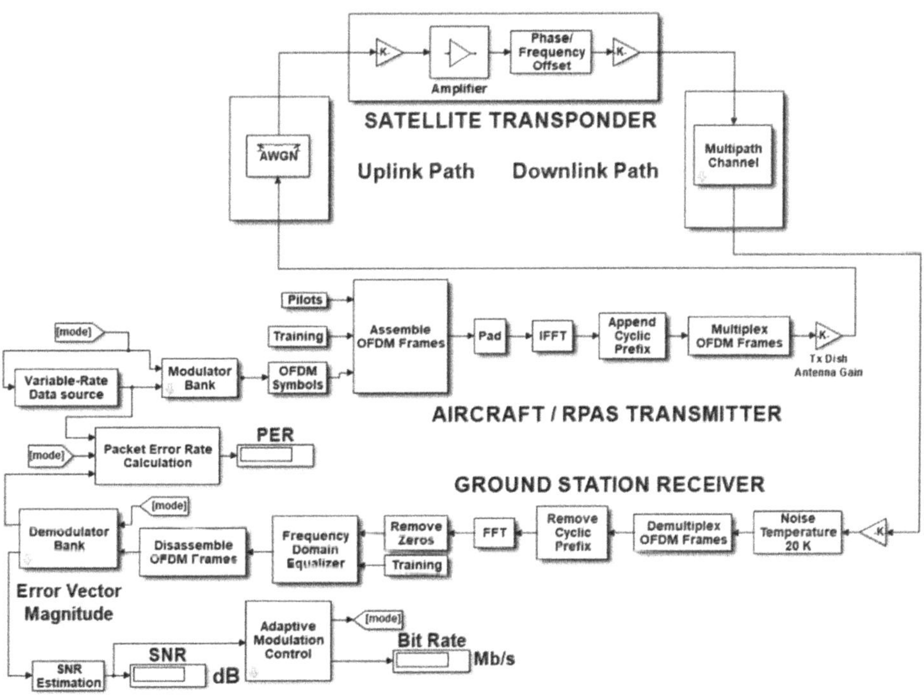

Satellite Communication Channel Simulation

For calculations the following parameters were set up: Aircraft/RPAS antenna gain was taken 12.4 (an antenna diameter ≈ 0.4 m at 4 GHz), ground station antenna gain – 62.2 (an antenna diameter ≈ 2.0 m at 4 GHz), satellite antennas gain – 31.1 (an antenna diameter ≈ 1.0 m at 4 GHz); phase/frequency offsets at a satellite transponder are equal to zero; noise temperatures of a satellite amplifier and a ground receiver are equal to 290 K.

SNR dependencies in a ground receiver on the SNR in downlink for the SNR in uplink 40 dB and different types of satellite amplifier nonlinearity and modulation modes are given in Figures 35–39. During modeling the value of a packet error rate was kept at zero by changing the type of modulation (using the SNR estimation in a ground receiver and adaptive rate control). In accordance with this, the SNR was changed.

The options for the modeling of satellite amplifier nonlinearity are Linear, Cubic Polynomial, Hyperbolic Tangent, Ghorbani and Rapp models.

For Cubic Polynomial and Hyperbolic Tangent Models the satellite Amplifier block applies the AM/AM nonlinearity by using the third-order input intercept point IIP3 = 100 dBm parameter and uses the AM/PM conversion (10 degrees per dB) parameter.

Figure 35. Dependencies of the SNR in a receiver on the SNR in downlink at the SNR in uplink 40 dB for Linear Amplifier model

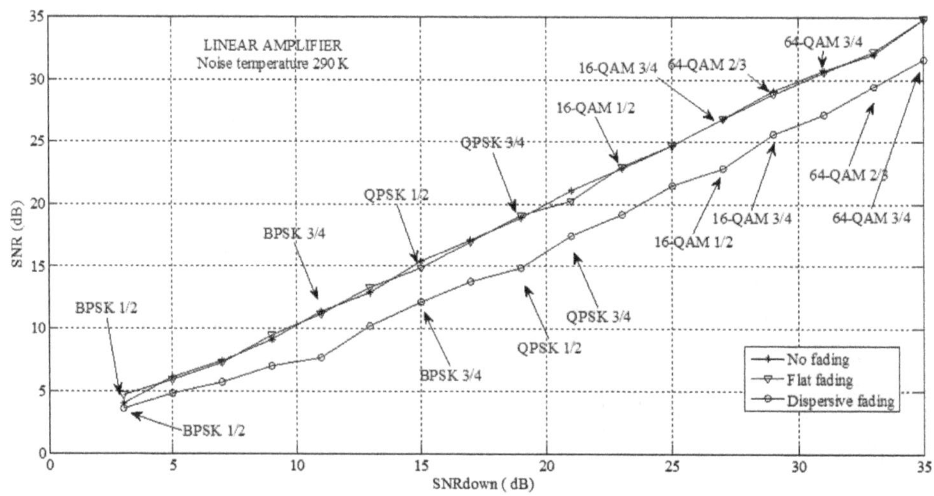

Figure 36. Dependencies of the SNR in a receiver on the SNR in downlink at the SNR in uplink 40 dB for Hyperbolic Tangent model

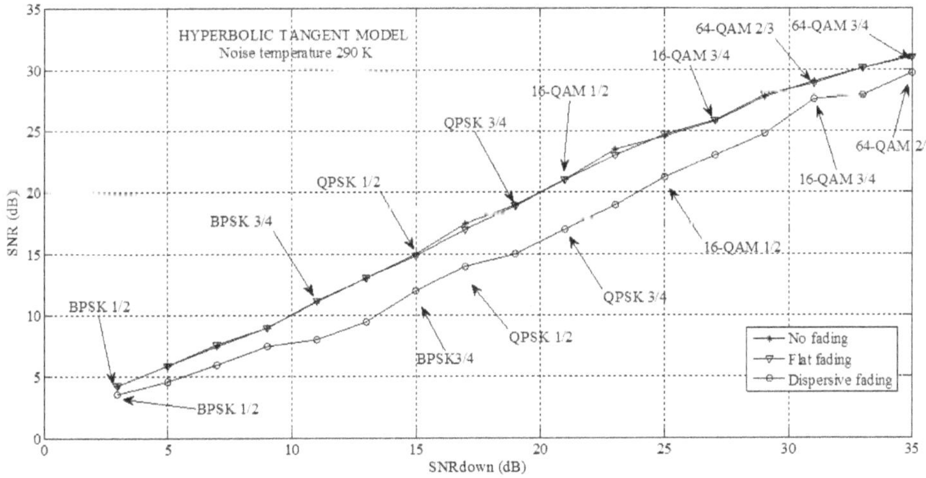

Figure 37. Dependencies of the SNR in a receiver on the SNR in downlink at the SNR in uplink 40 dB for Cubic Polynomial model

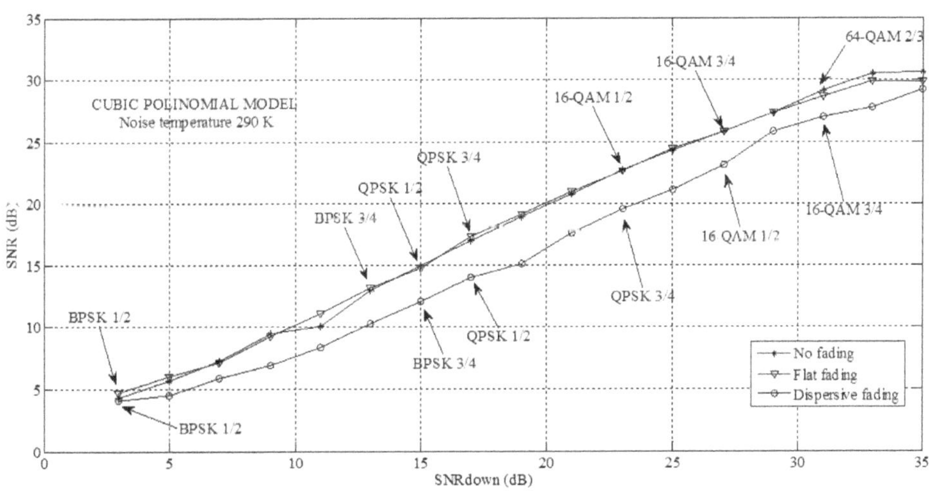

For Ghorbani Model the Input scaling (-1.5957 dB) parameter scales the input signal before the nonlinearity is applied. The AM/AM parameters [x1 = 8.1081, x2 = 1.5413, x3 = 6.5202, x4 = -0.0718] are used to compute the amplitude gain for an input signal. The AM/PM parameters [y1 = 4.6645,

Figure 38. Dependencies of the SNR in a receiver on the SNR in downlink at the SNR in uplink 40 dB for Ghorbani model

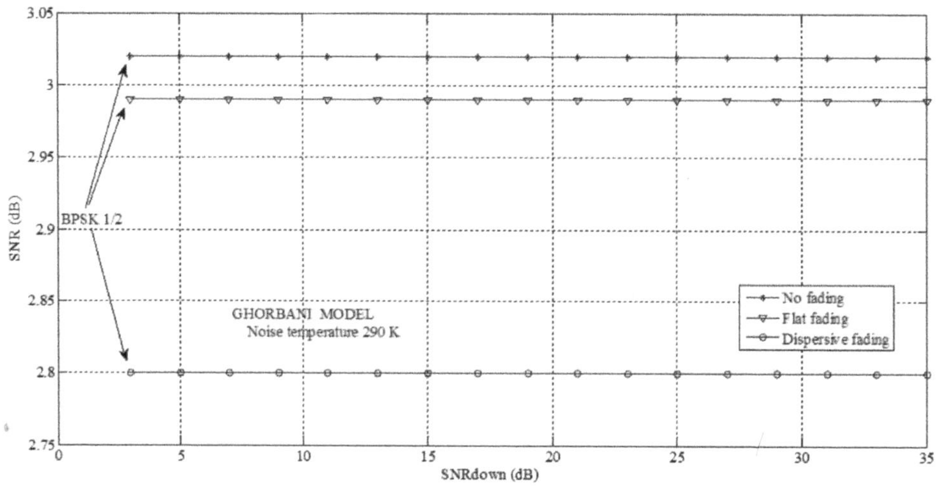

Figure 39. Dependencies of the SNR in a receiver on the SNR in downlink at the SNR in uplink 40 dB for Rapp model

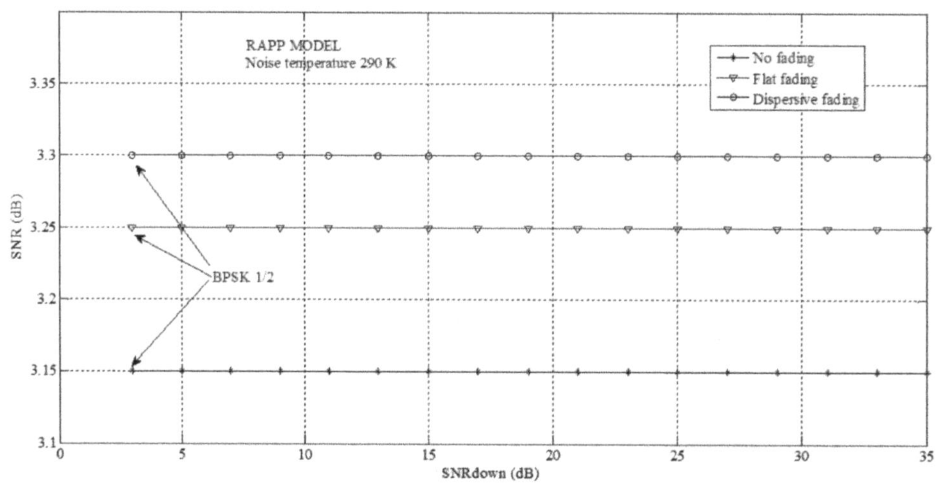

y2 = 2.0965, y3 = 10.88, y4 = -0.003] are used to compute the phase change for an input signal. For Rapp Model the amplitude gain for an input signal is computed with the Smoothness factor S=0.5 and the Output saturation level Q_{sat} = 1. The Rapp model does not apply a phase change to the input signal.

Doppler Frequency Shift

In the case of information exchange with mobile objects, there is an uncertainty of the frequency caused by the Doppler effect. In this connection, the impact of this effect on the OFDM channels of satellite telecommunications systems with Rayleigh fading is analyzed in this study.

Diagrams in Figures 40-44 allow to compare transmitted data T_x, SNR, bit rates, BER, signal constellations, equalized power spectra and bandlimited impulse responses in both downlink channels (Figure 22) for different types of satellite amplifier nonlinearities in case of dispersive fading. The numbers on the displays in the lower left corner show the values of a SNR and a bit rate at given moment of simulation time.

The delay time of the first path is 0.0 s and the gain is 0.0 dB, the delay time of the second path is $5 \cdot 10^{-8}$ s and the gain is -4 dB. Doppler spectrum type for all cases is Gaussian, the parameter MaxDoppler is 1000 Hz, a SNR in uplink is 40 dB and in downlink is 30 dB for all cases.

Figure 40. Dispersive fading in case of Linear satellite amplifier (MaxDoppler 1000 Hz)

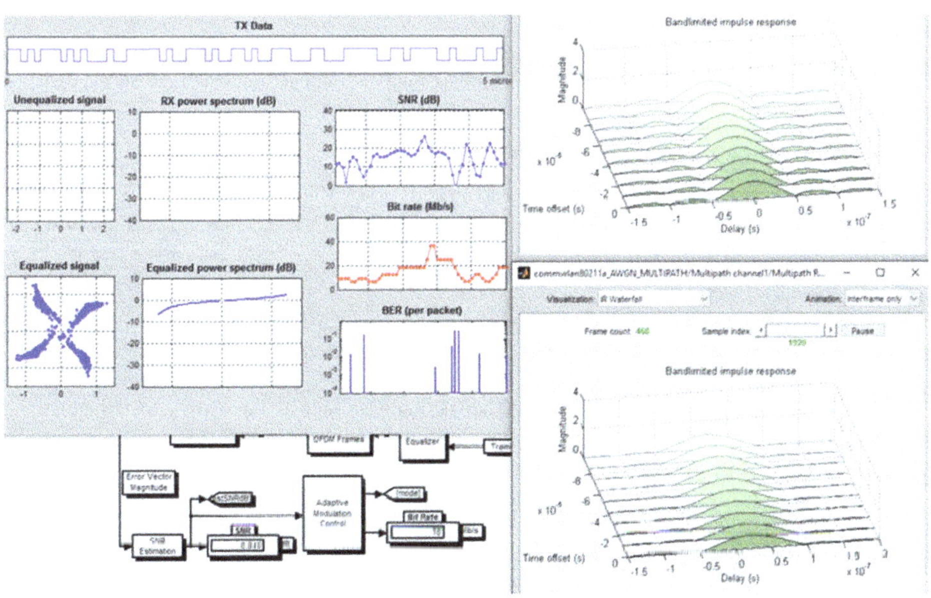

Figure 41. Dispersive fading in case of Cubic model (MaxDoppler 1000 Hz, IIP3= 100 dBm)

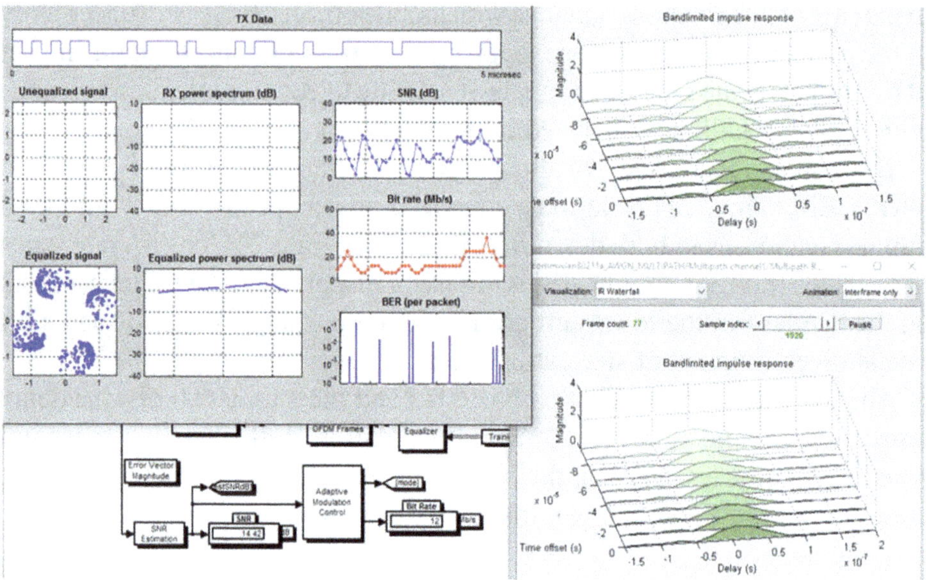

Figure 42. Dispersive fading in case of Hyperbolic model (MaxDoppler 1000 Hz, IIP3= 100 dBm)

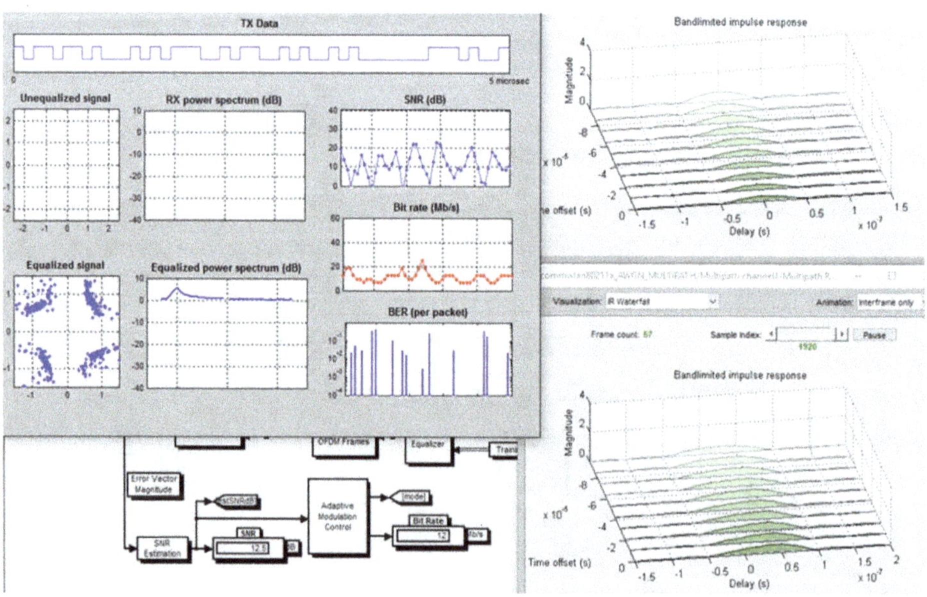

Figure 43. Dispersive fading in case of Ghorbani model (MaxDoppler 1000 Hz)

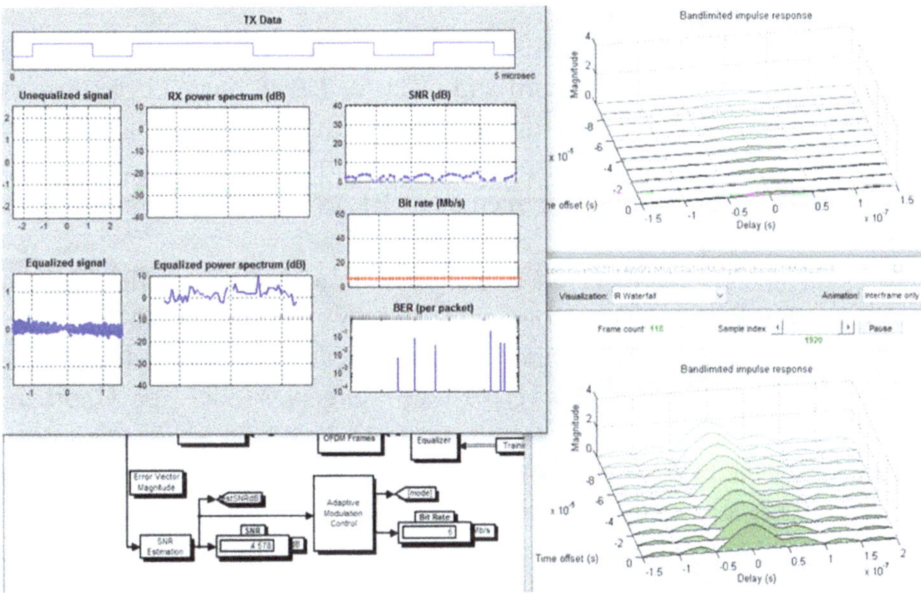

Figure 44. Dispersive fading in case of Rapp model (MaxDoppler 1000 Hz)

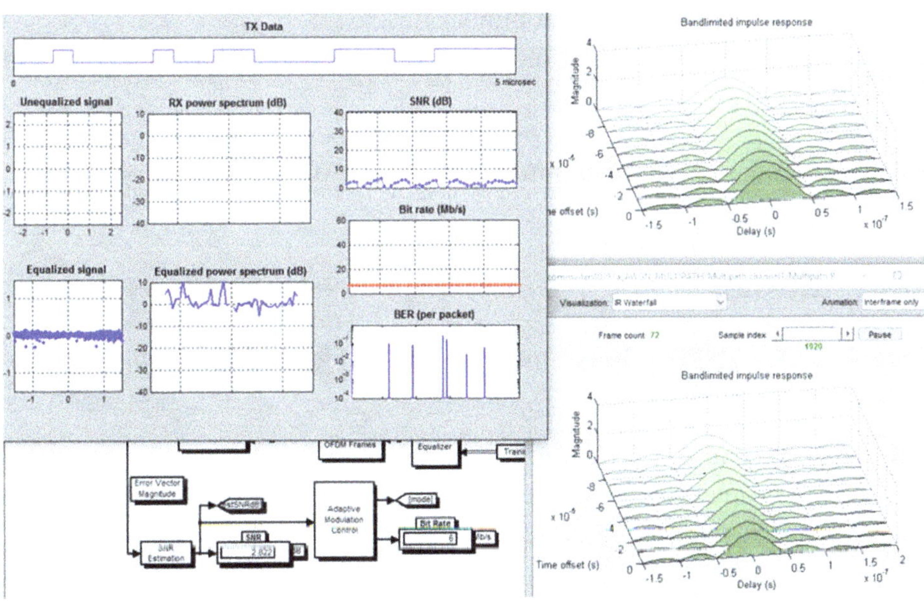

CONCLUSION

The model of RPAS satellite OFDM link is developed for the first time on a basis of IEEE 802.11a standard and used for RPAS channel parameters evaluation. Created model takes into account the Rayleigh fading in downlink and nonlinearities in satellite transponder. Proposed approach can be considered as a method for estimating parameters of RPAS satellite communication channel with fading.

Dependencies were received of the SNR in a receiver on the SNR in downlink for different types of a modulation (BPSK, QPSK, 16QAM, 64QAM) and bit rates for memoryless nonlinear satellite amplifiers (the Cubic Polynomial Model, the Hyperbolic Tangent Model, the Ghorbani Model and the Rapp Model).

For the selected values of a third-order intercept point results for the Cubic Polynomial Model and the Hyperbolic Tangent Model are similar to the Linear Model, but substantially differ from the Ghorbani Model and the Rapp Model. For the Ghorbani and Rapp Models data are observed at very low meanings of the SNR in a receiver and fluctuate. These data substantially differ for no fading, flat and dispersive fading cases and are given for average meanings.

A downlink with "Rayleigh Fading" has three modes: no fading, flat and dispersive fading. For dispersive fading, the results turn out to be the worst for all the types of considered nonlinearities.

It is shown how the type of modulation varies depending on the level of the SNR and the type of fading.

The developed model allows predicting the operation of the channel with Rayleigh fading and can be helpful for designing of communication systems.

REFERENCES

Baddeley, A. (2005). Going forward with JTRS. *Military Information Technology*, *9*(7), 8–13.

Cioni, S., Corazza, G. E., Neri, M., & Vanelli-Coralli, A. (2006). On the use of OFDM radio interface for satellite digital multimedia broadcasting systems. *International Journal of Satellite Communications and Networking*, *24*(2), 153–167. doi:10.1002at.836

Cioni, S., Corazza, G. E., Neri, M., & Vanelli-Coralli, A. (2006). On the use of OFDM radio interface for satellite digital multimedia broadcasting systems. *International Journal of Satellite Communications and Networking*, *24*(2), 153–167. doi:10.1002at.836

Csurgai-Horváth, L., & Bitó, J. (2007). Multipath propagation fade duration modeling of land mobile satellite radio channel. *Híradástechnika*, *LXII*(7), 22–26.

Dahlman, E., Parkvall, S., & Skold, J. (2011). *4G LTE/LTE-Advanced for Mobile Broadband*. Oxford, UK: Academic Press.

El-Khatib, Z., MacEachern, A., & Mahmoud, S. A. (2012). Distributed CMOS bidirectional amplifiers: broadbanding and linearization techniques. In Modulation schemes effect on RF power amplifier nonlinearity and RFPA linearization techniques. Springer, doi:10.1007/978-1-4614-0272-5

Fiebig, U. (1999). Modeling rain fading in satellite communications links. *Vehicular Technology Conference. IEEE VTS 50th*, 3, 1422 – 1426. 10.1109/VETECF.1999.801497

Ghorbani, A., & Sheikhan, M. (1991). The effect of solid state power amplifiers nonlinearities on MPSK and M-QAM signal transmission. *Sixth Int. Conference on Digital Processing of Signals in Comm.*, 193-197, 1991.

Global Air Navigation Plan for CNS/ATM Systems ICAO. (1998). *Doc. 9750*. Retrieved from http://www.icao.int/publications/Documents/9750_2ed_en.pdf

Gregorio, F. H. (2007). *Analysis and compensation of nonlinear power amplifier effects in multi-antenna OFDM systems*. Dissertation for the degree of Doctor of Science in Technology. Retrieved from http://lib.tkk.fi/Diss/2007/isbn9789512290017

Grekhov, A. M., & Kondratiuk, V. M. (2017). RPAS ADS-B and Trajectory Control Data Transmission via Satellite. *Proceedings of National Aviation University*, *72*(3), 26–32. doi:10.18372/2306-1472.72.11978

Hanzo, L., Akhtman, Y., & Wang, L. (2011). *MIMO-OFDM for LTE, WiFi and WiMax. Coherent versus Non-coherent and Cooperative Turbo-transceivers*. West Sussex, UK: John Wiley & Sons.

IEEE P802.11. (2004). Wireless LANs, "TGn Channel Models", IEEE 802.1103/940r4.

Islam, A. A., Kader, B. A., & Julkarnain, C. (2013). BER performance analysis of a real data communication through WiMAX-PHY layer over an AWGN and fading channels. *International Journal of Electrical & Computer Sciences*, *10*(2), 13–16.

Jantunen, P. (2004). *Modeling of nonlinear power amplifiers for wireless communications*. The thesis for the degree of Master of Science. Retrieved from http://www.researchgate.net/publication/224263342_Nonlinear_RF_power_amplifier_behaviouralanalysis_of_wireless_OFDM_systems

Jeon, W. G., Chang, K. H., & Cho, Y. S. (1999). An equalization technique for orthogonal frequency-division multiplexing systems in time-variant multipath channels. *IEEE Transactions on Communications*, *47*(1), 27–32. doi:10.1109/26.747810

Kharchenko, V. P., Grekhov, A. M., & Ali, I. (2015a). Method for Parameters Estimation of Aviation Satellite Communication Channel. *Proceedings of the National Aviation University*, *64*(3), 7–14. doi:10.18372/2306-1472.64.8927

Kharchenko, V. P., Grekhov, A. M., & Ali, I. (2015b). Influence of Nonlinearity on Aviation Satellite Communication Channel Parameters. *Proceedings of the National Aviation University*, *65*(4), 12–21. doi:10.18372/2306-1472.65.9815

Kharchenko, V. P., Grekhov, A. M., & Ali, I. (2015c). Method of Fading Impact Estimation on Parameters of Aeronautical Satellite Communication Channel. *Bulletin of Engineering Academy of Ukraine*, *3*, 50–56.

Kharchenko, V. P., Grekhov, A. M., Ali, I., & Udod, Y. M. (2016). Effects of Rician Fading on the Operation of Aeronautical Satellite OFDM Channel. *Proceedings of the National Aviation University*, *67*(2), 7–16. doi:10.18372/2306-1472.67.10426

Masud, M. A., Samsuzzaman, M., & Rahman, M. A. (2010). Bit Error Rate Performance Analysis on Modulation Techniques of Wideband Code Division Multiple Access. *Journal of Telecommunications*, *1*(2), 22–29.

Modiano, E. (2004). Satellite data networks. *Journal of Aerospace Computing, Information, and Communication*, *1*(10), 395–398. doi:10.2514/1.12800

O'Droma, M., Mgebrishvili, N., & Goacher, A. (2004). Theoretical analysis of intermodulation distortion in OFDM signals in the presence of nonlinear RF high power amplifiers. In *IEEE 59th VTC* (vol. 3, pp. 1295 – 1299). IEEE.

Park, D., & Song, H. (2007). A new PAPR reduction technique of OFDM system with nonlinear high power amplifier. *IEEE Transactions CE, 53*(2), 327 – 332.

Rapp, C. (1991). Effects of HPA-nonlinearity on a 4-DPSK/OFDM-signal for a digital sound broadcasting system. *Proceedings of the Second European Conference on Satellite Communications*, 179-184.

Recommendation, I. T. U.-R. F. 1093-1. (1997). *Effects of multipath propagation on the design and operation of line-of-sight digital radio-relay systems*. Retrieved from https://www.itu.int/dms_pubrec/itu-r/rec/f/R-REC-F.1093-1-199709-S!!PDF-E.pdf

Recommendation, I. T. U.-R. P. 530-16. (2015). *Propagation data and prediction methods required for the design of terrestrial line-of-sight systems*. Retrieved from https://www.itu.int/rec/R-REC-P.530-16-201507-I/en

Roque, D., & Siclet, C. (2013). Performances of weighted cyclic prefix OFDM with low-complexity equalization. *IEEE Communications Letters, 17*(3), 439–442. doi:10.1109/LCOMM.2013.011513.121997

Rumanek, J., & Šebesta, J. (2010). New channel coding methods for satellite communication. *Wuxiandian Gongcheng, 19*(1), 155–161.

Saleh, A. A. M. (1981). Frequency-independent and frequency-dependent nonlinear models of twt amplifiers. *IEEE Transactions on Communications, 29*(11), 1715–1720. doi:10.1109/TCOM.1981.1094911

Sharma, D., & Srivastava, P. (2013). OFDM simulator using MATLAB. *International Journal of Emerging Technology and Advanced Engineering, 3*(9), 493–496.

Sipon, M. M., Mahbubur, R. M., Godder, T. K., Chandra, S. B., & Parvin, M. T. (2011). Performance comparison of AWGN, flat fading and frequency selective fading channel for wireless communication system using 4QPSK. *International Journal of Computer and Information Technology, 1*(2), 82–90.

STANAG 4607/AEDP-7. (2010). NATO Ground Moving Target Indicator Format. (GMTIF).

STANAG 4609/AEDP-8. (2009). NATO Digital Motion Imagery Format.

STANAG 7023/AEDP-9. (2009). NATO Primary Image Format.

Varade, S., & Kulat, K. (2012). BER comparison of Rayleigh fading, Rician fading and AWGN channel using chaotic communication based MIMO-OFDM system. *International Journal of Soft Computing and Engineering*, *1*(6), 2231–2307.

Chapter 3
Satellite Channels Based on IEEE 802.16 Standard

ABSTRACT

This chapter considers the modeling of RPAS/Aircraft data transmission via channels based on IEEE 802.16 standard. RPAS communication channel with a fading was analyzed using original model. Dependencies of a SNR in ground receiver on a SNR in downlink for different types of RPAS amplifier nonlinearity were obtained. Signals constellations of received signals were compared for different Doppler shifts. The influence of the aircraft transmitter nonlinearity for different types of fading in the channel was studied using "80216dstbc Rayleigh," "80216dstbc Rician," "80216d Rayleigh," and "80216d Rician" models. The possibility of the nonlinearity correction using pre-distortion was revealed. The impact of space-time diversity (MISO 2x1) for different types of fading in the channels was investigated. The effect of the Doppler's frequency shift on the operation of communication channels was analyzed.

INTRODUCTION

WiMAX Standards IEEE 802.16

The standard WiMAX was published at the end of 2001. In accordance with the hierarchy of wireless access standards, it belongs to the class MAN (Metropolitan Area Network). For a number of factors, such as bandwidth, coverage and services, WiMAX outperforms the Wi-Fi standard (IEEE802.11)

DOI: 10.4018/978-1-5225-8214-4.ch003

of the LAN class (Local Area Network), allowing it to build regional, national and even global networks with a developed infrastructure. At the physical level, two fundamentally different technologies are used in the WiMAX standard. Data can be transmitted by modulating one carrier frequency (SC-Single Carrier) or multiple sub-carriers-OFDM technology. In the SC mode, the same requirements apply to radio channels as in radio relay networks: the use of direct rays only and the use of narrowly directional antennas, suppression of all reflected beams in order to eliminate intersymbol interference. In this regard, SC technology cannot be used in mass-use networks with multipath propagation of radio waves in communication channels.

The transition to OFDM technology occurred in 2004 with the advent of the new WiMAX standard: 802.16-2004. This technology eliminates intersymbol interference. In the next version of the standard, the OFDM parameters were significantly changed. In particular, they switched to scalable OFDM: the number of subcarriers used became dependent on the operating band (SOFDM), and the subscriber began to allocate a certain number of subchannels (SOFDMA - Scalable OFDM Access). In addition, there was an opportunity of realization of handovers. This version of the WiMAX standard was given the name of mobile WiMAX or 802.16e standard. The 802.16e option is the basic one in the existing WiMAX networks. The last few years the standard was constantly improved. For example, it was supplemented with 802.16i and 802.16j standards. The latter allows expanding existing networks by using repeaters. In 2011, a new version of the WiMAX standard, 802.16m, was approved. It is designed for building networks with a bandwidth above 100 Mb/s and for implementing a number of new promising services.

IEEE 802.16: Was released in 2001; frequency bands 11 – 66 GHz; mobility – no; technology – SC; channel width - 20, 25, 28 MHz.

IEEE 802.16-2004 (802.16d): Was released in 2004; frequency bands 2 – 11 GHz; mobility – no; technology – SC or OFDM (256); channel width - 1,75; 3,5; 7; 14; 1,25; 5; 10; 15; 8,75 MHz.

*IEEE 802.16***e:** Was released in 2005; frequency bands 11 – 66 GHz, 2 - 11 GHz (fixed), 2 – 6 GHz (mobile); mobility – yes; technology – SC or OFDM (256), or SOFDM (128, 512, 1024, 2048); channel width - 1,25; 5; 10; 20 MHz.

IEEE 802.16k: Was released in 2007; frequency bands 11 – 66 GHz, 2 – 11 GHz (mobile); mobility – yes; technology – SC or OFDM (256), or SOFDM (128, 512, 1024, 2048); channel width - 1,25; 5; 10; 20 MHz.

IEEE 802.16-2009: Was released in 2009; frequency bands 11 – 66 GHz, 2 – 11 GHz (mobile); mobility – yes; technology – SC or OFDM (256), or SOFDM (128, 512, 1024, 2048); channel width - 1,25; 5; 10; 20 MHz.

IEEE 802.16j: Was released in 2009; frequency bands 11 – 66 GHz, 2 – 11 GHz (mobile); mobility – yes; technology – SC or OFDM (256), or SOFDM (128, 512, 1024, 2048) + retranslation; channel width - 1,25; 5; 10; 20 MHz.

IEEE 802.16m: Was released in 2011; frequency bands below 3,6 GHz; mobility – yes; technology –SOFDMA; channel width - 1– 20 MHz.

WiMAX is a long-range system covering kilometers of space that typically uses licensed frequency spectra (although it is possible to use unlicensed frequencies) to provide a connection to the Internet, such as a point-to-point provider, to the end user. Different 802.16 family standards provide different types of access, from mobile (similar to data transfer from mobile phones) to fixed (alternative to wired access, where the user's wireless equipment is tied to a location).

Wi-Fi is a shorter-acting system, usually covering tens of meters, which uses unlicensed frequency bands to provide access to the network. Usually, Wi-Fi is used by users to access their own local network, which may or may not be connected to the Internet.

WiMAX and Wi-Fi have a completely different Quality of Service (QoS) mechanism. WiMAX uses a mechanism based on establishing a connection between the base station and the user's device. Each connection is based on a special scheduling algorithm that can guarantee the QoS parameter for each connection. Wi-Fi, in turn, uses a QoS mechanism similar to that used in Ethernet, in which packets receive a different priority. This approach does not guarantee the same QoS for each connection.

WiMAX technology has several advantages:

- WiMAX networks allow operators and service providers to cover not only new potential users economically, but also expand the range of information and communication technologies for users already having fixed access.
- The standard combines technologies of the operator level (to unify many subnets and give them access to the Internet), as well as the technology of the "last mile" (the final segment from the point of entry into the provider's network to the user's computer), which creates universality and, consequently, increases the reliability of the system.
- Wireless technologies are more flexible and, as a result, are easier to deploy, as they can be scaled as needed.

- Ease of installation as a factor in reducing the cost of deploying networks in developing countries, sparsely populated or remote areas.
- The coverage range is an essential indicator of the radio communication system. Now, most wireless broadband technologies require direct visibility between network objects. WiMAX, thanks to the use of OFDM technology, creates coverage areas in the absence of line of sight from the client equipment to the base station, with distances of kilometers.
- WiMAX technology initially contains the IP protocol, which allows it to be easily and transparently integrated into local networks.
- WiMAX technology is suitable for fixed, moving and mobile network objects on a single infrastructure.

Channel Coding

Multipath propagation of the radio signal can lead to a weakening and even complete suppression of some subcarriers due to interference of the direct and delayed signals. To solve this problem, channel coding is used. The standard IEEE 802.16-2004 provides traditional coding technologies and relatively new methods. Conventional includes convolutional encoding with Viterbi decoding and Reed-Solomon codes. To relatively new - block and convolutional turbo codes. Interleaving data is used to increase coding efficiency without reducing the code rate. Interleaving increases coding efficiency, because error packets are fragmented into small fragments, with which the coding system manages.

Flexibility

An important feature of the physical layer is the ability to select the bandwidth for the channel bandwidth. The standard provides the choice of bandwidth in steps of 1.25 MHz to 20 MHz with many intermediate options, which allows more efficient use of the radio-frequency spectrum. In addition, the standard contains an adaptive signal-code construction, that is, the system is adjusted to the characteristics of the channel at each moment of time. In accordance with the standard, depending on the signal-to-noise ratio, the system selects a modulation method in which stable operation can be provided.

WiMAX Adaptive Modulation and Coding

WiMAX modulation and coding is adaptive, enabling it to vary these parameters according to prevailing conditions. WiMAX modulation and coding can be changed on a burst by burst basis per link. To determine the required WiMAX modulation and coding scheme the channel quality feedback indicator is used.

DOWNLINK

Modulation: BPSK, QPSK, 16 QAM, 64 QAM; BPSK optional for OFDMA-PHY.

Coding:

Mandatory - convolutional codes at rate 1/2, 2/3, 3/4, 5/6.

Optional - convolutional turbo codes at rate 1/2, 2/3, 3/4, 5/6; repetition codes at rate 1/2, 1/3, 1/6, LDPC, RS-Codes for OFDM-PHY.

UPLINK

Modulation: BPSK, QPSK, 16 QAM; 64 QAM optional.

Coding:

Mandatory - convolutional codes at rate 1/2, 2/3, 3/4, 5/6.

Optional - convolutional turbo codes at rate 1/2, 2/3, 3/4, 5/6; repetition codes at rate 1/2, 1/3, 1/6, LDPC.

The Main Achievements of the Mobile Mode

- Stability to multipath signal propagation and intrinsic interference.
- Scalable bandwidth of the channel.
- Time Division Duplex (TDD) technology that allows efficient processing of asymmetric traffic and simplifies the management of complex antenna systems by relaying the session between channels.
- Hybrid-Automatic Repeat Request (H-ARQ) technology, which allows maintaining a stable connection with a sharp change in the direction of the client equipment.
- The allocation of frequencies and the use of subchannels with a high load makes it possible to optimize the data transmission taking into account the signal strength of the client equipment.
- Encrgy management allows optimizing the power consumption for maintaining the communication of portable devices in standby or idle mode.

- Network-Optimized Hard Handoff (HHO) technology, which allows up to 50 milliseconds or less to change the client switching time between channels.
- Technology Multicast and Broadcast Service (MBS), which combines the functions of DVB-H, MediaFLO and 3GPP E-UTRA for:
  - Achieve high data rates using a single-frequency network;
  - Flexible distribution of radio frequencies;
  - Low power consumption of portable devices:
  - Quick switching between channels.
- Smart Antenna technology, supporting subchannels and relay session between channels, which allows using complex antenna systems, including beamforming, space-time marking, spatial multiplexing.
- Technology Fractional Frequency Reuse, which allows to control the overlap/intersection of channels to re-engage frequencies with minimal loss.
- The frame size of 5 milliseconds creates the best compromise between reliability of data transmission due to the use of small packets and overhead due to the increase in the number of packets.

Orthogonal Space-Time Block Codes (OSTBCs)

In the modern satellite communications, the requirements to increase the data transfer rate, expand the coverage area and reliability of the connection are constantly increasing. There are exist two problems: the phenomenon of signal fading and inter-channel interference. Fading is associated with the re-reflection of electromagnetic waves from obstacles in the path of their propagation. Many copies of the transmitted signal arrive at the antenna of the receiver at a time, which, when combined in antiphase, are mutually suppressed. This effect reduces the received power to zero and creates so-called "dead zones" in the coverage area of the network. Inter-channel interference is associated with signal distortions in the network cell under the influence of neighboring cell signals, in which the same frequency range is used for data transmission.

The solution to these problems is the use of MIMO technology, which generally implies that each radio equipment participating in the data exchange will have multiple antennas.

The simplest example is a system consisting of a two-antenna transmitter and a single-antenna receiver. It was called MISO (Multiple Input - Single

Output). The most common scheme to date, in which each transceiver has a single antenna, is called SISO (Single Input - Single Output). The effectiveness of MIMO in the conditions of signal fading and channel interference is caused by the following reasons:

1. Growth of the mean value of the signal-to-noise ratio due to the coherent addition of signals emitted by the transmitter antennas. The degree of increase is proportional to the number of receiving antennas.
2. The effect of cochannel interference on the noise immunity of reception can be interpreted as an increase in noise in a given frequency band. Therefore, an improvement in reception against noise background when used MIMO means, in particular, a weakening of the effect of channel interference.
3. To counteract fading, the diversity of signals over time, frequency and in space is used. Time or frequency diversity can be realized in the classical SISO system, but the combined diversity in space is a hallmark of MIMO technology.

In addition, the diversity of signals on the transmitting and receiving sides is distinguished. The first view is implemented only in systems with a transmitter having two or more antennas. Signals, independently radiated by each of them and arriving at the receiver, will have different propagation paths. We can assume that a set of independent radio channels appears between the devices. This allows us to hope that, at least in one of them, fading does not lead to a complete suppression of signal power.

Diversity of signals on the receiving side means that a receiver must have several antennas. Then, each of them receives simultaneously signals emitted by the transmitting device. The likelihood that fading will lead to mutual suppression of signals on all antennas at once is much less than in SISO systems.

With the simultaneous use of a multi-antenna receiver and a multi-antenna transmitter, the advantages described above are combined. A set of formed radio communication channels is called a MIMO channel.

The *fundamental problem* of MIMO systems is the development of algorithms for the distribution of information bits between transmit antennas; generation, emission and reception of signals. Engineers and researchers working in this field focus on three tasks: providing a high information transfer rate, low error probability per bit, and/or relative simplicity of the device implementing a particular algorithm. In practice, three transmission schemes over MIMO channels are used: Space-Time Coding (STC); Spatial Multiplexing (SM); and method of beamforming.

The principle of the STC scheme is that signals are generated from the bits of the initial information sequence in accordance with the rules defined by this code. The transmitter antennas emit these signals simultaneously and for a predetermined time interval. Moreover, from the same bit sequences, several different signals are formed, that is, redundancy is introduced, which can be reduced by using error correction codes.

The STC encoding method is divided into: STTC (Space-Time Trellis Codes); STBC (Space-Time Block Codes); STTTC (Space-Time Turbo Trellis Codes); and LSTC (Layered Space-Time Codes).

In the case of using the spatial multiplexing scheme, the original data stream is distributed (multiplexed) between transmission channels, the number of which is equal to the number of transmission antennas. After the formation of signals from the initial information sequences, their independent, but simultaneous emission by the transmitter antennas occurs for a predetermined time interval. That is, redundancy is not introduced in this case.

Thus, in an ideal case, the transmission rate of information increases as many times as the number of antennas is present in the transmitting device. The specific scheme SM is determined by the way in which the data stream is multiplexed and signals based on it are generated. SM is used in situations where the primary task is to increase the transmission speed of information, and the effect of fading and noise in the MIMO channel is not enough. Otherwise, the gain from independent and parallel data transmission will be leveled by the need for frequent retransmissions from errors during the reception and decoding of the signal.

The STC and SM schemes are used in situations where the MIMO channel parameters are not known exactly, otherwise the beamforming method is applied. In this case, information is transferred in a single stream. The advantage of this method is the ability to significantly improve the spectral efficiency and reduce the probability of error per bit in comparison with SISO systems, but the need for an accurate knowledge of the parameters of the MIMO channel significantly limits the scope of its application.

The features of the problem being solved condition the implementation of this or that transmission scheme via MIMO channels in the system under development. Compared to the rest of the schemes, the STC provides relatively easy implementation of the encoding/decoding devices and low error probability, but the information transfer rate remains relatively low. In SM methods, the transmission rate increases, but the probability of error per bit also increases. The method of forming a radio beam provides the lowest of all three schemes the probability of error, but its application in real conditions is not always possible.

In addition to the three considered MIMO channel transmission schemes, there are three main reception strategies.

1. The maximum likelihood strategy provides a minimum average error probability in comparison with other strategies, however, it requires computationally complex algorithms to implement it. The receiver calculates all possible signals in the system in the absence of noise. One of them, whose Euclidean distance between him and the received signal is the smallest, is taken as the transmitted signal.
2. The work of linear receivers used in practice is based on the zeroing strategy. Such receivers nullify the effect of interference between signals received from different transmitter antennas and perform detection independently for each channel. At the same time, the complexity of reception devices decreases, but in cases where interference really has a significant effect on the type of signals received, the performance falls sharply.
3. BLAST receivers (Bell Labs Layered Space-Time) implement the Nulling and Canceling algorithm based on the Decision Feedback strategy. At the same time, the performance is average between the receivers using the maximum likelihood criterion and linear receivers.

The WirelessMAN-OFDM and WirelessMAN-OFDMA specification includes the option of using Space Time Coding in combination with MIMO and Adaptive Antenna System (AAS) technologies. These specifications define the orthogonal space-time block coding scheme proposed in 1998 by Siavash Alamouti (Alamouti, 1998).

INFLUENCE OF TRANSMITTER NONLINEARITIES ON DATA TRANSMISSION FROM RPAS

The aviation authority and people in general wants partially cover the aviation transportation system with the help of unmanned machines. RPAS are a new way of using flying machines. This is only the first step in transition from fully controlled manned vehicles into fully guided unmanned vehicles. This is impossible without information exchange through wireless communication systems. Nevertheless, even perfect in researchers minds wireless systems in practice confront with quality of data transfer (EUROCAE WG73 UAS, 2008).

Currently there are many methods and technics via which this imperfection in receiving and sending of data can be reduced. One of them is OFDM technology (Faezah & Sabira, 2009). This technology is a method of digital data encoding on multiple carrier frequencies. The main feature and advantage of this technology is ability to resist when severe difficulties in channel are take place. For example, confront with fading at high frequencies or narrow band interferences caused by multipath propagation.

The information transmitting by means of OFDM signals became the standard for many modern radio systems in connection with a number of advantages - high spectral efficiency, low level of an intersymbol interference, high quality of transmitting in the conditions of frequency-selective fading.

It is especially important to provide power efficiency for an information transmitting in aviation complexes with rigid restriction of spatially-frequency parameters for onboard radio-electronic equipment. For this, simulations are mandatory to infer the performance of RPAS communication systems. However, issues related to the RPAS channel nonlinearities still are not investigated in detail.

Basic requirements for RPAS data rate are stated in the NATO standards (STANAG 4609/AEDP-8, 2009); STANAG 7023/AEDP-9, 2009; STANAG 4607/AEDP-7, 2010).

Nonlinear distortion is a source of major degradation of modulation fidelity in multicarrier systems with OFDM signals. OFDM signals significantly improve spectrum efficiency, reduce frequency-selective fading problems, but are sensitive to nonlinear distortion (O'Droma, Mgebrishvili, & Goacher, 2004). The primary source of this nonlinear distortion is the radio frequency transmitter power amplifier. Nonlinear power amplifiers for wireless communications were modeled (Jantunen, 2004) and nonlinear power amplifier effects in multi-antenna OFDM systems were analyzed (Gregorio, 2007). Modulation schemes effect on radio frequency power amplifier nonlinearity were considered in book (El-Khatib, MacEachern, & Mahmoud, 2012). A new reduction technique of OFDM system with nonlinear high power amplifier was proposed (Park & Song, 2007). The use of OFDM radio interface for satellite digital multimedia broadcasting systems (Cioni, Corazza, Neri, & Vanelli-Coralli, 2006), a BER for MIMO-OFDM systems (Varade & Kulat, 2012), performances of weighted cyclic prefix OFDM with equalization (Roque & Siclet, 2013) were studied.

The aim of this study is: 1) to design model of RPAS OFDM communication channel "RPAS–Ground Station" with adaptive modulation using MATLAB Simulink software; 2) to calculate parameters of a channel with different fading types for several nonlinearities level (Grekhov & Kondratiuk, 2017).

Model "OFDM_RPAS_Multlpath_802.16d" for "RPAS-Ground Station" Channel

RPAS communication channel was analyzed using demo model *commwman80216d* designed on a basis of IEEE 802.16 standard. The model (Figure 1) consists of "RPAS Transmitter", "Ground Station", and "Downlink" with Rician fading (Figure 2).

The tasks performed in the communication system include: generation of random bit data that models a downlink burst consisting of an integer number of OFDM symbols; Forward Error Correction (FEC), consisting of a Reed-Solomon (RS) outer code concatenated with a rate-compatible inner Convolutional Code (CC); data interleaving; modulation, using one of the BPSK, QPSK, 16-QAM or 64-QAM constellations specified; OFDM transmission using 192 sub-carriers, 8 pilots, 256-point FFTs, and a variable cyclic prefix length.

Figure 1. Model of OFDM_RPAS_Multipath_802.16d for RPAS link

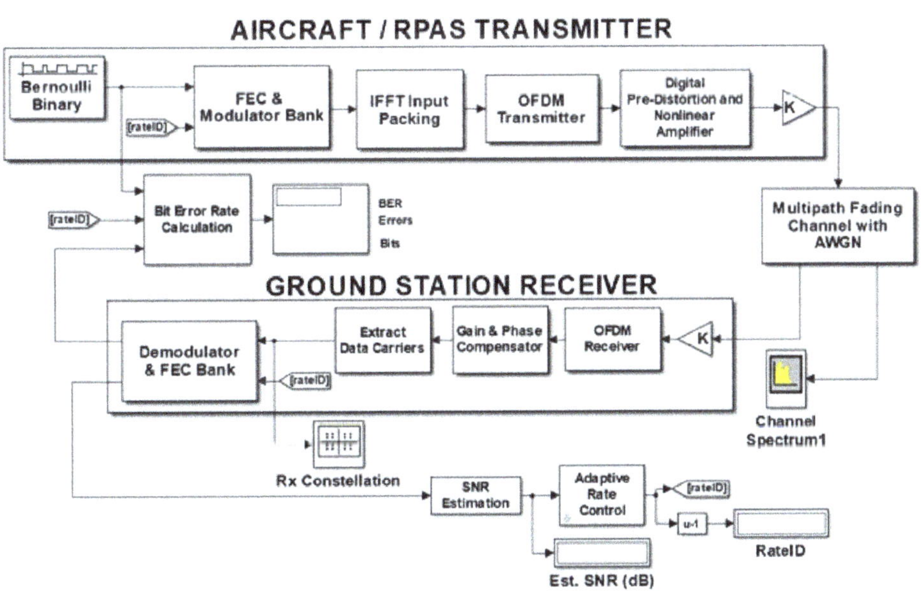

Figure 2. RPAS downlink

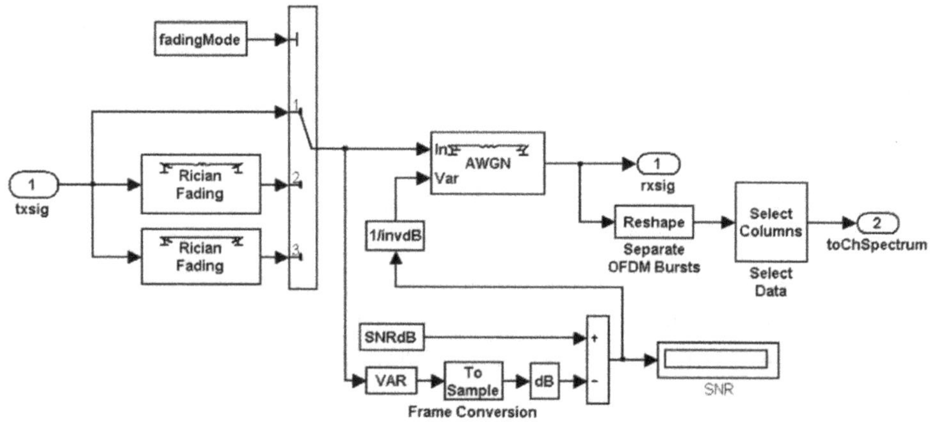

An optional memoryless nonlinearity can be driven at several backoff levels. An optional digital pre-distortion capability can correct the nonlinearity.

It is possible to choose a non-fading, flat-fading, or dispersive multipath fading channel.

Ground Station OFDM receiver includes channel estimation using the inserted preambles. Hard-decision demodulation followed by deinterleaving, Viterbi decoding, and Reed-Solomon decoding. Both models also use an adaptive-rate control scheme based on SNR estimates at the receiver to vary the data rate dynamically based on the channel conditions. The model uses the standard-specified set of seven rates, each corresponding to a specific modulation and RS-CC code rate.

The model includes blocks for measuring and displaying the bit error rate after FEC, the channel SNR and the rate. Spectrum Scope blocks display the spectra of both the OFDM transmitter output and the faded AWGN channel output. A Scatter Plot scope displays the AM/AM and AM/PM characteristics of the signal at the output of the memoryless nonlinearity. A Scatter Plot scope displays the received signal, helping to visualize channel impairments and modulation adaptation as the simulation runs.

Model Parameters configuration block allows choosing and specifying system parameters, such as channel bandwidth, number of OFDM symbols per burst and the cyclic prefix factor. Varying these parameter values allows to experiment with the different WiMAX profiles.

It is possible to vary the state of the nonlinearity and the digital pre-distortion via the Amplifier nonlinearity and Digital pre-distortion parameters. A Saleh model implements the nonlinearity, with three different backoff options. The digital pre-distortion function fits polynomials to the empirically determined AM/AM and AM/PM characteristics of the nonlinearity, and then creates a lookup table by which to pre-distort the signal. Since the nonlinearity induces a gain compression on the input signal, the pre-distortion applies a "gain expansion" on the signal, such that the composite gain is linear.

The Low SNR thresholds [4 10 12 19 22 28] (dB) for rate control parameter directly affects the adaptive-rate control. This parameter is a six-element vector representing the boundaries between the adjoining seven SNR ranges that correspond to the seven rates. Ideally, the simulation should use the highest throughput mode that achieves the desired bit error rate.

The following blocks display numerical results: the Bit Error Rate Display block shows the bit error rate, number of errors and the total number of bits processed; the Est. SNR (dB) display block at the top level shows an estimate of the SNR based on error vector magnitude; the SNR block in the Channel subsystem shows the SNR based on received signal power; the Rate ID corresponds to the specific modulation RS-CC rate currently in use.

The model allows varying the fading parameters and the AWGN variance (in SNR mode) added to the signal. As a result, it is possible to examine how well the receiver performs with different fading characteristics (choosing the appropriate K factor, maximum Doppler's shift, number of paths, path gains) and generate BER curves for varying SNR values.

The Multipath Rician Fading Channel block implements a baseband simulation of a multipath Rician fading propagation channel. This block models mobile wireless communication systems when the transmitted signal can travel to the receiver along a dominant line-of-sight or direct path.

Relative motion between the transmitter and receiver causes Doppler's shifts in the signal frequency. It is possible to specify the Doppler's spectrum of the Rician process. For channels with multiple paths, it is possible to assign each path a different Doppler's spectrum, by entering a vector of Doppler's objects in the Doppler's spectrum field.

Because a multipath channel reflects signals at multiple places, a transmitted signal travels to the receiver along several paths, each of which may have differing lengths and associated time delays. In the block's parameter dialog box, the discrete path delay vector specifies the time delay for each path. The number of paths indicates the length of discrete path delay vector.

Fading causes the signal to become diffuse. The K-factor parameter, which is part of the statistical description of the Rician distribution, represents the ratio between the power in the line-of-sight component and the power in the diffuse component. The ratio is expressed linearly, not in decibels. The K-factor parameter controls the gain's partition into line-of-sight and diffuse components.

It is possible to specify the K-factor parameter as a scalar or a vector.

RPAS Communication Channel Nonlinearity Simulation

The following parameters in the model were set up: RPAS antenna gain was taken 12.4 (an antenna diameter ≈ 0.4 m at 4 GHz), ground station antenna gain – 62.1 (an antenna diameter ≈ 2.0 m at 4 GHz).

In the Model Parameters configuration block channel bandwidth was taken 3,5 MHz and 7 MHz, number of OFDM symbols per burst was taken 1, 20 and 100. Obtained results were similar and are given on figures for channel bandwidth 7 MHz and OFDM symbols per burst 20.

Modelling of RPAS communication channel was realized for no fading, frequency-flat fading and frequency-dispersive fading when RPAS amplifier nonlinearity is disabled (Figure 3), moderate and digital pre-distortion is

Figure 3. Dependencies of a SNR in ground receiver on SNR in downlink when RPAS amplifier nonlinearity is disabled: no fading (points), frequency-flat fading (circles), and frequency-dispersive fading (squares); Max diffusive Doppler's shift is 0,5 Hz

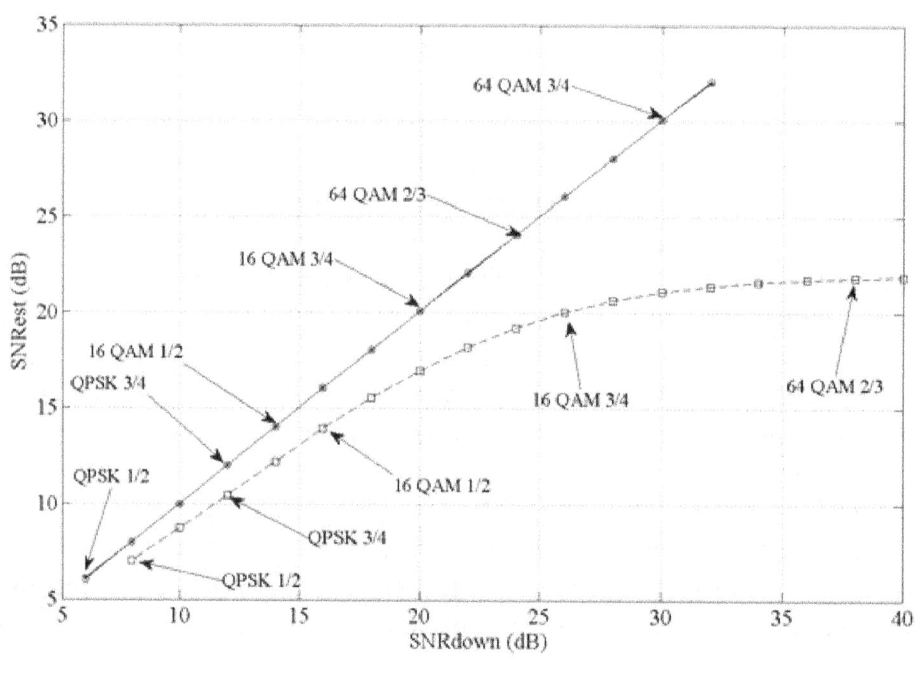

enabled (Figure 4), moderate and digital pre-distortion is disabled (Figure 5), severe and digital pre-distortion is disabled (Figure 6).

The downlink (Figure 2) contains two channels with Rician fading. If the K-factor parameter is a scalar, then the first discrete path of the channel is a Rician fading process (it contains a line-of-sight component) with the specified K-factor, while the remaining discrete paths indicate independent Rayleigh fading processes (with no line-of-sight component).

For the first Rician channel (Figure 2) K = 0,5; the Doppler's frequency shift of the line-of-sight component is 0 Hz; an initial phase of the line-of-sight component is 0 rad; maximal Doppler's frequency shift of the diffusive component is 0,5 Hz; the Doppler's spectrum type is Rounded; the discrete path delay is 0 s; the average path gain is 0 dB.

For the second Rician channel (Figure 2) K = 0,5; the Doppler's frequency shift of the line-of-sight component is 0 Hz; an initial phase of the line-of-sight component is 0 rad; maximal Doppler's frequency shift of diffusive

Figure 4. Dependencies of SNR in ground receiver on SNR in downlink when RPAS amplifier nonlinearity is moderate and digital pre-distortion is enabled: no fading (points), frequency-flat fading (circles), and frequency-dispersive fading (squares); Max diffusive Doppler's shift is 0,5 Hz

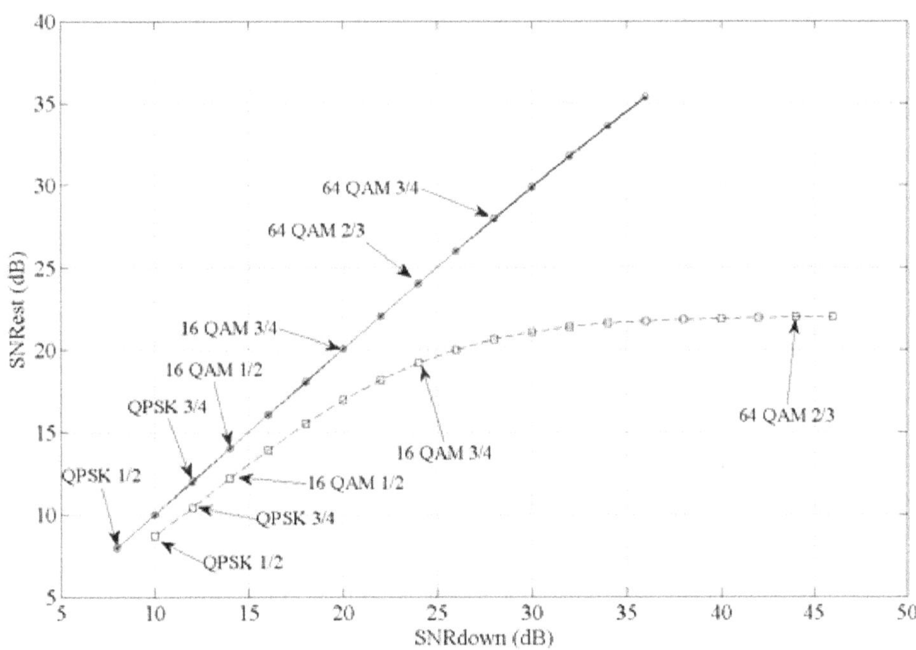

Figure 5. Dependences of SNR in ground receiver on SNR in downlink when RPAS amplifier nonlinearity is moderate and digital pre-distortion is disabled: no fading (points), frequency-flat fading (circles), and frequency-dispersive fading (squares); Max diffusive Doppler's shift is 0,5 Hz

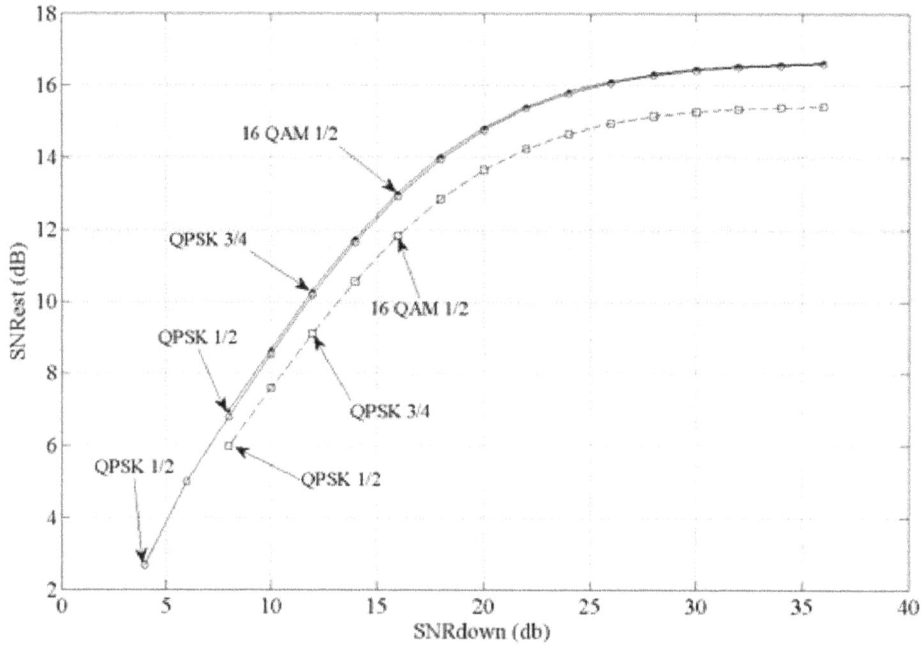

components is 0,5 Hz; the Doppler's spectrum type is Rounded; the discrete path delay vector is [0 0,4 0,9]•10^{-6} s; the average path gain vector is [0 -5 -10] dB. In this case, the line-of-sight component is Rician and two dispersive components is Rayleigh.

Dependences of a SNR in ground receiver on SNR in downlink when RPAS amplifier nonlinearity is disabled are shown in Figure 3. It can be seen that the curve for the frequency-flat fading coincides with the curve for no fading curve and grows linearly with grows of a SNR in downlink. The frequency-dispersive component has a nonlinear character and goes to "saturation" at a $SNR_{est} \approx 22$ dB. A transition to higher types of modulation occurs at large values of a SNR in downlink. The highest modulation for the frequency-flat component is 64QAM3/4 and for the frequency-dispersive component 64QAM2/3.

Figure 6. Dependences of SNR in ground receiver on SNR in downlink when RPAS amplifier nonlinearity is severe and digital pre-distortion is disabled: no fading (points), frequency-flat fading (circles), and frequency-dispersive fading (squares); Max diffusive Doppler's shift is 0,5 Hz

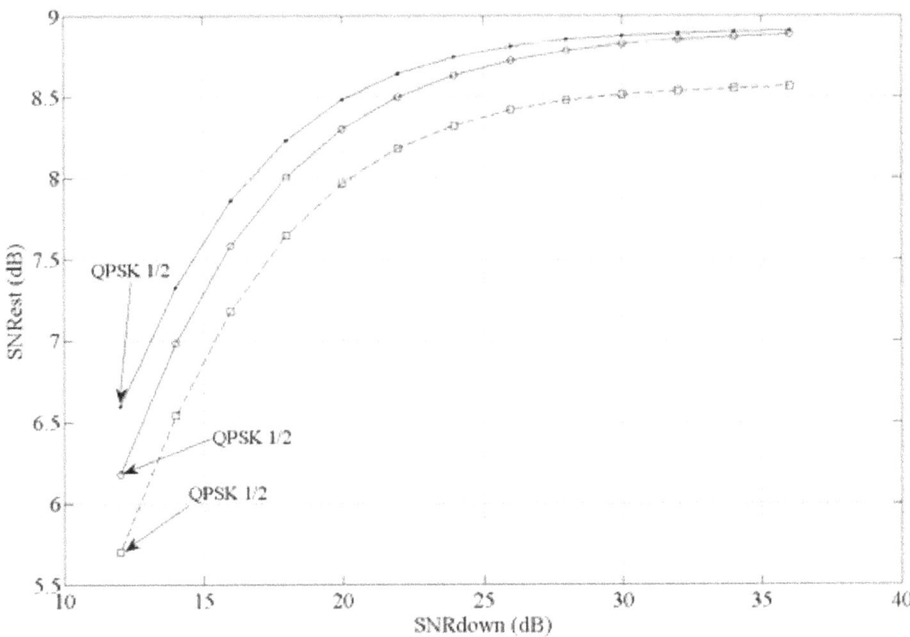

Dependencies of a SNR in ground receiver on a SNR in downlink when RPAS amplifier nonlinearity is moderate and digital pre-distortion is enabled are shown in Figure 4. It can be seen that the curve for the frequency-flat fading coincides with the curve for no fading curve and grows linearly with grows of SNR in downlink. For the frequency-dispersive component, the channel begins to work without errors at SNR in downlink ≈10 dB (and not at 8 dB as in the previous case), but the transition to higher modulations occurs at lower values of SNR in downlink. The curve goes to "saturation" at the same value of ≈ 22 db. The highest modulation for the frequency-flat component is 64QAM3/4 and for the frequency-dispersive component 64QAM2/3.

Dependencies of a SNR in ground receiver on a SNR in downlink when RPAS amplifier nonlinearity is moderate and digital pre-distortion is disabled are shown in Figure 5. It can be seen that the curve for the frequency-flat fading coincides with the curve for no fading curve and both has non-linear character. Both curves go to "saturation" at SNR_{est} ≈ 16,5 dB and have the highest modulation 16QAM1/2.

Dependencies of a SNR in ground receiver on a SNR in downlink when RPAS amplifier nonlinearity is severe and digital pre-distortion is disabled are shown in Figure 6. All curves are non-linear and "split". Only one type of modulation is observed QPSK1/2. "Saturation" in the absence of fading and frequency-flat fading takes place at $SNR_{est} \approx 8{,}9$ dB, and for the frequency-dispersive fading at $SNR_{est} \approx 8{,}5$ dB.

The data in Figure 3-6 are obtained for the diffuse component Doppler's shift of 0,5 Hz. Changes in time of the channel impulse response and signal constellations for the Doppler's shift of 100 Hz are shown in Figure 7.

CONCLUSION

For investigation the influence of RPAS amplifier nonlinearity the model was developed with adaptive modulation for RPAS OFDM communication channel.

Dependencies of a SNR in ground receiver on a SNR in downlink on the type of a modulation (BPSK, QPSK, 16QAM, 64QAM), a bit rate for different nonlinearities levels and types of a fading were received. Signals constellations were compared for different Doppler's shifts.

On the basis of received data (under the given conditions: a type of nonlinearity, a number of OFDM symbols, gains of antenna dishes and the RPAS amplifier nonlinearity type) the channel parameters were estimated: the level of a SNR, for which the RPAS communication channel is "open"; the type of a modulation and data transfer rate, which are possible under the given conditions.

Created model and received results can be used for further development and improvement of communication channel integrity and efficiency as well as for modeling and investigation of channel characteristics under other parameters and conditions of information transmission.

ADS-B SATELLITE COMMUNICATION CHANNEL

On 20 June 2012 satellite operator Iridium has decided that from 2015 they will be putting ADS-B receivers on its next-generation satellite constellation, aimed at bringing global, real-time aircraft surveillance for air navigation service providers (Iridium, 2012).

Figure 7. Time variations in the impulse response of RPAS communication channel and signal constellations for Max diffusive Doppler's shift 100 Hz; a nonlinearity is moderate and a pre-distortion enabled for the frequency-selective fading; SNR_{down} *= 30 dB*

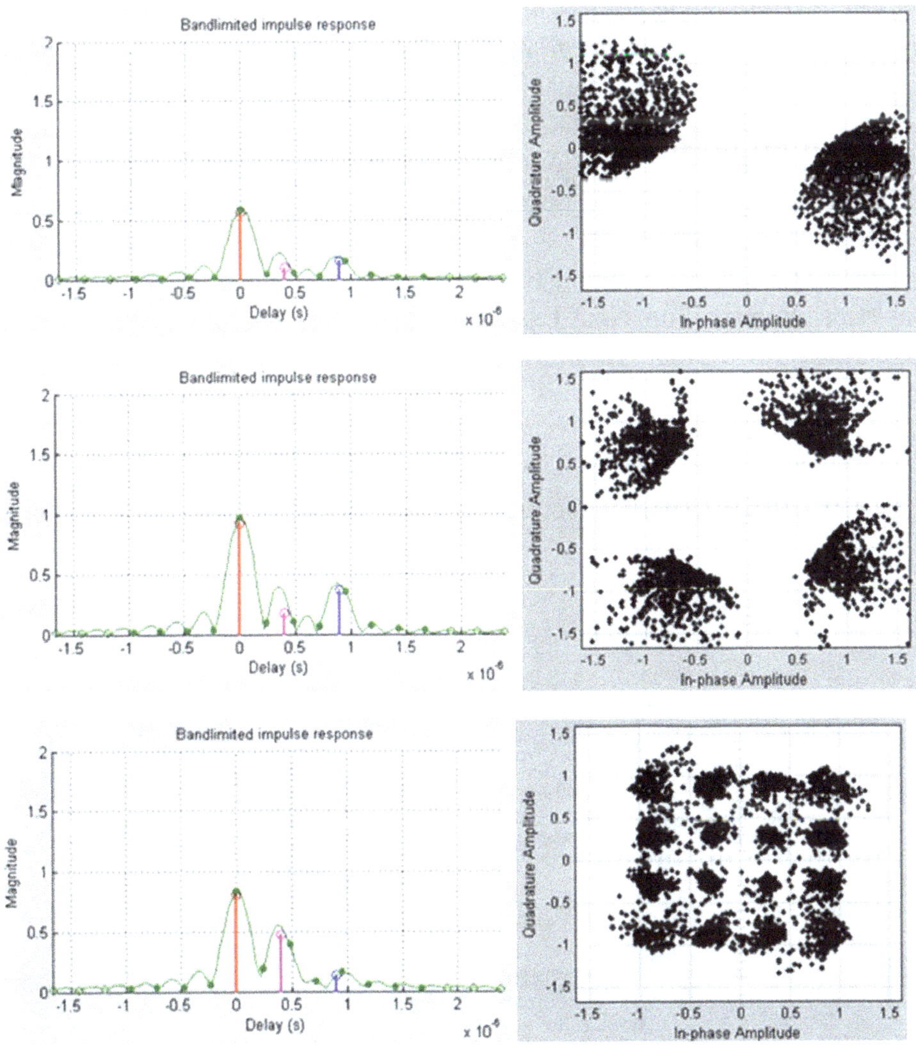

The model for Iridium satellite communication channel "Aircraft-Satellite-Ground Station" was built using MATLAB Simulink software (Kharchenko, Barabanov, & Grekhov, 2012; 2013). Dependencies of a BER on free space path losses, antennas diameter, phase/frequency offsets, satellite transponder linear gain, aircraft and satellite transponder high power amplifier backoff level, and phase noise were received and analyzed.

In 2017, Iridium started new satellite network with just 20 satellites of the Iridium Next constellation in orbit. Iridium Communications' six remaining launches with SpaceX will place their entire 75-satellite network into polar orbit (NASA, 2017).

Iridium, through its joint venture Aireon LLC, deploys ADS-B receivers on all Iridium-NEXT constellation members to establish a global, real-time aircraft surveillance capability through ADS-B signals sent by every commercial aircraft (EUROCONTROL, 2012; Aireon, 2017). ADS-B represents periodic transmissions of data by an aircraft's Mode-S transponder at the 1090 MHz frequency, containing the aircraft ID, its position, altitude and intent (Figure 8).

Harris Corporation built 81 ADS-B 1090 Extended Squitter receiver payloads for the Iridium NEXT satellites. Tracking over 10000 aircraft at any given time, the system will deliver data to air traffic control centers with a latency of less than 1,5 seconds at an update rate of less than eight seconds (Aireon, 2017).

ADS-B message travelling time and average downlink utilization for different Iridium link architectures were estimated in paper (Kharchenko,

Figure 8. Communication links (Aireon, 2017)

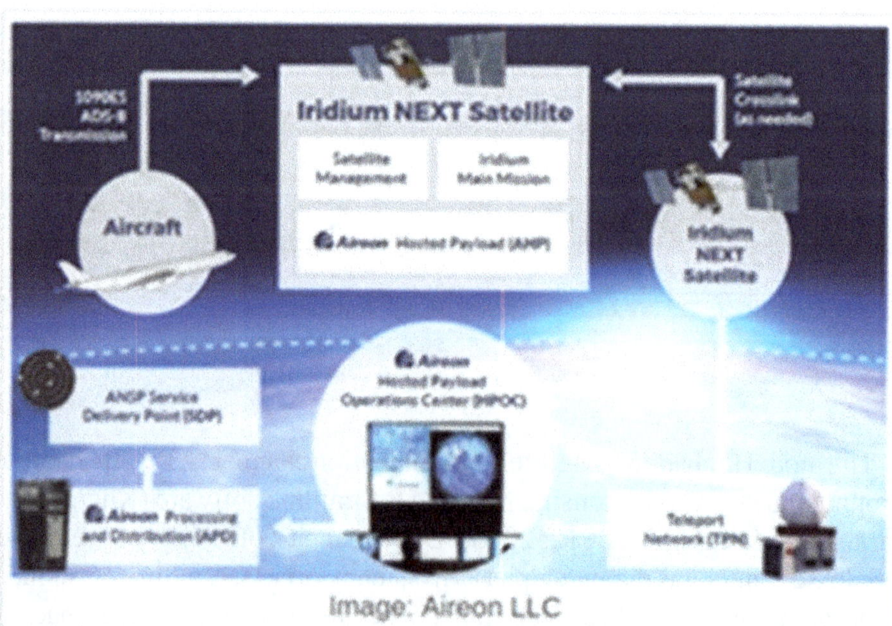

Wang Bo, Grekhov, & Kovalenko, 2014). It was found that the delay is about 1,4-1,5 seconds. Dependences of message travelling time on different number of satellites (N = 1-10) for several aircraft (n = 1-3) were obtained on the base of original models.

The WiMAX Forum defines two system profiles based on (IEEE Standard 802.16d, 2004) and (IEEE 802.16e, 2005), called fixed and mobile system profiles, respectively. The IEEE 802.16 standards provide a large set of fundamentally different options for designing a wireless broadband system, including, for example, multiple options for Physical (PHY) layer implementation, Media Access Control (MAC) architecture, frequency bands, and duplexing. So many options lead to several possible system variants, which are all compatible with the IEEE standards (Pinola & Pentikousis, 2008).

The advantage of WiMAX comes from combining OFDM with smart antenna technologies. Modern mobile satellite telecommunication technologies combine the spatial diversity and OFDM (Dahlman, Parkvall, & Skold, 2011). Comparison of Rayleigh Fading, Rician Fading and AWGN Channel using Chaotic Communication based on MIMO - OFDM System was provided in a paper (Varade & Kulat, 2012).

This study provides an approximation for modelling, which is based on the standards IEEE 802.16d and 802.16dstbc with STBC. The aim of a study is (Kharchenko, Grekhov, Kondratiuk, & Nagorna, 2017):

1. To simulate the operation of the aeronautical satellite channel for ADS-B data transmission on a basis of the IEEE 802.16 standard;
2. To design models of OFDM communication channel "Aircraft-Satellite-Ground Station" with adaptive modulation and error-control coding using MATLAB Simulink software;
3. To study the influence of the aircraft transmitter nonlinearity for different types of fading in the channel (Rayleigh and Rician) and to reveal the possibility of the correction using pre-distortion;
4. To investigate the impact of space-time diversity (MISO 2x1) for different types of fading in the channels;
5. To analyze the effect of the Doppler's frequency shift on the operation of communication channels.

"Aircraft-Satellite-Ground Station" Link Models

Proposed for ADS-B satellite channel models (Table 1) were designed using MATLAB demo examples *commwman80216dstbc* and *commwman80216d.*

The "Aircraft Transmitter" (Figure 9) performs the following tasks: generation of random bit data that models an uplink burst consisting of an integer number of OFDM symbols; FEC, consisting of a Reed-Solomon outer code concatenated with a rate-compatible inner convolutional code; data interleaving; modulation, using one of the BPSK, QPSK, 16-QAM or 64-QAM constellations; OFDM transmission using 192 sub-carriers, 8 pilots, 256-point FFTs, and a variable cyclic prefix length; space-time block coding using an Alamouti code (Alamouti, 1998); an optional memoryless nonlinearity that can be driven at several backoff levels; an optional digital pre-distortion capability that corrects the nonlinearity.

The "Ground Station Receiver" (Figure 10) includes channel estimation using the inserted preambles; for the STBC model, this implies diversity combining; hard-decision demodulation followed by deinterleaving, Viterbi (Viterbi, 1971) decoding, and Reed-Solomon decoding.

The MISO fading uplink (Figure 11a) is used for models 80216dstbc Rayleigh and 80216dstbc Rician (Table 1). The STBC link model uses a MISO fading channel to model a two transmitter, one receiver (2x1) system. The fading parameters are identical for two links. The Space-Time Diversity Combiner block uses the channel estimates for each link and combines the received signals. The combining operation performs simple linear processing using the orthogonal signaling employed by the encoder.

For models 80216d Rayleigh and 80216d Rician (Table 1), it is possible to choose a non-fading, flat-fading, or dispersive uplink multipath fading channel.

Figure 9. "Aircraft Transmitter"

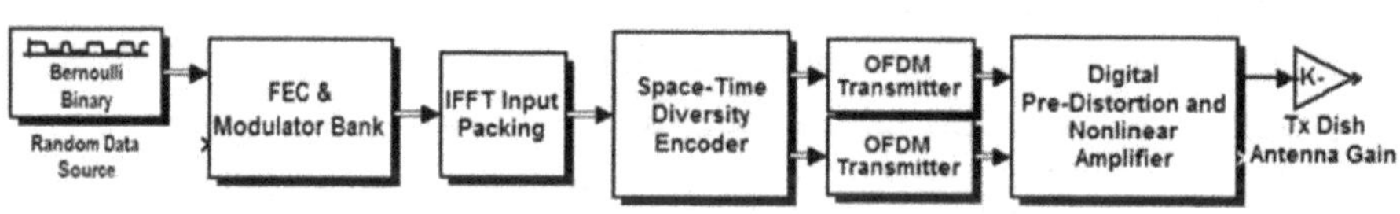

Figure 10. "Ground Station Receiver"

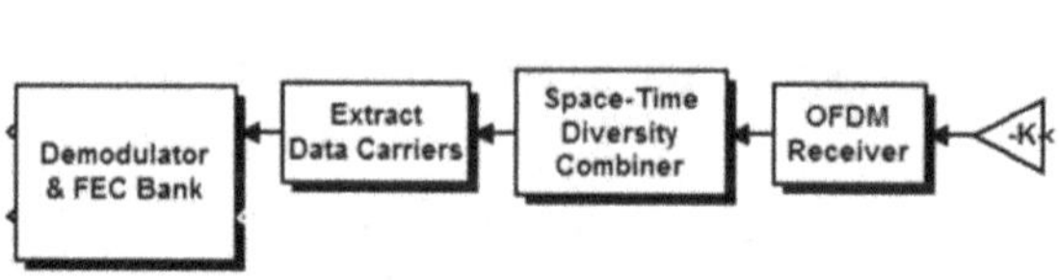

Table 1. Models for aviation satellite link based on IEEE 80216 standard

N	Model	Uplink	Downlink
1	*80216dstbc Rayleigh*	MISO channel (2x1) uses multipath Rician fading with same parameters for each link (Figure 11a): K-factor: 0.5 Delay vector (s): [0 0,4 0,9]*10^{-6} . Gain vector (dB): [0 -5 -10]. The first discrete path is a Rician fading process (it contains a line-of-sight component) with the specified K-factor, while the remaining discrete paths are independent Rayleigh fading processes (with no line-of-sight component).	Multipath Rayleigh Fading Channel (Figure 11b) Delay vector (s): [0 0,4 0,9]*10^{-6}. Gain vector (dB): [0 -5 -10]. Discrete paths are independent Rayleigh fading processes. Doppler's spectrum type: Flat.
2	*80216dstbc Rician*	The same as for Model 1 (Figure 11a).	Multipath Rician Fading Channel (Figure 11c) K-factor: 0.5 Delay vector (s): [0 0,4 0,9]*10^{-6} . Gain vector (dB): [0 -5 -10]. The first discrete path is a Rician fading process (it contains a line-of-sight component) with the specified K-factor, while the remaining discrete paths are independent Rayleigh fading processes (with no line-of-sight component). Doppler's spectrum type: Flat.
3	*80216d Rayleigh*	Multipath Rician fading channel. K-factor: 0.5 Delay vector (s): [0 0,4 0,9]*10^{-6}. Gain vector (dB): [0 -5 -10].	The same as for Model 1 (Figure 11b).
4	*80216d Rician*	The same as for Model 3.	The same as for Model 2 (Figure 4c).

Figure 11. MISO Uplink, Rayleigh and Rician Downlinks

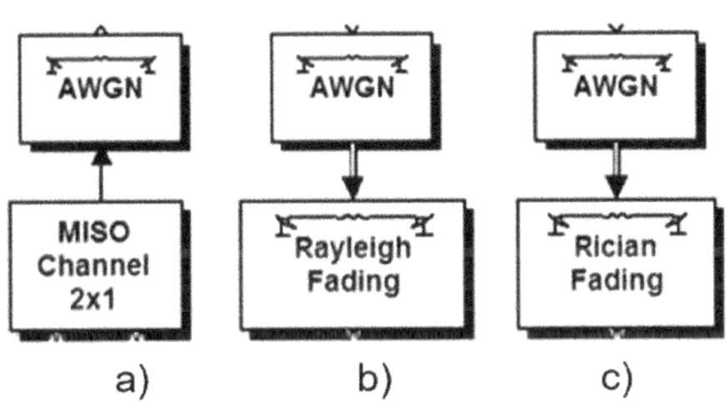

All models use an adaptive-rate control scheme based on SNR estimates at the receiver to vary the data rate dynamically based on the channel conditions. The models use the standard-specified set of seven rates for OFDM-PHY, each corresponding to a specific modulation and RS-CC code rate.

In the "Satellite Transponder" Figure 12, the Satellite Receiver System Temperature block simulates the effects of thermal noise on a complex, baseband signal. Modeling was provided for effective satellite noise temperatures 290 K (typical noise level). The Complex Baseband Amplifier block models linear amplifier with gain 10 dB.

The Model Parameters configuration block allows to choose for all models and specify system parameters, such as channel bandwidth, number of OFDM symbols per burst and the cyclic prefix factor.

It is possible to vary the state of the nonlinearity and the digital pre-distortion via the Amplifier nonlinearity and Digital pre-distortion parameters. A Saleh (Saleh, 1981) model implements the nonlinearity with three different backoff options. The digital pre-distortion function fits polynomials to the empirically determined AM/AM and AM/PM characteristics of the nonlinearity, and then creates a lookup table by which to pre-distort the signal. Since the nonlinearity induces a gain compression on the input signal, the pre-distortion applies a "gain expansion" on the signal, such that the composite gain is linear.

The Aircraft Transmitter, Satellite Transponder and Ground Station Receiver Dish Antenna Gain blocks multiply the input by a constant value (gain).

The following blocks display numerical results: the Bit Error Rate Display block shows the bit error rate, number of errors and the total number of bits processed; the Est. SNR (dB) display block at the top level shows an estimate of the SNR based on error vector magnitude; the SNR block in the Channel subsystem shows the SNR based on received signal power; the Rate ID corresponds to the specific modulation RS-CC rate currently in use.

Figure 12. "Satellite Transponder"

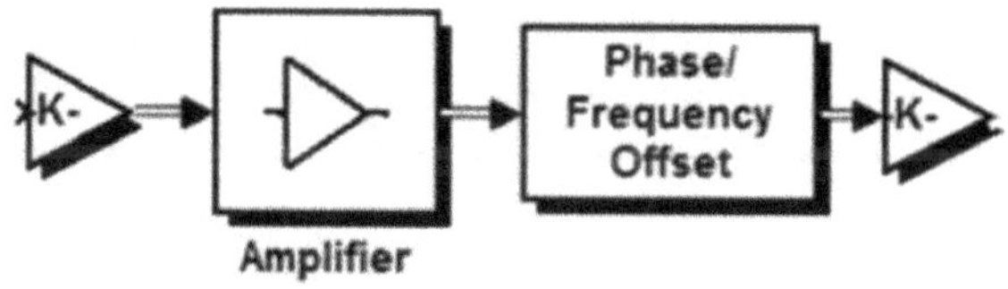

Nonlinearity Impact on Aeronautical Satellite Channels Operation

For calculations, the following parameters in the models were set up: "Aircraft Transmitter" antenna gain was taken 3.1 (an antenna diameter ≈ 0.4 m at 1 GHz), "Satellite Transponder" antennas gains were taken 7.8 (an antenna diameter ≈ 1.0 m at 1 GHz), "Ground Station Receiver" antenna gain – 15.5 (an antenna diameter ≈ 2.0 m at 1 GHz).

In the Model Parameters configuration block channel bandwidth was taken 3,5 MHz and 7 MHz, number of OFDM symbols per burst was taken 1, 20 and 100. Obtained results were similar and are given on figures for channel bandwidth 3.5 MHz and OFDM symbols per burst 20.

Dependencies of estimated SNR_{est} in "Ground Station Receiver" on SNR_{up} in uplink for different levels of "Aircraft Transmitter" nonlinearity are given in Figures 13 and 14 for models *80216dstbc Rayleigh* and *80216dstbc Rician* (Table 1) respectively.

The arrows in the figures show the moment of transition to higher types of modulation. The common for both types of fading is the possibility of correcting the moderate nonlinearity by introducing a pre-distortion (plots for points and circles coincide).

All curves go to "saturation" but in case of disabled nonlinearity this happens for Rayleigh fading (Figure 13) at a value of $SNR_{est} \approx 16$ dB and have the highest modulation 16QAM1/2, while for Rician fading (Figure 14) at a value of $SNR_{est} \approx 21$ dB and have the highest modulation 16QAM3/4.

Both models operate almost identically up to values of $SNR_{up} \approx 10$ dB, but for large values of SNR_{up} the deviations increase (with the exception of the case of strong transmitter nonlinearity). In general, it should be noted that the channel with the Rician fading has better performance for data transmission.

Space-Time Diversity (MISO 2x1) for Different Types of Fading in Channels

Requirements for the throughput of satellite communication networks are very high and are constantly growing. The obvious options for increasing the bandwidth are increasing the channel width and using higher-order modulations. However, they do not completely solve the problem of providing high throughput because the frequency range is still limited.

Figure 13. Dependencies of SNR_{est} in "Ground Station Receiver" on SNR_{up} in uplink when "Aircraft Transmitter" nonlinearity is: disabled (points), moderate nonlinearity (stars), moderate nonlinearity with correctible pre-distortion (circles), severe nonlinearity (squares) for 80216dstbc Rayleigh model; Max Doppler's shift is 0,5 Hz

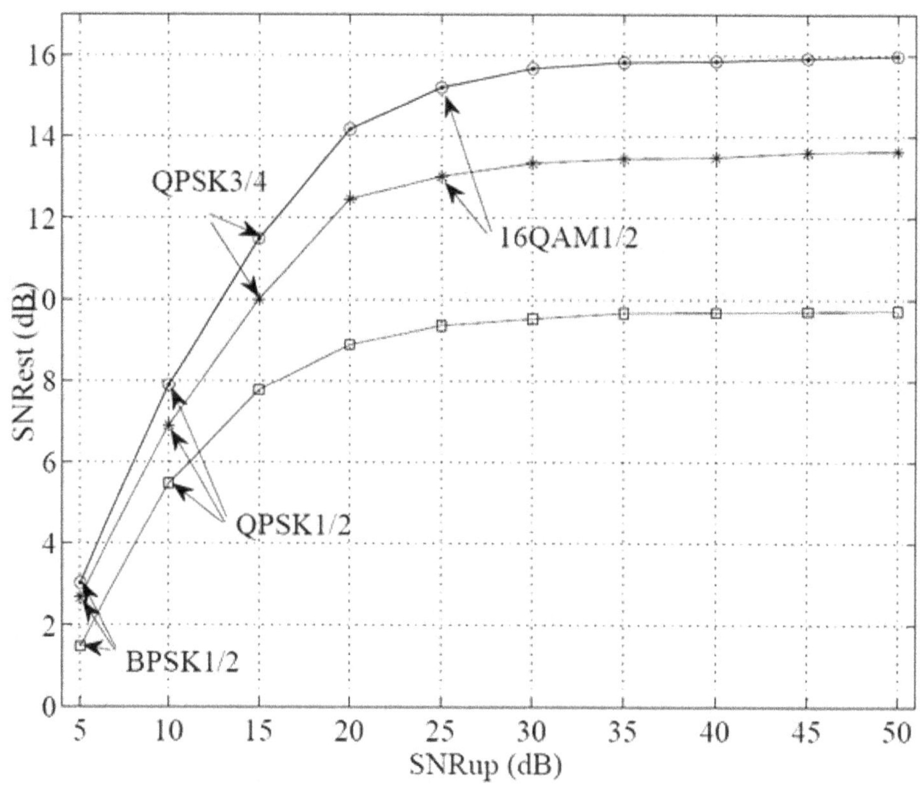

Moreover, the use of higher order modulation implies an increase in SINR (Signal to Interference plus Noise Ratio), which also has its limit. Another way to increase the throughput of wireless systems is to use multiple transmit and receive antennas (MIMO) and special signal processing in this case.

The diversity transmission (Tx Diversity) is the case of using more antennas on the transmitting side than at the receiving end. From the MIMO point of view, such a system is called MISO. The simplest case of such a system, where the transmitting antennas are two, and the receiving one is called MISO 2x1 (models 80216dstbc Rayleigh and 80216dstbc Rician in Table 1).

MISO does not increase the bandwidth of the channel, but increases the reliability of the transmission. At the same time, the use of MISO allows to

Figure 14. Dependencies of SNR_{est} in "Ground Station Receiver" on SNR_{up} in uplink when "Aircraft Transmitter" nonlinearity is: disabled (points), moderate nonlinearity (stars), moderate nonlinearity with correctible pre-distortion (circles), severe nonlinearity (squares) for 80216dstbc Rician model; Max Doppler's shift is 0,5 Hz

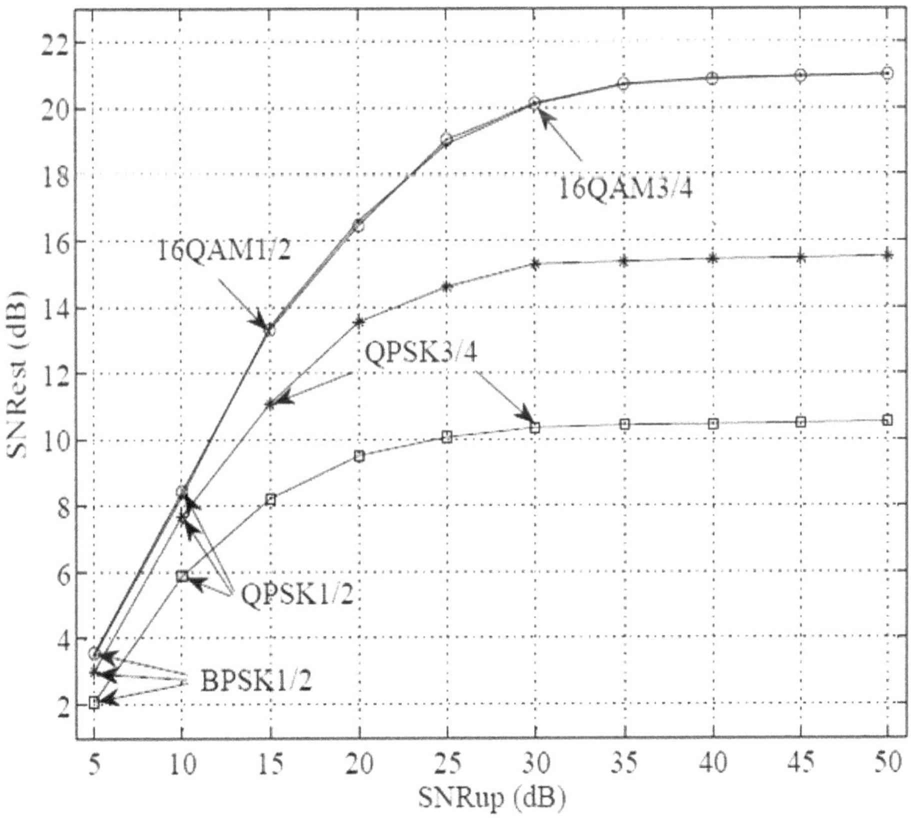

transfer the necessary additional signal processing from the receiving side (mobile station) to the transmitting (base station). Space-time coding is used to generate a reliable signal. The effect of space-time coding is demonstrated in Figure 15, where the graphs for different types of fading are compared. Calculations were carried out for the case of moderate transmitter nonlinearity with correctible pre-distortion. The arrows in the figure show the moment of transition to higher types of modulation and demonstrate how the parameters of data transmission vary depending on the signal propagation conditions.

In the case of Rayleigh fading, the use of space-time coding (Figure 15, points) improves the data transfer (higher values of the SNR_{est} are observed)

Figure 15. The impact of space-time diversity (MISO 2x1) for different types of fading in the channels: 80216d Rayleigh model (squares), 80216d Rician model (stars), 80216dstbc Rayleigh model (points), 80216dstbc Rician model (circles); Max Doppler's shift is 0,5 Hz

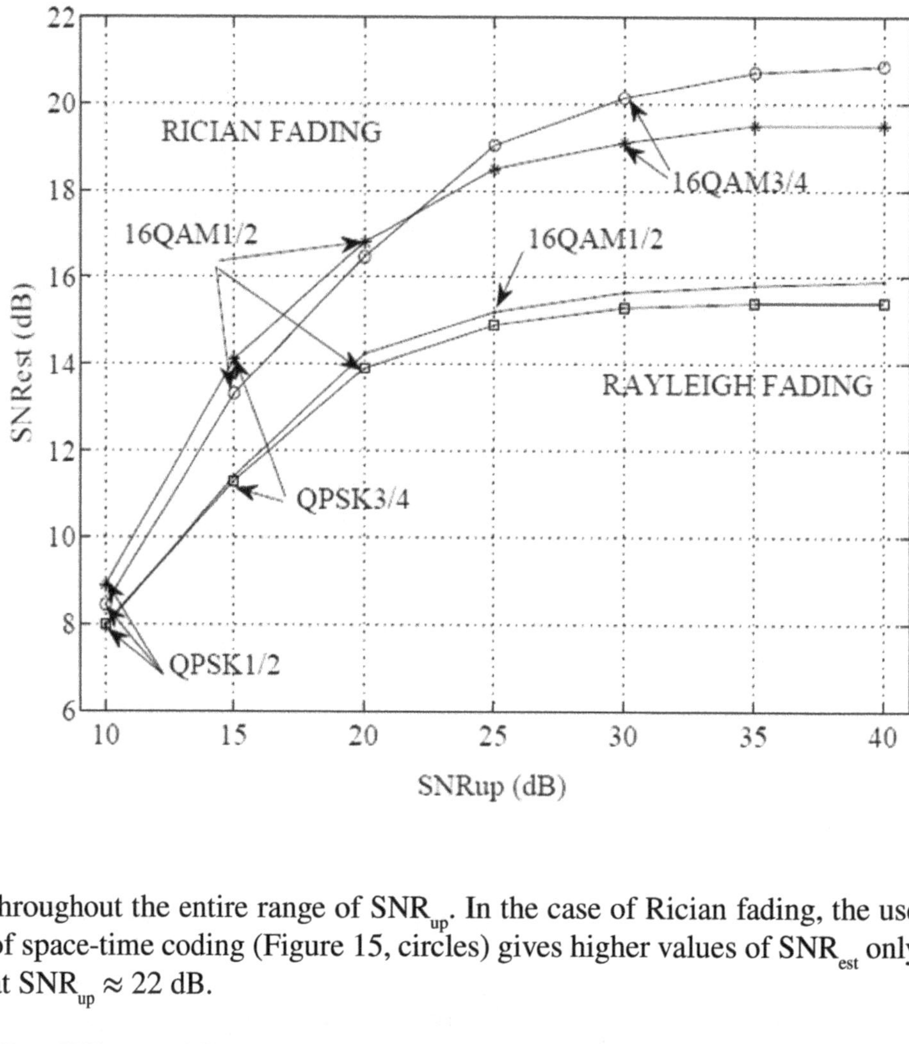

throughout the entire range of SNR_{up}. In the case of Rician fading, the use of space-time coding (Figure 15, circles) gives higher values of SNR_{est} only at $SNR_{up} \approx 22$ dB.

The Effect of Doppler's Frequency Shift on the Operation of Communication Channels

The Doppler's frequency shift occurs in the case of a relative motion of the transmitter and receiver. The aircraft and the satellite move relative to each other, and the satellite moves relative to the ground station. Therefore, it is necessary to take into account the Doppler's shift of the frequency both in the communication channel "Aircraft-Satellite" and in the channel "Satellite-Ground Station".

Doppler's shifts in all models were set symmetrically in the channel up and down. BER values and modulation types for different Doppler's frequencies are given in Table 2.

As follows from Table 2 the BER performance of all models is degraded with the increasing the Doppler's frequency shift. The modulation order is decreasing at the same time. The change in the behavior of multipath components with Doppler's frequency increasing is shown in Figure 16 and is typical for all models considered. Figure17 depicts the signal constellations changing with Doppler's frequency shift increasing, which is similar for all models under given conditions.

CONCLUSION

This study deals with ADS-B satellites channel (Figure 8) and original models (Table 1) of the communication channel "Aircraft-Satellite-Ground Station" link based on IEEE 802.16 standard.

The influence of the aircraft transmitter nonlinearity for different types of fading was studied and the possibility of the nonlinearity correction using pre-distortion was revealed.

The combination of OFDM with spectral efficient multiple antenna techniques is used to overcome the frequency selective problems (Bannour & Mohammad, 2015). Benefits and shortcomings of various MIMO technologies like spatial multiplexing, spacc-timc coding, spatial modulation and transmit antenna selection for performance optimization are considered in (Kumbhani

Table 2. BER values and modulation types for different Doppler's frequencies

Models	1 Hz		10 Hz		100 Hz		500 Hz	
	BER	Modulation	BER	Modulation	BER	Modulation	BER	Modulation
80216d Rayleigh	0.0	16QAM1/2	0.0037	16QAM1/2	0.13	BPSK1/2	0.37	BPSK1/2
80216d Rician	0.0	16QAM1/2	0.0085	16QAM1/2	0.15	BPSK1/2	0.40	BPSK1/2
80216dstbc Rayleigh	0.0	16QAM1/2	0.0021	16QAM1/2	0.10	BPSK1/2	0.34	BPSK1/2
80216dstbc Rician	0.0	16QAM1/2	0.00091	16QAM1/2	0.14	BPSK1/2	0.39	BPSK1/2

Note: SNR_{up} = 20 dB; the case of moderate aircraft transmitter nonlinearity with correctible pre-distortion; in models *80216d Rayleigh* and *80216d Rician* the uplink was taken with frequency-selective fading; in models *80216d Rician* and 80216dstbc Rician in the down link Doppler's shift of line-of-sight component and Doppler's shifts of diffusive components were the same.

Figure 16. Typical dependencies for all models of multipath fading components in case of Doppler's shift 1 Hz (a) and 500 Hz (b); SNR_{up} = 20 dB; the case of moderate aircraft transmitter nonlinearity with correctible pre-distortion

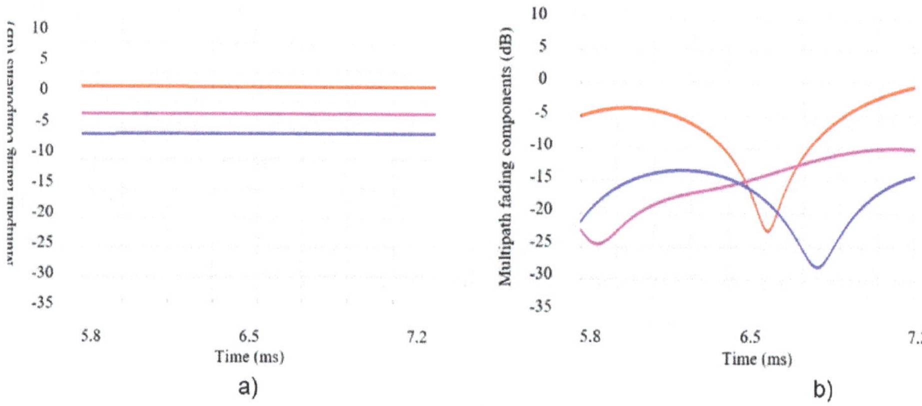

& Kshetrimayum, 2017). In this study the effect of space-time diversity for different types of fading was investigated and graphs were compared.

Joint iterative channel and data estimation in high mobility MIMO-OFDM systems set forth in the book (Sand, 2010). Doppler's shift (1, 10, 100, 500 Hz) impact on the MIMO OFDM system in vehicular channel condition is analyzed with the change in m value of Nakagami channel and with the variation in the modulation schemes (Sur & Bera, 2012). The effect of Doppler's shift on the MIMO-OFDM systems in troposcatter fading channels is that a slight Doppler's shift influences the system performance significantly (Xie, Chen,

Figure 17. Typical received signal constellations for all models in case of Doppler's shift 1 Hz (a) and 500 Hz (b); SNR_{up} = 20 dB; the case of moderate aircraft transmitter nonlinearity with correctible pre-distortion

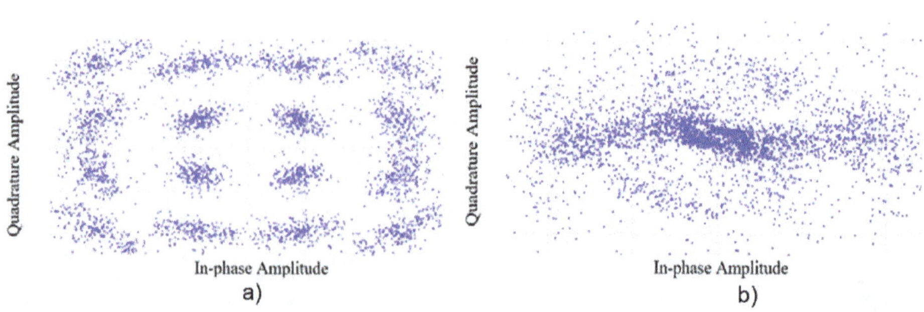

& Liu, 2016). The effect of Doppler's shift on the performance of the 2x2 MIMO system (Hussein & Qaradaghi, 2017) has been studied for 4-QAM, 8-QAM, 16-QAM, and 64-QAM modulations and for different values of Doppler's shift (5, 50, 100 Hz).

In this study the analysis of models performance in mobility condition, i.e. the effect of the Doppler's shift on the system (Table 2) showed that with increasing of Doppler's frequency the BER increases and the higher modulation order is the system become more sensitive towards Doppler's change.

The significance of the obtained results is that with the help of calculations and modelling (non-linearity, MISO 2x1, Doppler's shift) it is possible not only to identify problems in the early stages of the design of aeronautical satellite communication channels, but also to minimize design errors, reduce the time and cost of setting up equipment, assembling a laboratory measuring system, and ensure maximum performance and scalability in accordance with future needs. As a result, the data of such calculations quickly turns into the necessary tool of the researcher and developer engaged in the design and integration of MIMO components and systems in the satellite communication network.

REFERENCES

Aireon. (2017). Retrieved from https://aireon.com/

Alamouti, S. M. (1998). A simple transmit diversity technique for wireless communications. *IEEE Journal on Selected Areas in Communications*, *16*(8), 1451–1458. doi:10.1109/49.730453

Bannour, A., & Mohammad, A. M. (2015). *Coding for MIMO-OFDM in future wireless systems*. Springer.

Cioni, S., Corazza, G. E., Neri, M., & Vanelli-Coralli, A. (2006). On the use of OFDM radio interface for satellite digital multimedia broadcasting systems. *International Journal of Satellite Communications and Networking*, *24*(2), 153–167. doi:10.1002at.836

Dahlman, E., Parkvall, S., & Skold, J. (2011). *4G LTE/LTE-Advanced for mobile broadband*. Oxford, UK: Academic Press.

El-Khatib, Z., MacEachern, S., & Mahmoud, S. A. (2012). Distributed CMOS bidirectional amplifiers: broadbanding and linearization techniques. In Modulation schemes effect on RF power amplifier nonlinearity and RFPA linearization techniques. Springer. doi:10.1007/978-1-4614-0272-5

EUROCAE WG73 UAS. (2008). *Concept of RPAS required communication performance methodology for the command, control and communication link.* Retrieved from Internet: https://www.uavdach.org/News/WG73_CClink_RRCPDraftforWG73CommentV0%2010.pdf

EUROCONTROL. (2012). *ADS-B Sites*. Retrieved from http://www.eurocontrol.int/ads_b_sites

Faezah, J., & Sabira, K. (2009). Adaptive modulation for OFDM systems. *International Journal of Communication Networks and Information Security, 1*(2), 1–8.

Gregorio, F. H. (2007). *Analysis and compensation of nonlinear power amplifier effects in multi-antenna OFDM systems*. Dissertation for the degree of Doctor of Science in Technology. Retrieved from http://www.gwg.nga.mil/misb/docs/nato_docs/STANAG_4609_Ed3.pdf

Grekhov, A. M., & Kondratiuk, V. M. (2017). RPAS ADS-B and Trajectory Control Data Transmission via Satellite. *Proceedings of National Aviation University, 72*(3), 26–32. doi:10.18372/2306-1472.72.11978

Hussein, D. H., & Qaradaghi, T. M. (2017). Effect of Doppler's shift frequency on the performance of 2x2 OSTBC-OFDM system. *QalaaiZanist Scientific Journal, 2*(2), 230–238.

IEEE Standard 802.16d-2004. (2004). IEEE standard for local and metropolitan area networks, Part 16: Air interface for fixed broadband wireless access systems.

IEEE Standard 802.16e-2005. (2005). IEEE standard for local and metropolitan area networks, Part 16: Air interface for fixed and mobile broadband wireless access systems, Amendment 2: Physical and medium access control layers for combined fixed and mobile operation in licensed bands.

Iridium-Adds-ADS-B-to-its-Iridium-NEXT-Constellation. (2012). Retrieved from http://www.aviationtoday.com/2012/06/20/iridium-adds-ads-b-to-its-iridium-next-constellation/

Jantunen, P. (2004). *Modeling of nonlinear power amplifiers for wireless communications.* The thesis for the degree of Master of Science. Retrieved from <http://www.researchgate.net/publication/224263342_Nonlinear_RF_power_amplifier_behavioural_analysis_of_wireless_OFDM_systems>

Kharchenko, V. P., Barabanov, Y. M., & Grekhov, A. M. (2012). Modeling of aviation telecommunications. *Proceedings of the National Aviation University, 50*(1), 5–13.

Kharchenko, V. P., Barabanov, Y. M., & Grekhov, A. M. (2013). Modeling of ADS-B data transmission via satellite. *Aviation, 17*(3), 119–127. doi:10.3846/16487788.2013.840057

Kharchenko, V. P., Bo, W., Grekhov, A. M., & Kovalenko, M. A. (2014). Investigation of ADS-B messages traffic via satellite communication channel. *Proceedings of the National Aviation University, 61*(4), 7–14. doi:10.18372/2306-1472.61.7580

Kharchenko, V. P., Grekhov, A. M., Kondratiuk, V. M., & Nagorna, K. M. (2017). ADS-B Satellite Communication Channel Based on IEEE 802.16 Standard. *Transport and Aerospace Engineering, 5*(1), 18–27. doi:10.1515/tae-2017-0014

Kumbhani, B., & Kshetrimayum, R. S. (2017). *MIMO wireless communications over generalized fading channels.* CRC Press. doi:10.1201/9781315116778

NASA. (2017). Retrieved from https://www.nasaspaceflight.com/2017/07/iridium-satellite-network-55-more/

O'Droma, M., Mgebrishvili, N., & Goacher, A. (2004). Theoretical analysis of intermodulation distortion in OFDM signals in the presence of nonlinear RF high power amplifiers. In IEEE 59th VTC (vol. 3, pp. 1295 – 1299). IEEE. doi:10.1109/VETECS.2004.1390462

Park, D., & Song, H. (2007). A new PAPR reduction technique of OFDM system with nonlinear high power amplifier. *IEEE Transactions CE, 53*(2), 327 – 332.

Pinola, J., & Pentikousis, K. (2008). Mobile WiMAX. *The Internet Protocol Journal, 11*(2), 20-32. Retrieved from https://www.cisco.com/c/en/us/about/press/internet-protocol-journal/back-issues/table-contents-40/112-wimax.html

Roque, D., & Siclet, C. (2013). Performances of weighted cyclic prefix OFDM with low-complexity equalization. *IEEE Communications Letters*, *17*(3), 439–442. doi:10.1109/LCOMM.2013.011513.121997

Saleh, A. A. M. (1981). Frequency-independent and frequency-dependent nonlinear models of TWT amplifiers. *IEEE Transactions on Communications*, *29*(11), 1715–1720. doi:10.1109/TCOM.1981.1094911

Sand, S. (2010). *Joint iterative channel and data estimation in high mobility MIMO-OFDM systems*. Logos Verlag Berlin GmbH.

STANAG 4607/AEDP-7. (2010). *NATO ground moving target indicator format*. Retrieved from http://standards.globalspec.com/std/1300603/nato-stanag-4607

STANAG 4609/AEDP-8. (2009). NATO digital motion imagery format.

STANAG 7023/AEDP-9. (2009). *NATO primary image format*. Retrieved from https://booksmovie.org/similar-pdf-stanag-7023-nato.html

Sur, S. M., & Bera, R. (2012). Doppler's shift impact on the MIMO OFDM system in vehicular channel condition. *International Journal of Information Technology and Computer Science*, *8*(8), 57–62. doi:10.5815/ijitcs.2012.08.07

Varade, S., & Kulat, K. (2012). BER comparison of Rayleigh fading, Rician fading and AWGN channel using chaotic communication based MIMO-OFDM system. *International Journal of Soft Computing and Engineering*, *1*(6), 2231–2307.

Viterbi, A. (1971). Convolutional codes and their performance in communications systems. *IEEE Transactions on Communication Technology*, *5*(5), 751–772. doi:10.1109/TCOM.1971.1090700

Xie, Z., Chen, X., & Liu, X. (2016). Effect of Doppler's shift on the MIMO-OFDM systems in troposcatter fading channels. *Proceedings of the Fifth International Conference on Network, Communication and Computing*, 246-251.

Chapter 4
Investigation of Aircraft and RPAS Data Traffic via Satellite Communication Channel

ABSTRACT

This chapter considers simulation of traffic in the transmission of ADS-B messages with the help of low-orbit satellite complex Iridium. Different models of communication channel "Aircraft-Satellites-Ground Stations" were built using NetCracker Professional 4.1 software. Influence of aircraft and satellites amount on average link utilization and message travel time was studied for telecommunication channels with intersatellite link and bent-pipe architecture. The effect of communication channel "saturation" during simultaneous data transmission through a satellite communication channel from many aircraft was investigated. Influence of protocol type, size of transactions, time between transactions, and channel latency on traffic was studied. A method for estimation of traffic losses was proposed and dependencies of the data loss coefficient on the size of transactions were received.

INTRODUCTION

Data Exchange in Aviation Telecommunications

New times have come for aviation communications. Soon it will be basically a data exchange, and voice communication will be used in non-standard and emergency situations. The main emphasis in the development of aeronautical communications is now being made on functioning on a global scale.

DOI: 10.4018/978-1-5225-8214-4.ch004

Globalization makes it possible to use channels more efficiently and provides sharing of opportunities by many users.

There are two main categories of aeronautical communications:

- Safety-related modes of communication that must have high integrity and speed: communication between the air traffic management bodies and the aircraft to provide air traffic control, the transfer of flight information, warning messages, and communication for the purposes of aeronautical operational control, implemented by aircraft operators to address issues related to safety, regularity and efficiency of operations;
- Non-safety-related modes of communication: aeronautical administrative communication, carried out by aviation personnel and/or aviation organizations for administrative and private matters; aviation communication for passengers.

In addition, for the transmission of surveillance data and data that provide better knowledge of the air situation, direct communication and communication in the broadcasting mode are used.

Communication "Air – Ground"

It is assumed that the regular air-to-ground communication during the flight phase along the route will basically be the exchange of digital data. In this case, the user selects a specific message from a pre-compiled list using the menu on the monitor screen, adds some specific parameters (or arbitrary text), and then sends the message. In some cases, data is transferred between automated airborne and ground systems without the need for manual intervention. Such an exchange of data, according to experts, will greatly reduce the amount of voice communication and, therefore, will reduce the workload of pilots and controllers. However, this does not mean that the voice communication will completely disappear. In busy, the use of voice communication is likely to continue to be preferred. In emergency or non-standard situations, voice communication will be maintained as the main air-ground connection.

Air-ground messages are transmitted over one of the following radio links:

- **AMSS (Aeronautical Mobile Satellite Service):** Geostationary communication satellites designed specifically for mobile communications provide near-global coverage and voice and data communication channels. AMSS is used for aircraft operating in oceanic and remote areas of continental airspace. However, it is possible that next-generation satellite systems will be used for aircraft operating in high-density airspace.

- **VHF (Very High Frequency 30-300 MHz, 10-1m):** Communication link: existing VHF equipment has high operational reliability and will continue to be used for voice communication in busy areas, as well as for the transmission of irregular general messages in its coverage areas. Where saturation of the VHF band for aeronautical communications can occur, there is a reduction in channel spacing from 25 kHz to 8.33 kHz, which increases the number of available channels in such areas.
- **HF (High Frequency 3-30 MHz, 100-10m):** Communication line: radio communication in the HF band for transmitting messages over long distances has limited availability, mainly because of the variability in the propagation characteristics of radio signals. It is assumed that as the use of AMSS in oceanic/remote areas increases, the congestion of HF channels will decrease. Until a new satellite constellation, acceptable for aviation use and covering the entire globe, is established for flight maintenance over the polar regions, HF communication will remain the only possible mode of communication in these areas.
- **VDL Mode 2 (Very-High Frequency Digital Link):** Provides an air-to-ground data link compatible with ATN and uses digital radio communication methods. The nominal data rate of 31.5 kbit/s is compatible with the 25 kHz channel diversity characteristics used for analog VHF and VDL Mode 3 (combined voice and data) mode. Mode 2 allows the use of ATN protocol sets for various operational application processes, thereby providing a significant increase in the efficiency of the VHF channel use.
- **VDL Mode 3:** Uses Time Division Multiple Access (TDMA) and provides the application of both voice communication systems and data transmission systems. Improved use of the VHF spectrum is achieved by providing four separate radio channels on a single carrier (channel spacing of 25 kHz).
- **VDL Mode 4:** Uses the method of Self-Organizing Time Division Multiple Access (STDMA) and is designed to provide the use of surveillance applications (for example, ADS B). This mode is considered for use in other data link applications.
- **SSR Mode S (Secondary Surveillance Radar Mode Select):** Data link and is an air-ground data link, which is specifically designed to transmit limited data. It can be used in mixed conditions, which are

characterized by different levels of the possibility of applying a data link in mode S.

The HF data link provides an air-to-ground data link that is compatible with ATN, and is primarily designed to complement AMSS in the ocean/remote areas and provide the basic connectivity in the polar regions. The data links of AMSS, VDL, SSR Mode S and HF band ensure their interconnection with terrestrial networks so that the onboard terminal can be connected to any terrestrial system. ATN communication allows ground-based data transmission subnets, air-ground data subnets and on-board data subnetworks to interact with each other, providing specific aviation application processes.

The above air-ground data links are ATN-compatible and therefore can be used as subnets of the ATN. In the ATN structure, the subnets are interconnected using ATN tracers, which select the "best" route for each information message. In this regard, the choice of a particular air-ground data link is often inconspicuous for the end user. The data link radio links used to communicate with aircraft in flight are of exceptional importance for safety, regularity and economy of operations.

Communication "Ground-Ground"

It is assumed that the bulk of regular communication between terrestrial aviation users and systems, as in the previous case, will be the exchange of data. Such data exchange between, for example, meteorological offices, NOTAM (Notice to Airmen) services, aviation data banks, ATC units, etc., can be a) messaging with arbitrary text; b) exchange of pre-selected information messages (with some parts added manually); c) automated data exchange between computerized systems.

Various terrestrial networks that have already been implemented by States, a group of States or commercial service providers will continue to serve the exchange of data between aviation users. However, only those networks that use packet switching methods and are compatible with the basic ISO OSI model will be able to use ATN's interworking services. As the ATN progressively deploys, the use of the Aeronautical Fixed Telecommunications Network (AFTN) will be reduced. However, during the transitional period, it will be possible to interconnect the AFTN terminals with the ATN using special internetworking transitions. Voice communication between ATC units will continue to be required in emergency or normal situations. Given the relatively small use of voice communication, dedicated direct voice communications will be gradually replaced by aeronautical switched networks capable of processing voice messages and digital data. There is also a trend towards the use of completely digital methods of switching and voice communications, as more flexible and less expensive leased line digital communications become widely available.

Flight Data Recorder in On-Line Mode

Investigation of information flows in the system based on Iridium satellite communication system for provision of the flight data recorder in on-line mode was provided in a report (Final report on research work No. 721-DB11, 2013).

On-board recorder is a device for recording the basic parameters of flight, internal indicators of aircraft systems, crew negotiations, etc. Information from on-board recordings is usually used to find out the causes of aviation events. The on-board recorder is part of the system of objective control of the aircraft, which collects information about the state of the material part (fuel pressure at the engine inlet, hydraulic pressure, engine speed, air temperature behind the turbine, etc.), about the actions of the crew (the degree of deviation of the control bodies, aircraft wing mechanization during take-off/landing), navigation (flight speed and altitude, course, passing of traffic lights) and other data.

Usually on the aircraft two flight data recorders are installed: the voice, recording the crew's negotiations, and the parametric, fixing parameters of the flight. In addition, many modern aircraft have two sets of recordings: operational (not having a protective case and designed to control the operation of systems and crew after the flight) and emergency (in a strong hermetic body). Information recording can be done on optical or magnetic (metal wire or magnetic tape) media. Recently, a flash memory is widely used.

Emergency recorders with magnetic carriers withstand shock overload in 1000 g and can store information with full fire coverage for 15 minutes. According to the requirements of the current TSO-C124 standard, data storage should be provided within 30 minutes of full fire exposure, shock overloads of 3400 g for 6 ms, static overloads over 2 tonnes for 5 minutes, and immersion at a depth of 6000 m during the month. To facilitate the search for recordings, they are built into radio beacons and acoustic pingers, which are automatically switched on after an accident.

Ground personnel reads information from the operational storage after each flight. The information is deciphered and analyzed in order to determine whether the crew during the flight made unacceptable actions or evolutions

- whether the maximum roll or pitch allowed by the manufacturer was not exceeded, whether there was an excessive overload at landing, not exceeding the set time of work on afterburst or take-off modes, etc. This data also allows to monitor the development of aircraft resources and timely work schedules, thereby reducing the frequency of failures and improving the reliability of aircraft and flight safety.

Parametric and voice recorders were structurally separated: the first was located in the tail of the aircraft, and the second - in the cabin. However, since during a disaster the cabin most often collapses stronger than the tail part of the fuselage, the voice recorder subsequently also moved to the tail.

Each commercial aircraft in the world is equipped with one of these strong, reinforced, waterproof boxes - on-board recorders. When analyzing hundreds of crashes in the sky, they discovered the details - speed, altitude, state of the aircraft engines and mechanization control. This information allowed analysts to conclude on the causes of the accident and to find ways to save thousands of lives in the future.

However, sometimes a flight data recorder can not be found. Without the important data that the record holds, analysts can not say exactly what the cause of the disaster was.

The flight data recorder is perhaps the most ambitious invention in the history of safety. However, technology must move on and improve it. Instead of storing all the data in the on-board recorder, which may not be found in an emergency, it would be much better to obtain all the necessary information from the aircraft in real time with the continuous transmission of data directly to the ground facilities or via satellite. In case of receiving non-standard data, this system could request additional information from the on-board computer and also record it. This could greatly simplify the collection of information about the event and allowed to store data for a long time.

Flight Data Transmission in Real Time

The use of the telecommunication channels "aircraft-LEO satellite-Gground station" will enable the creation of a "cloud" without borders using networks of servers and databases. This system will not have any restrictions for real-time data transmission: over water and land, in thunder and sunny weather, in the mountains and in the plain, from the ground and up to maximum flight levels.

The invisible "cloud" could be used not only for the disclosure of the causes of air crashes, but also for the effective management of air traffic controllers. With remote monitoring of land data using onboard systems, it is much faster and more likely to answer the question whether there is an emergency situation in the sky due to a mistake in piloting or a technical problem with an airliner.

Currently, most flight recorders can record *256* different streams of digital data or settings *per second* and store them for 25 hours before new data is recorded over them. You can also save 180 minutes of talk, while old models could only save 30 minutes of talk. Information is recorded on media with different sampling rates. Manufacturers provide hardware and software that is needed for decoding and analyzing data, and sometimes they can provide their representatives who can assist with data inspection. They can help extract information if the flight recorder is badly damaged by high temperature, shock, or damaged data cables. In these cases, representatives are required to find other ways to extract all the necessary data. This operation can take time from weeks to several months.

Data recorded in the flight varies depending on the flight stage: take-off, landing, flight at the declared flight level. The US Federal Aviation Administration says about 88 parameters to be recorded. One of the typical parameters is the change in altitude. Other data, such as flight time parameters, speed, vertical acceleration, course relative to the magnetic north, fuel consumption, data on mechanization and engine parameters. Most of the parameters are written in the size of four 12-bit counters per second, others can be written less frequently. Airlines may collect additional information for their own use.

The satellite-based aviation system assumes that the aircraft will transmit information either directly to the ground, or through low-orbit satellites to remote ground stations. When flying, for example, over water or in the mountains, the information will be transmitted to the ground through a network of LEO satellites. Thus, the data transmission network will cover even the polar regions. Since the throughput of satellites is a limited resource, it is possible to transfer data of only a few flight indicators. Today, most aircraft already send some information to ground stations. Continuously received data relates to the flight path and speed, as well as information about failures and other technical issues for repair services. This system basically uses VHF, which can handle only 16 bits per second.

One of the main problems is the lack of a homogeneous medium for transmitting the signal. The earth is covered by various wireless communication systems - some of which are for cities, some for the countryside, others for use at sea. To stay in touch with each aircraft, network systems will have to be reconfigured between all these communication channels. For example, aircraft flying above ground at low altitudes can access high-bandwidth networks by tapping on a mobile network receiver using VHF and UHF. When flying at high altitudes or over water, aircraft will need to work with satellite systems that provide less bandwidth.

Today, satellites that provide the required bandwidth, operate on the Ku-band (microwave) and use protocols known as MPLS VLANs (multi-protocol switching over labels of a virtual LAN). These channels provide data transfer through the Internet Protocol servers on the ground. This may be necessary for the rapid change in the amount of data transmitted in accordance with the flight status. For example, additional data should be transmitted during take-off and landing, when some parameters change faster than during a flight at a given flight level. Similarly, when terrestrial monitoring systems notice something unusual, they may ask for additional data to make it clear. To solve the problem associated with the rapid change in the volume of requested data, the "invisible cloud" should include dynamic scheduling and channels with different bandwidth for the needs of the aircraft. It is necessary to organize data transfer taking into account the limited throughput of the satellite network. The satellite network can transmit only those parameters that are marked by significant deviations from the previous sampling. In other words, it is necessary to stop the transmission of some data when the allocated traffic is completely busy. But it is necessary to restore the full transfer using the free channel that has appeared. After the data gets to the server, expert systems can begin processing this information by comparing metrics with previous data to identify catastrophic metrics.

Formation of Events List Registered by the Flight Data Recorder

Since the amount of data to be saved is hundreds of gigabytes per day, only certain parts of the data must be stored. Armed with such a system of concise data, expert systems and specialists working in tandem can detect repetitive errors associated with a variety of problems: problems of service, pilot training, weather conditions, airport runway, etc. The obtained knowledge can be used for training of pilots and air traffic controllers in order to avoid recurrence of accidents.

Indicative list of registered events:

1. The roll angle is more than permissible.
2. The speed is greater than the maximum allowable when flying with an open cargo hatch.
3. The speed is greater than the maximum allowable when the chassis is released.
4. The speed is greater than the maximum permissible when removing the chassis.
5. The vertical overload is greater than the maximum allowable for different configurations.
6. The Mach number is greater than the maximum permissible.
7. The angle of attack is greater than the maximum allowable for different configurations.
8. The speed is greater than the maximum allowed for different configurations.
9. The angle of attack at the moment of takeoff.
10. Height at the beginning of wing flaps removing.
11. Height of passing of lighthouses on the glide path.
12. The angle of attack at the moment of touching the ground. Maximum overload at landing.
13. Overload is more than permissible.
14. Overflow of the counter of extreme overloads.
15. The brake system is under pressure.
16. Angle sensor failure.
17. Failure of the amplifier channel.
18. Failure of strain gauge sensor.
19. Failure of weight load meter.
20. No data is entered for the cargo.
21. The fuel measurement system is not turned on.
22. Weight and alignment are defined and are normal.
23. Take-off weight is more than the maximum permissible.
24. Take-off alignment is not normal.
25. Flight centering is not normal.
26. Carry out autocontrol readiness for takeoff. The parameters are not normal.
27. Manual control of readiness for takeoff. The parameters are not normal.
28. Signal on the engine's reverse position.
29. The chassis is not ready.
30. The brakes are faulty.
31. A fault in the fire alarm system has been detected.

32. There is no readiness of power supply.
33. The signal "door-hatches checked".
34. The angular position of the flaps does not correspond to the takeoff.
35. The hydraulic system is faulty.
36. The emergency recorder is not turned on.
37. Failure of the emergency registrar.
38. Signal to the retracted position of the spoilers.
39. The signal "XXX is not prepared".
40. There is a malfunction of XXX.
41. Not all tumblers "Air selection" are turned on.
42. There is no signal of automatic fuel supply.
43. Engine malfunction is detected.
44. No time is entered in the service information.
45. Not all tumblers "XXX" are turned on.
46. The signal "door-hatches checked".
47. The parameters "XXX" are not normal.
48. There is a malfunction in the hydraulic system.
49. Malfunction in the emergency recorder.
50. Failure of the emergency recorder.
51. Signal of the uncleared position of the spoilers.
52. There is no readiness for the chassis.
53. No readiness on the power supply system.
54. Engine ON/OFF switch signal.
55. The airborne shut-off damper of the engine is not closed.
56. Prepared for landing control. The parameters are not normal.
57. Flaps not released.
58. The chassis has not been released.
59. Brake interceptors are not prepared.
60. When the marker is over, the chassis and flaps are not released.
61. Flaps are released.
62. Centering beyond tolerance.
63. Failure of the calculator of the rate of presure change in the cabin.
64. Cabin pressure sensor failure.
65. Failure of the cabin power amplifier.
66. Failure of cabin outlet valve.
67. Control channel failure.
68. The voltage of the generator phase A is not normal.
69. The voltage of the generator phase B is not normal.
70. The voltage of the generator phase C is not normal.

71. The generator frequency is not normal.
72. The bus voltage is not normal.
73. Generator No. ... not turned on.
74. The rectifier is not turned on.
75. The transformer is not turned on.
76. Switch on the circuit stimulus signals.
77. No fire extinguishing system control.
78. No control of the ailerons of one of the sides.
79. Asymmetry of slats subchannel.
80. Asymmetry of the end flap subchannel.
81. Non-release of one of the flap sections.
82. Asymmetry of internal flaps subchannel.
83. Full failure of the transverse channel.
84. Slats/flaps – check.
85. Slats – failure.
86. Full failure of the longitudinal channel.
87. Ailerons - refusal.
88. There is no control of the ailerons of one of the sides.
89. Asymmetry of subchannels of flaps.
90. Asymmetry of subchannels of finite flaps.
91. Asymmetry of subchannels of inner flaps.
92. The filter in the pump discharge line is clogged.
93. Slats/flaps - checked.
94. Flaps - failure.
95. Above the norm, the temperature of the liquid in the pump.
96. Below the norm the fluid temperature in the pump.
97. Maximum temperature of the liquid during the flight in the pump.
98. Total operating time (under heating) of the left and right main pump.
99. The filter in the common drain line is clogged.
100. The filter in the drain line from the pumps is clogged.
101. The filter in a delivery line of the hydrotransformer is clogged.
102. The filter is clogged in the discharge line of the torque converter.
103. Does not work/pumps the pump.
104. Below the minimum permissible liquid level in the supply tank.
105. Failure of the transfer pump.
106. Calculated operation mode of air intake.
107. There is no power on the slat-adjusting valve.
108. The air temperature in front of the slat-adjusting valve is not normal.
109. Low air pressure in the crane-slat regulator.

110. Estimated flaps operation mode.
111. Estimated mode of operation of air intake dents for fuel tanks.
112. Failure of air intake flap for fuel tank drainage.
113. Dangerous overheating of wheel brake.
114. Malfunction of parking braking on the port side.
115. Parking brake failure on the starboard side.
116. Braking failure on the port side.
117. Malfunction of braking on the starboard side.
118. "Stuck" button "CONTROL" when checking the brakes.
119. Fault on the left side of the port underlining braking.
120. Malfunction on the starboard side of the deceleration braking.
121. Failure of the final switch of the open position of the main support pillar.
122. Failure of the final switch of the open position of the front support pillar.
123. No failure when checking the brakes for pressure.
124. Violation of the automatic operation of fans.
125. Check the brakes before take-off or landing.
126. Absence of pressure on the rack when the brakes warm up.
127. Absence of failures when checking brakes.
128. Operating time of the engine in different modes.
129. Number of flight cycles, run-out time.
130. Failure of low-pressure fuel pump.
131. The deviation of the rotational speed of the high-pressure rotor from the mean more than tolerance.
132. The temperature of the air cooling the engine is higher than permissible.
133. The temperature of the air cooling the engine is normal after going beyond the limit.
134. The oil pressure in the central engine is not normal.
135. The oil temperature at the engine inlet is higher than permissible.
136. The oil temperature at the engine inlet is above the permissible value for more than 10 minutes.
137. The amount of oil in the engine tank is ... below the norm and there is an appropriate signal.
138. The amount of oil in the engine tank above the maximum permissible level.
139. Changing the amount of oil in engine tank above the permissible level.

140. When starting engine in flight, temperature restrictions on re-starting are initiated.
141. When starting engine in flight, a temporary restriction on re-starting is initiated.
142. The engine was started at an altitude of over 8000 meters.
143. Engine was launched more than 3 times in flight.
144. At engine start in flight the maximum permissible value of the gas temperature is exceeded.
145. The time between engine starts is less than 30 seconds.
146. The time between the series of engine starts is less than 30 minutes.
147. The cold scrolling of engine has not been carried out.
148. Open engine valve.
149. The emergency shutter of the engine.
150. When engine starts, the unit is manually disabled.
151. The engine was launched in flight from the autorotation mode.
152. The engine was launched in flight with a twist.
153. The engine was launched in flight with a twist and a manual fuel feed switch.
154. Registration of an operating time on engine takeoff mode.
155. Engine stop signal.
156. Engine malfunction when working on the control program.
157. Fault in the fuel regulator of the engine.
158. There are conditions for processing information on the parameters of the engine in this control mode.
159. It is necessary to check the input from the panel of the engine's formulary values.
160. Calculation of the individual characteristics of the fuel-regulating equipment of the engine is completed.
161. The total degree of pressure increase in engine outside the tolerance.
162. Fuel consumption of engine outside the operational tolerance.
163. Frequency of rotation of the high-pressure rotor of the engine outside the operational tolerance.
164. Rotation frequency of the medium-pressure rotor of engine outside the tolerance.
165. Engine control.
166. The position of the throttle of engine does not meet the requirements.
167. The position of the engine blades does not meet the requirements.
168. Vibration on the rear engine sensor above the local tolerance limit.

169. The value of the vibration of engine exceeds the level of the global maximum.
170. Increased vibration of the engine.
171. Dangerous vibration of the engine.
172. Engine failure.
173. Malfunction of the engine thrust reverser.
174. Failure of engine oil thrust reverser.
175. Engine overheating.
176. Engine surging.
177. The lack of power supply for engine.
178. The failure of the engine power supply system.
179. The work of the reserve power supply channel of the engine (in case of failure of the main channel).
180. Limiting parameters of the engine.
181. The temperature of the gases of engine is below the limit.
182. The temperature of the gases of engine is higher than the maximum.
183. The rotational speed of the engine rotor fan above the limit.
184. The main pumps of hydraulic system are not switched off.
185. The air pressure in the left launch system is not normal.
186. The air pressure in the right launch system is not normal.
187. The "Air" switch is not set to the "Closed" position.
188. The temperature of the gases of engine is higher than permissible before restarting.
189. Engine start-up time does not meet the requirements.
190. Autocontrol ready for landing. The parameters are normal.
191. Failure of the calculator of the pressure change rate in the cabin.
192. Failure of the cabin pressure sensor.
193. Failure of the control unit in the channel for outputting the signal to the power amplifier.
194. Cylinder outlet valve failure.
195. Failure of the cabin servicing program.
196. The voltage of the generator phase is not normal.
197. The frequency of the generator is not normal.
198. The generator is not turned on.
199. Power amplifier failure.
200. The generator failure.

Let it be necessary to transfer from the aircraft to the satellite 200 parameters that describe the coordinates, speed, height, course, angular speed, overload, data from the control channels and other characteristics of the current state of the aircraft systems, given above. Each of the parameters must contain 4 digits. Then it is necessary to transmit 800 characters and according to the Hartley's formula ($2^i = N$, where i - the amount of information in bits, N - the number of characters in the alphabet, $2^{10} = 1024$) to represent 800 characters using 10-bit binary code, that is transmit 8000 bits of information.

If the rate of transfer of information (as in the case of the Iridium constellation) is 2400 bits/s, then the time t of 8000 bits file transfer is calculated by the formula

$$t = Q / q,$$

where Q is the file size, q is the data rate.

$$t = (8000 \text{ bits}) / (2400 \text{ bits/s}) \approx 3.3 \text{ seconds}.$$

The same time will be spent on the satellite transmission to the ground and, taking into account the delays, it can be assumed that approximately every 10 seconds some information from the flight data recorder can be updated on the ground and be accessible to the operators in the on-line mode.

BUILDING CHANNEL MODELS USING NETCRACKER PROFESSIONAL 4.1

Description of Models Design

NetCracker Professional 4.1 software was used to simulate the transmission of data via the aeronautical satellite communication channel (Kharchenko, Barabanov, Grekhov, Ivanenko, & Lobanov, 2012; Kharchenko, Wang Bo, Grekhov, & Kovalenko, 2014; Wang Bo, Kharchenko, Grekhov, & Ali, 2015; 2016). The following designations are used for channel models: ***AkSmGn***, where ***k*** is the number of aircraft ***A***, ***m*** is the number of satellites ***S***, ***n*** is the number of ground stations ***G***. In Figure 1 a model ***A1S1G1*** is given, which consists of airborne station, satellite and air traffic controller center. The model simulates packet switching of circuits with different data transfer rate. The packet latency, the probability of packets missing, and BER parameter are zero except in specially stipulated cases. InterLAN traffic is selected in the traffic profile.

Average Workload (AW) is the quantity of messages transmitted by communication link for given time interval. Average Utilization (AU) is the percentage of link capacity involved in load transmission for given time interval. Travel Time (TT) is the time that is needed for message to be delivered from transmitting device to receiving one.

Data streams have stochastic nature and are described using distribution laws for random values. Traffic parameters in NetCracker Professional 4.1 software were specified similar to queuing theory (Akimaru & Kawashima, 1999; Flood, 1998; Kennedy, 2003; Penttinen, 1999).

"Aircraft–Satellite–Ground Station" communication channel consists of separate links: A-A link is intended for aircraft data transfer to the transceiver antenna (Figure 2), A-S link is wireless data link to a satellite (Figure 3), S-G wireless link for transmitting data from a satellite to ground station antenna (Figure 4), G-G link between ground transceiver antenna and ATC center equipment, S-S link between satellites in models with several satellites.

Figure 1. Model A1S1G1 of satellite communication channel
(TS = 112 bytes, Constant PDF; TBT = 0.5 s, Constant PDF)

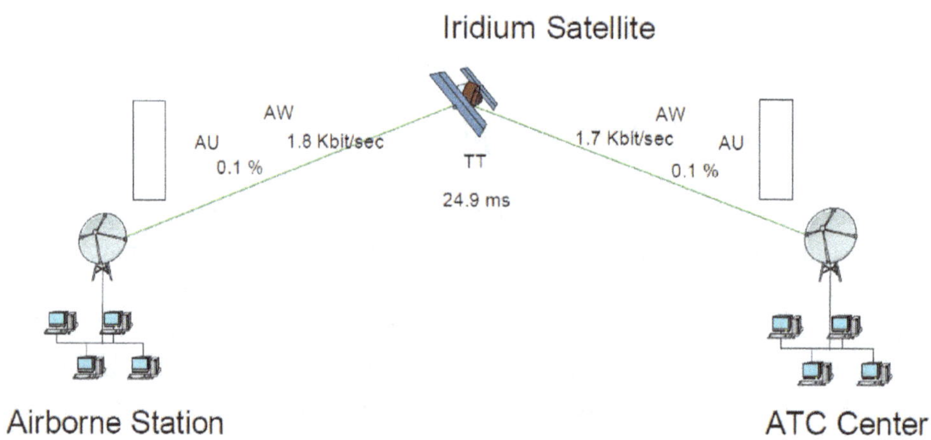

Figure 2. Aircraft equipment – ADS-B transceiver link (A-A link)

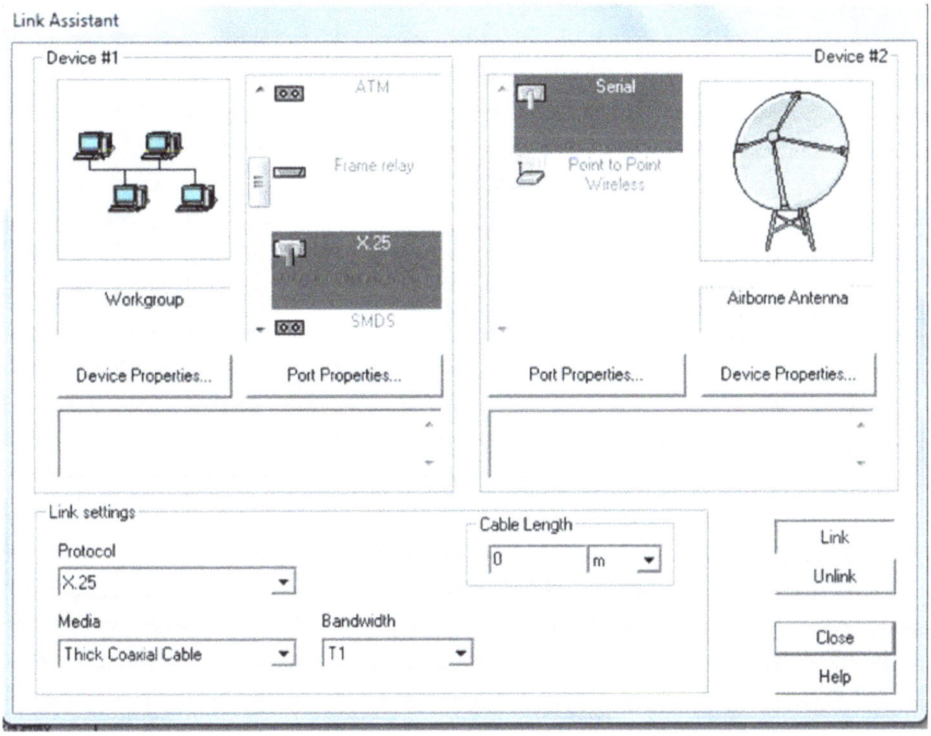

TRAFFIC IN COMMUNICATION CHANNELS "AIRCRAFT/ RPAS-SATELLITES-GROUND STATION"

Investigated Parameters

Regarding the capabilities of NetCracker Professional 4.1 the next parameters of "Aircraft- Satellite–Ground station" communication channel have been investigated (Kharchenko, Barabanov, Grekhov, Ivanenko, & Lobanov, 2012):

- **Channel Architecture:** Number of satellites in communication channel, number of aircraft in communication channel, and cable length of communication link;
- **Traffic Capabilities:** Transaction Size (TS) of transmitted message, Time Between Transactions (TBT);
- **Link Characteristics:** Communication link bandwidth, BER of communication link, communication link latency;

Figure 3. ADS-B transceiver – Satellite transponder link (A-S link)

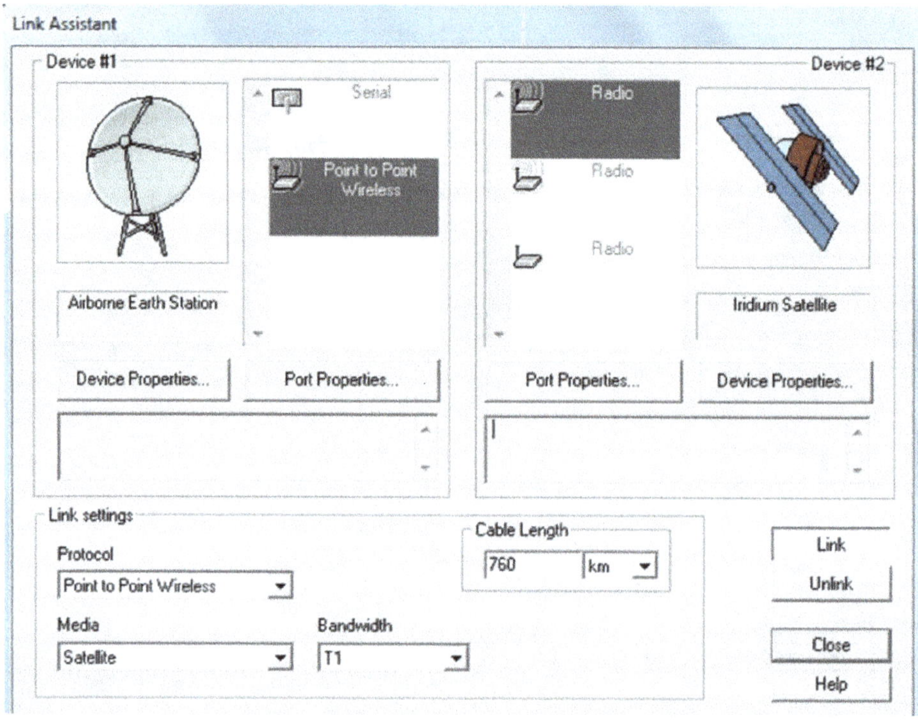

- **Equipment Properties:** Network protocol of the model, Packet Fail Chance (PFC) of telecom device.

The reaction of the system on the changes of abovementioned parameters is analyzed with the help of communication channel indicators.

Reference Model Parameters

For standardization of all obtained results, the reference model for investigations was created. The bandwidth, length and latency of communication link and its initial BER are represented in Figure 7 and Figure 8 for space segment and airborne/ground segments of the channel correspondingly.

Figure 4. Satellite transponder – Ground transceiver link (S-G link)

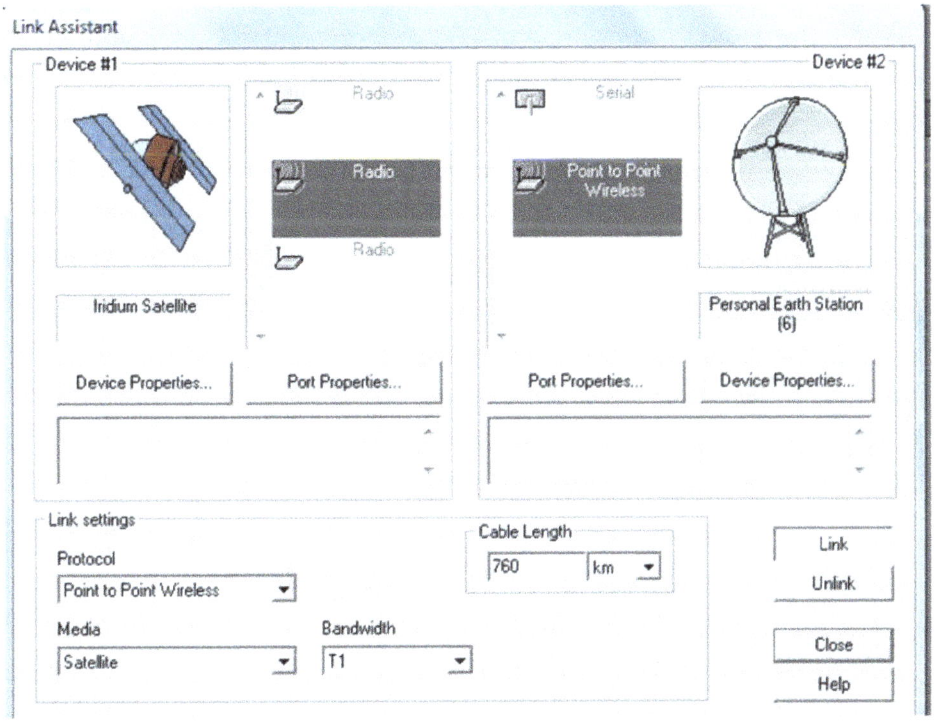

Transaction Size and Time Between Transactions as Well as Application Layer

Protocol of traffic are set in accordance to real parameters of ADS-B system (Figure 9).

The satellite characteristics of the model satisfy the Iridium system requirements. In the model with two satellites and more the transmission of data between aircraft – satellite node is organized as it is shown in Figure 10.

During the changes of one of parameter all other parameters remain the same that helps to reveal how exactly the setting influences on communication channel indicators (AW, AU and TT). Reference parameters of initial model are given in Table 1.

These data are used for the next analysis. Regarding these data potentially 1000 aircraft can be connected to the link. Further analysis is aimed to investigate how to increase the capacity of communication channel.

Figure 5. Ground transceiver – ATC center link (G-G link)

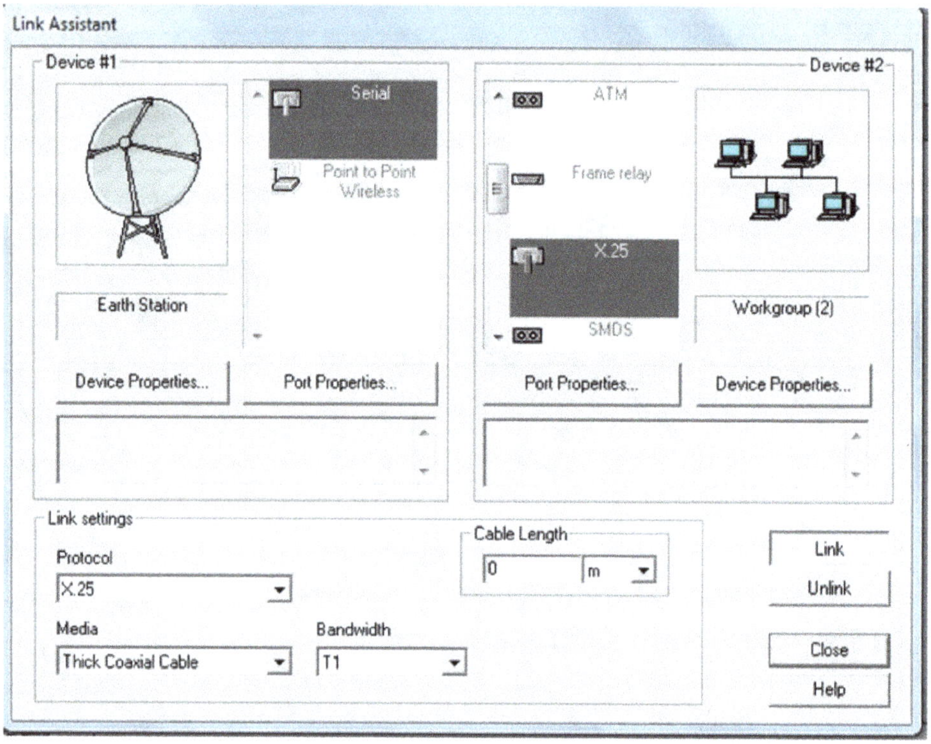

Impact of Satellites Number

The study of the effect of satellites number on data traffic used models ***A1S1G1*** (Figure 1), ***A1S2G1*** (Figure 11) and ***A1S5G1*** (Figure 12). The parameters of transmitted data, devices and communication channels in all models are set in accordance with the reference model. The satellite-to-satellite link parameters are determined identically to the real parameters of Iridium communication system. Results of traffic modelling are summarized in Table 2.

As follows from Table 2, the value of AW and AU parameters for all models stay the same and equal to initial value in reference model despite increasing number of satellites in the communication channel. AW and AU parameters of the satellite-to-satellite link possess the same value as at the aircraft–satellite link. AW and AU parameters in ***A1S5G1*** model are the same for all four satellite-to-satellite links.

Obtained data demonstrate that number of satellites does not affect average workload and average utilization neither aircraft–satellite link nor satellite–

Figure 6. Satellite - Satellite (S-S link)

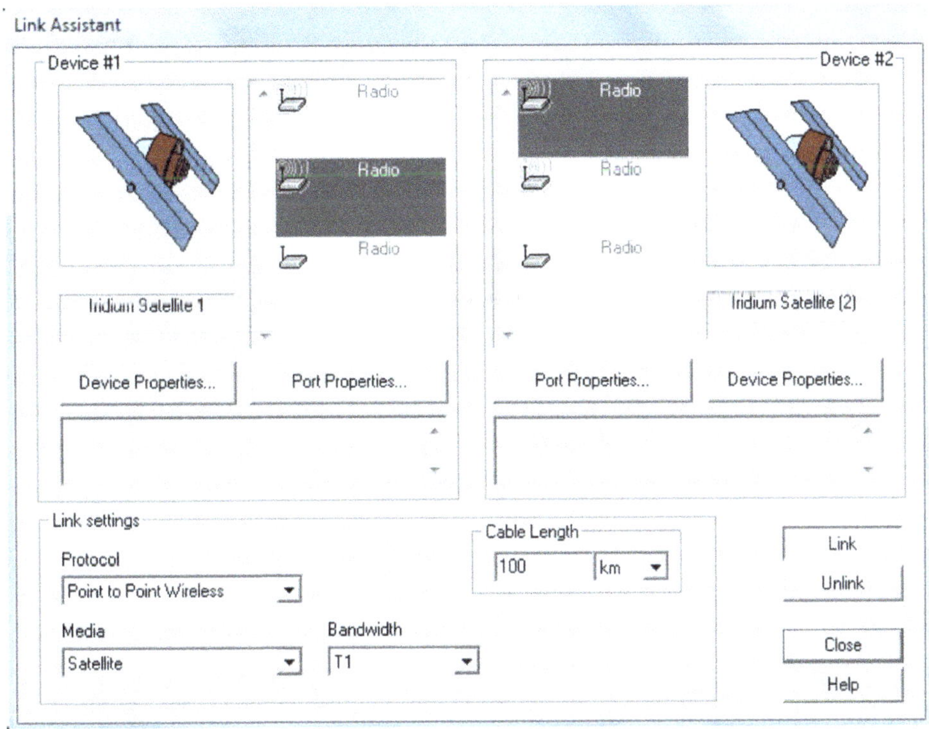

Figure 7. A–S, S–G links parameters (TS = 112 bytes, Constant PDF; TBT = 0.5 s, Constant PDF)

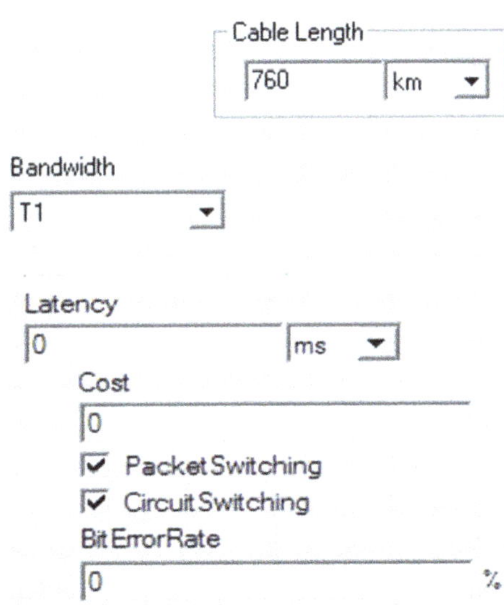

Figure 8. A–A, G–G links parameters (TS = 112 bytes, Constant PDF; TBT = 0.5 s, Constant PDF)

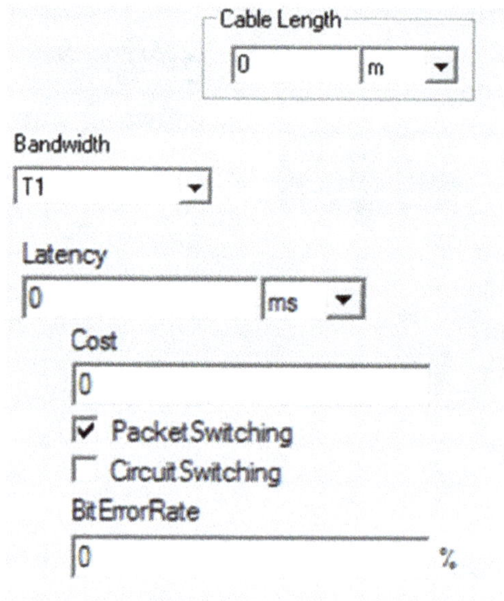

Figure 9. Traffic parameters (TS = 112 bytes, Constant PDF; TBT = 0.5 s, Constant PDF)

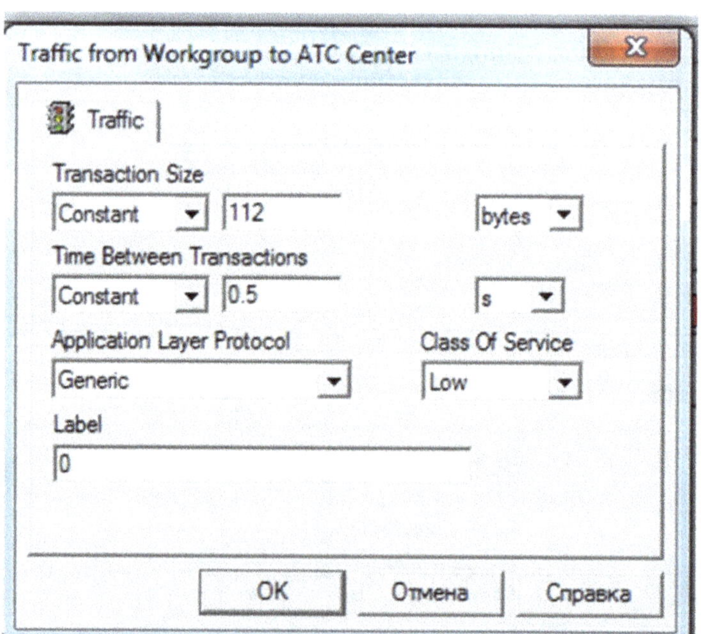

Figure 10. Satellite characteristics

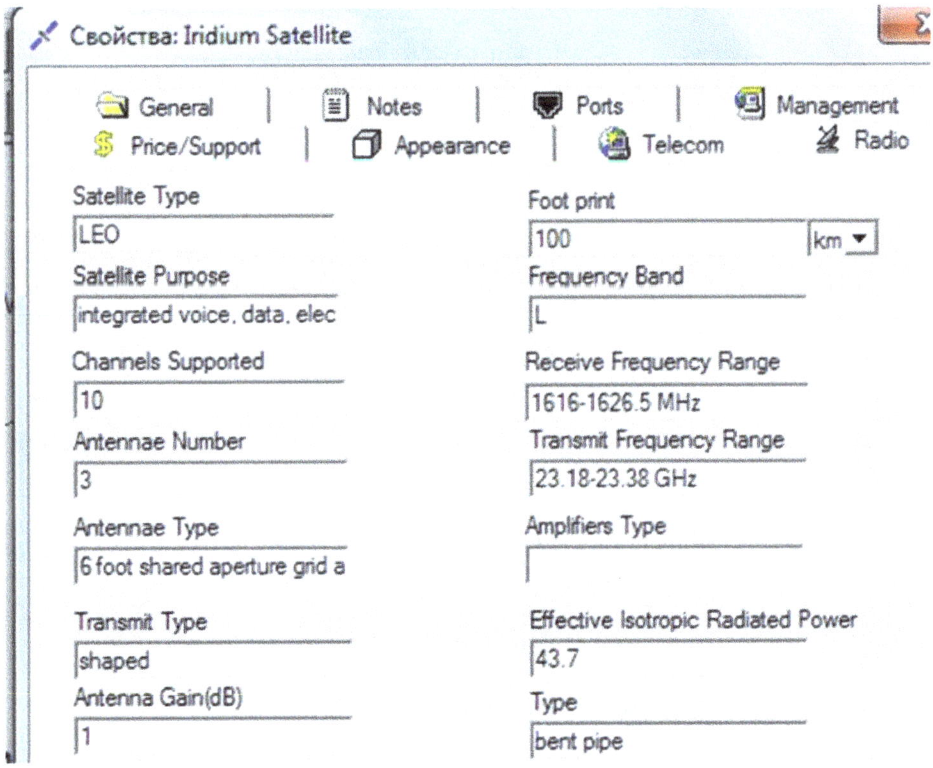

Table 1. Reference data of initial model

	A-S link	S-G link
AW (Kbit/s)	1,8	1,7
AU (%)	0,1	0,1
TT (ms)	24,9	

(TS = 112 bytes, Constant PDF; TBT = 0.5 s, Constant PDF)

ground station link. The only parameter that is affected by number of satellites is TT parameter. Its value rises with complication of communication link architecture.

Figure 11. A1S2G1 model (TS = 112 bytes, Constant PDF; TBT = 0.5 s, Constant PDF)

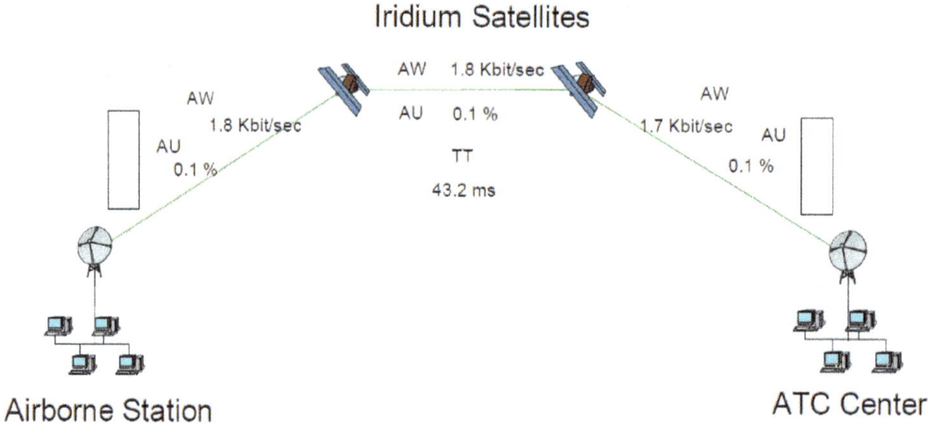

Figure 12. A1S5G1 model (TS = 112 bytes, Constant PDF; TBT = 0.5 s, Constant PDF)

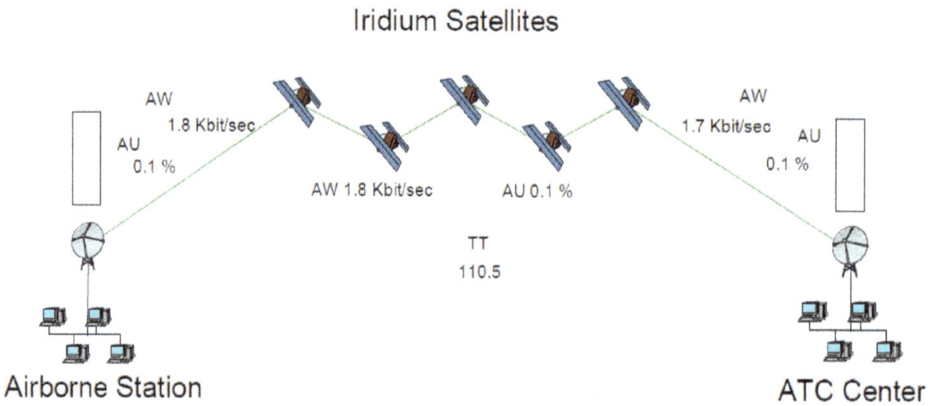

Table 2. Influence of satellites number on data traffic

	A1S1G1 model		*A1S2G1* model			*A1S5G1* model		
	A-S	**S-G**	**A-S**	**S-S**	**S-G**	**A-S**	**S-S**	**S-G**
AW (Kbit/s)	1,8	1,7	1,8	1,8	1,7	1,8	1,8	1,7
AU (%)	0,1	0,1	0,1	0,1	0,1	0,1	0,1	0,1
TT (ms)	24,9		43,2			110,5		

(TS = 112 bytes, Constant PDF; TBT = 0.5 s, Constant PDF)

Figure 13. A2S1G1 model (TS = 112 bytes, Constant PDF; TBT = 0.5 s, Constant PDF)

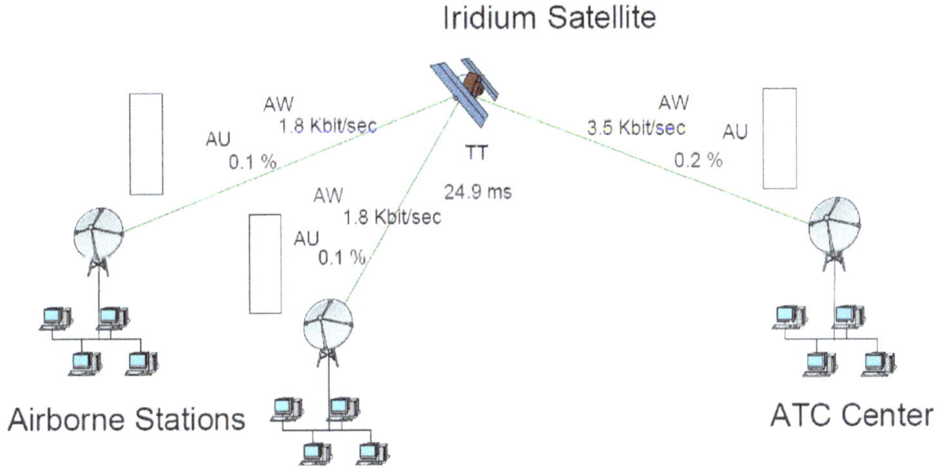

Figure 14. A3S1G1 model (TS = 112 bytes, Constant PDF; TBT = 0.5 s, Constant PDF)

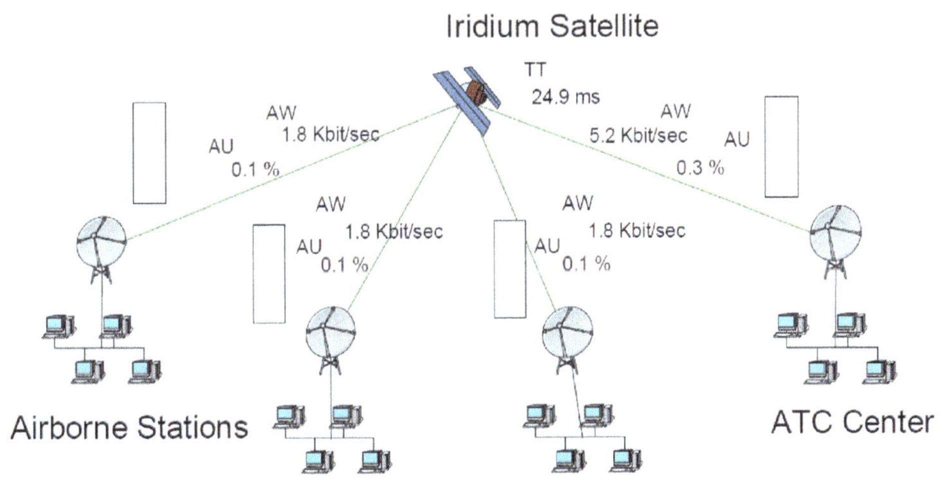

Impact of Aircraft Number

Hundreds of aircraft equipped with ADS-B system are simultaneously operating in airspace. That is why the channel model with several numbers of aircraft should be investigated. The study of the effect of aircraft number

Table 3. Influence of aircraft number on data traffic

	A1S1G1 **Model**		*A2S1G1* **Model**		*A3S1G1* **Model**	
	A-S	**S-G**	**A-S (1-2)**	**S-G**	**A-S (1-3)**	**S-G**
AW (Kbit/s)	1,8	1,7	1,8	3,5	1,8	5,2
AU (%)	0,1	0,1	0,1	0,2	0,1	0,3
TT (ms)	24,9		24,9		24,9	

(TS = 112 bytes, Constant PDF; TBT = 0.5 s, Constant PDF)

on data traffic used models A1S1G1 (Figure 1), A2S1G1 (Figure 13) and A3S1G1 (Figure 14).

Table 3 illustrates the behavior of the communication channel with additional ADS-B systems connected: AW and AU parameters of satellite-ground link equal to the sum of AW and AU parameters of all connected aircraft-satellite links.

Taking into consideration the fact that average utilization of satellite-ground link is proportionally growing with the number of aircraft connected to the channel, it is possible to calculate what amount of aircraft in service could

Figure 15. A3S1G1 model (TS = 500-600 Kbits, Uniform PDF; TBT = 1 s, LogNormal PDF)

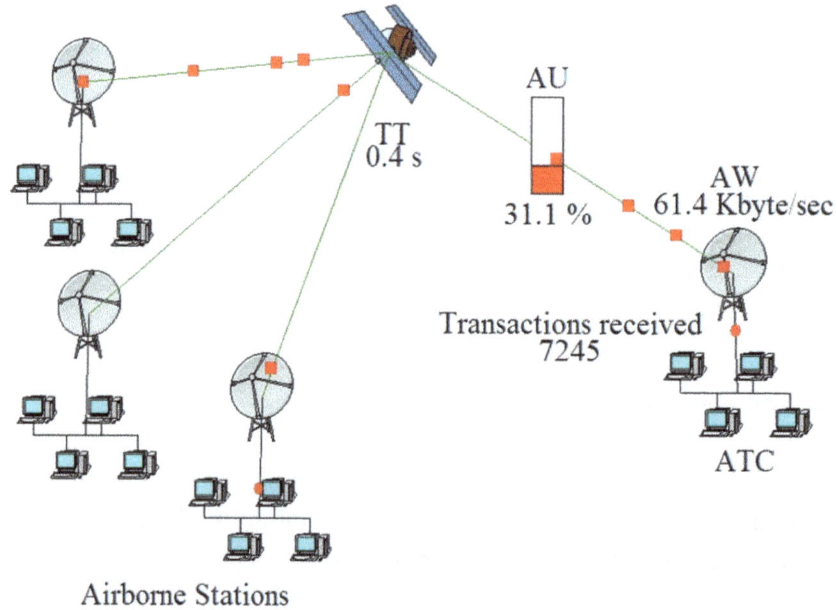

Figure 16. A4S1G1 model (TS = 500-600 Kbits, Uniform PDF; TBT = 1 s, LogNormal PDF)

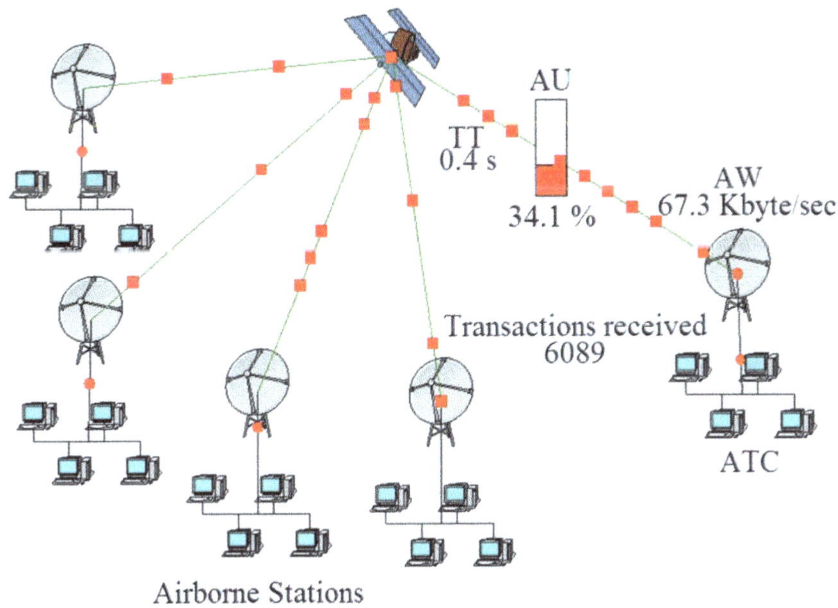

Figure 17. A5S1G1 model (TS = 500-600 Kbits, Uniform PDF; TBT = 1 s, LogNormal PDF)

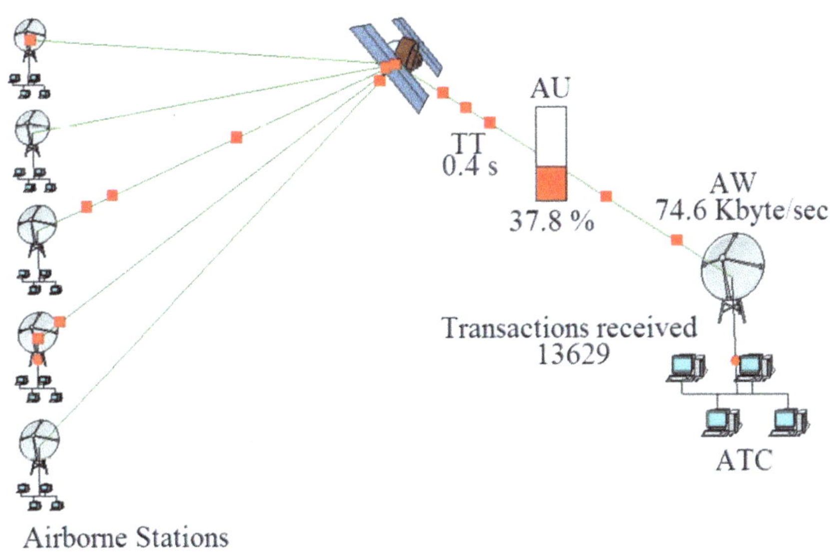

Figure 18. A6S1G1 model (TS = 500-600 Kbits, Uniform PDF; TBT = 1 s, LogNormal PDF)

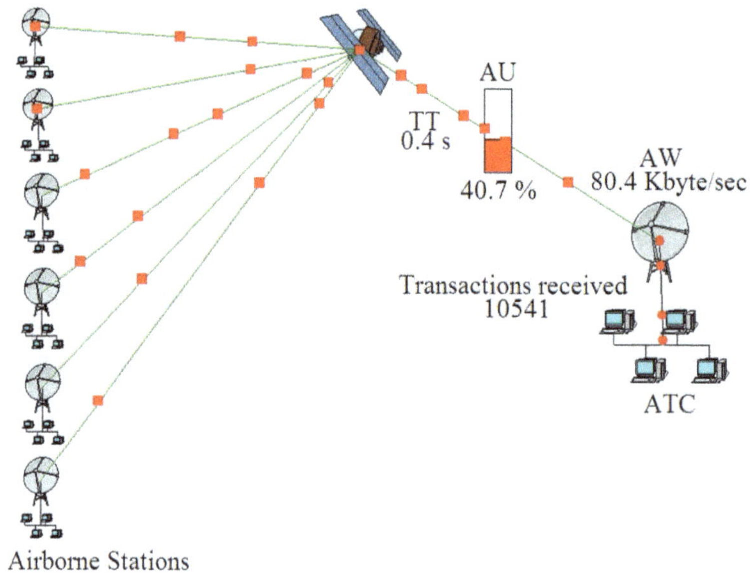

Figure 19. A7S1G1 model (TS = 500-600 Kbits, Uniform PDF; TBT = 1 s, LogNormal PDF)

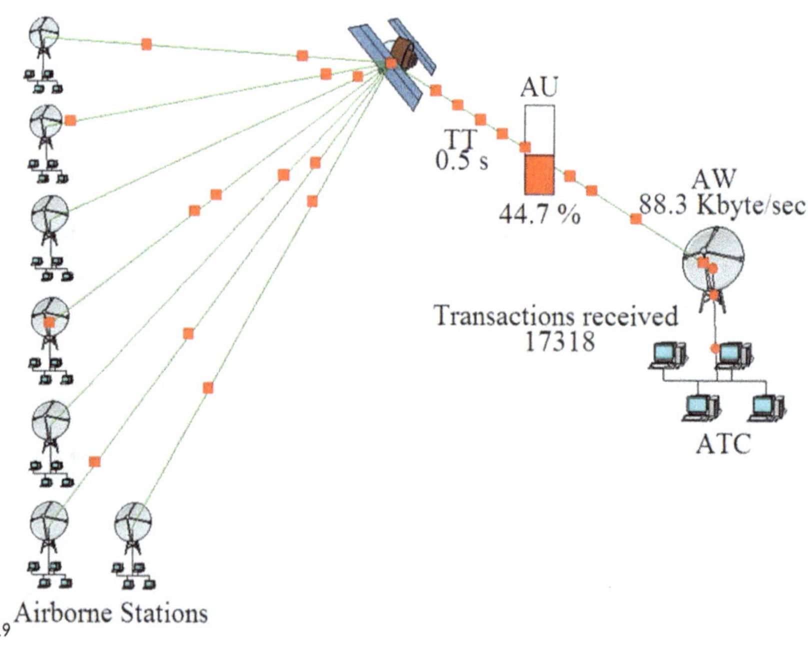

Fig19

Figure 20. A8S1G1 model (TS = 500-600 Kbits, Uniform PDF; TBT = 1 s, LogNormal PDF)

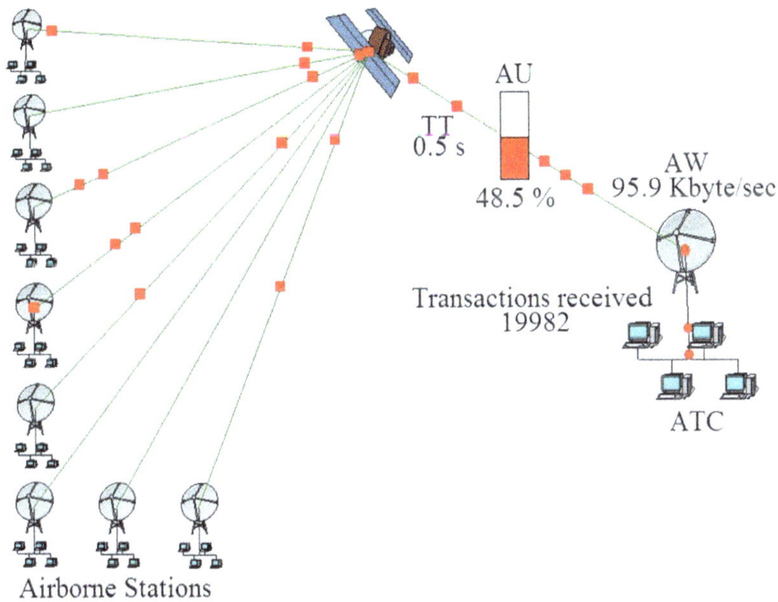

Figure 21. A9S1G1 model (TS = 500-600 Kbits, Uniform PDF; TBT = 1 s, LogNormal PDF)

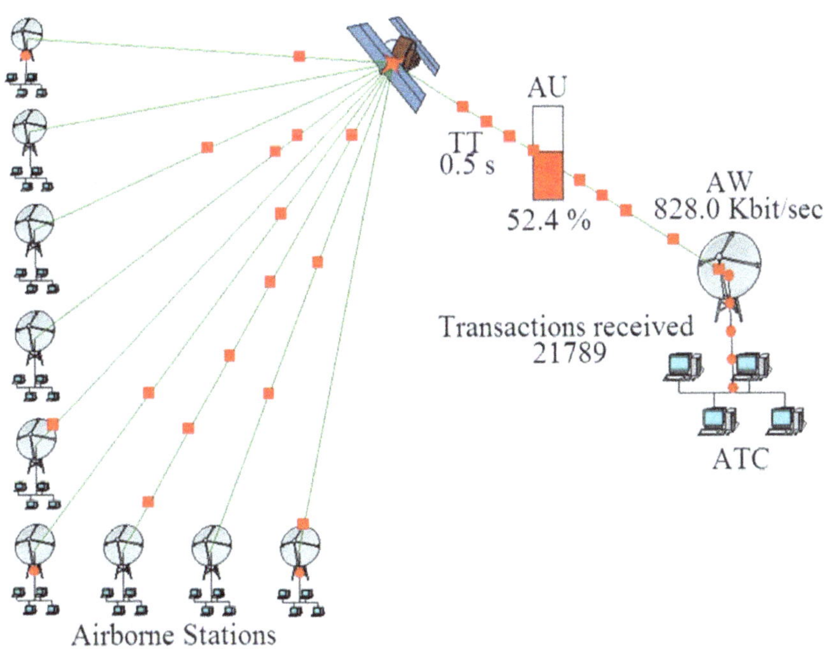

Figure 22. A10S1G1 model (TS = 500-600 Kbits, Uniform PDF; TBT = 1 s, LogNormal PDF)

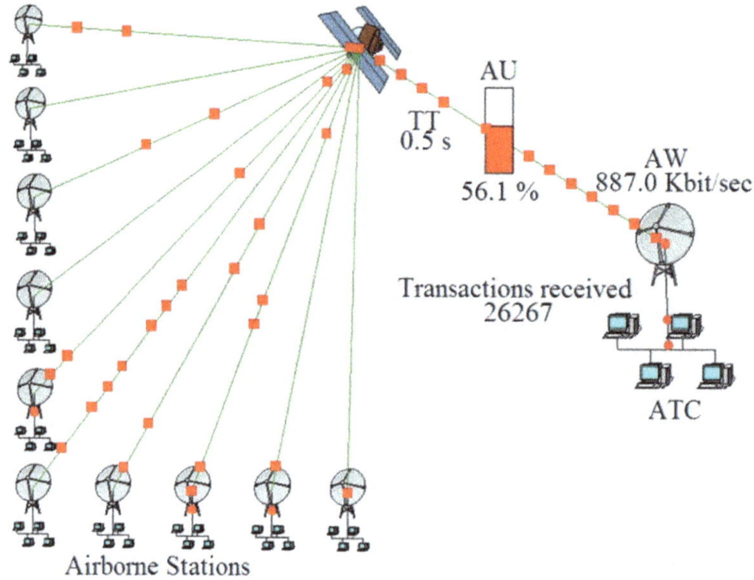

Figure 23. A11S1G1 model (TS = 500-600 Kbits, Uniform PDF; TBT = 1 s, LogNormal PDF)

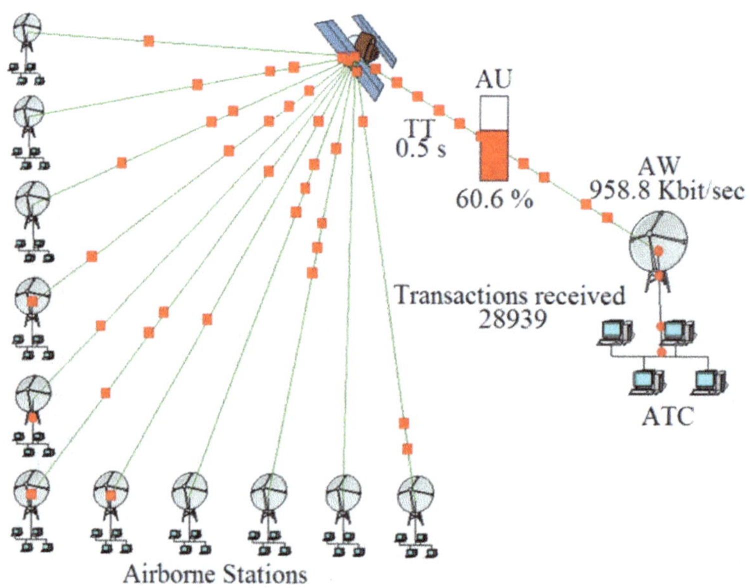

Figure 24. A1S10G1 model (TS = 500-600 Kbits, Uniform PDF; TBT = 0.03 s, Exponential PDF)

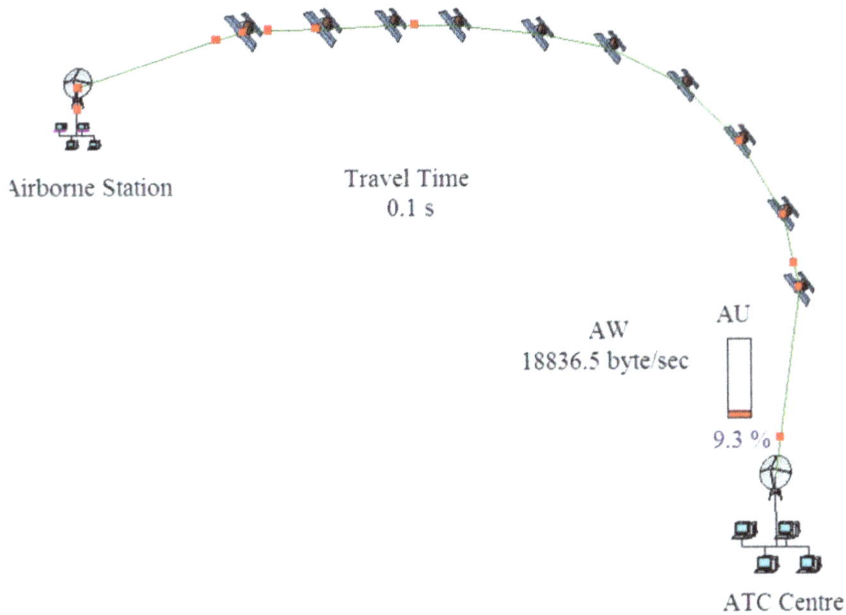

Figure 25. A2S10G1 model (TS = 500-600 Kbits, Uniform PDF; TBT = 0.03 s, Exponential PDF)

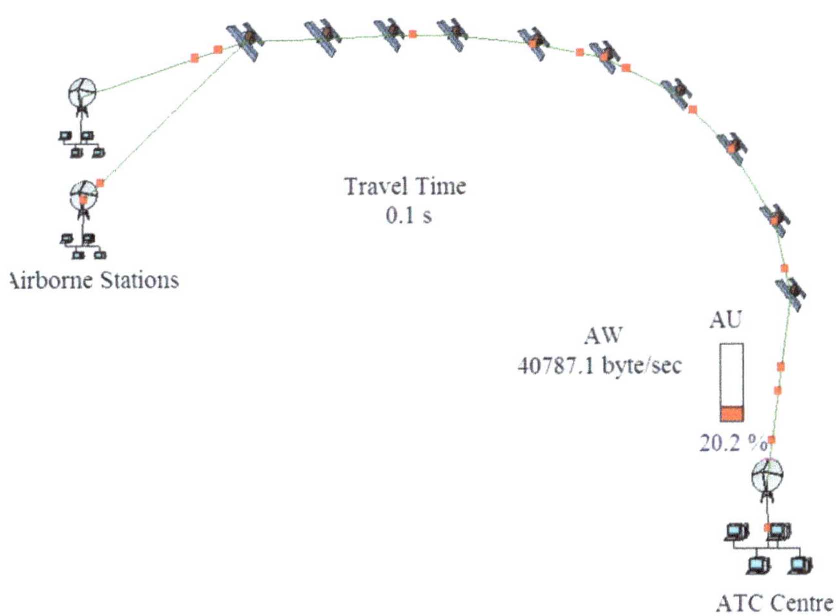

Figure 26. A3S10G1 model (TS = 500-600 Kbits, Uniform PDF; TBT = 0.03 s, Exponential PDF)

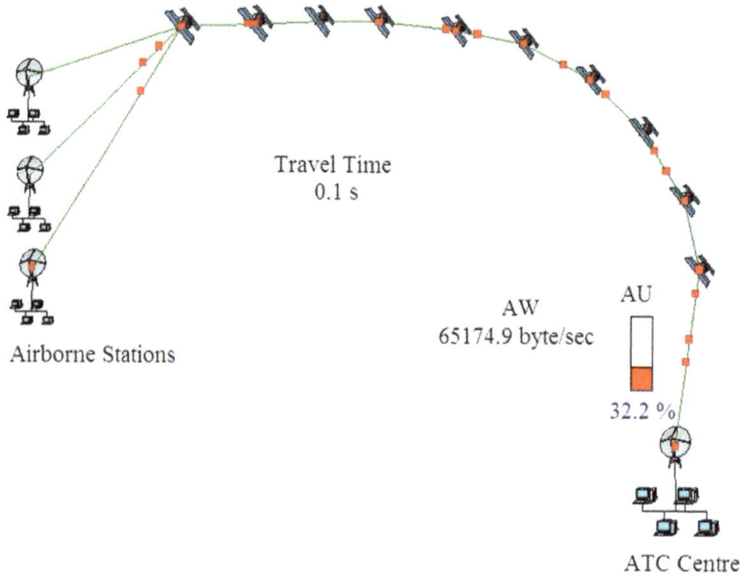

Figure 27. A4S10G1 model (TS = 500-600 Kbits, Uniform PDF; TBT = 0.03 s, Exponential PDF)

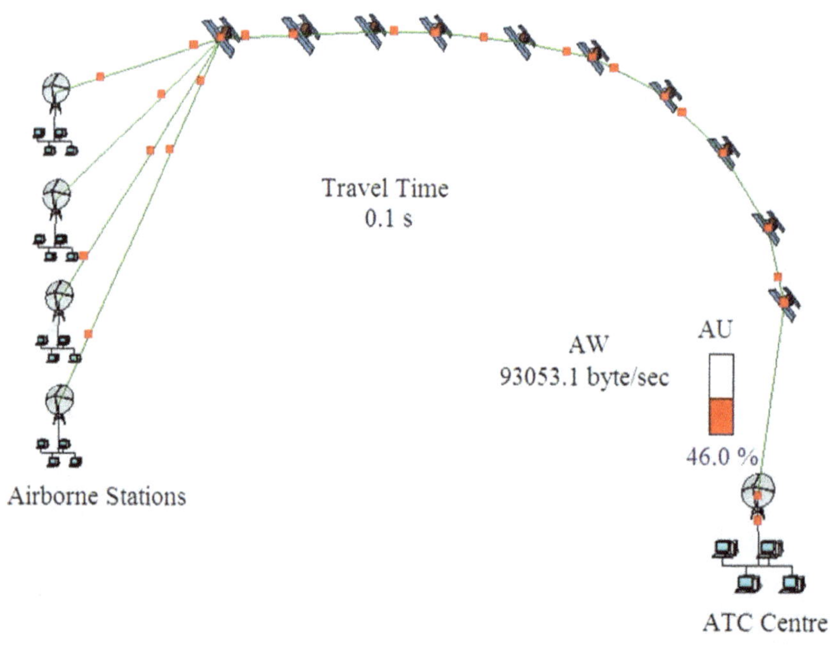

Figure 28. A5S10G1 model (TS = 500-600 Kbits, Uniform PDF; TBT = 0.03 s, Exponential PDF)

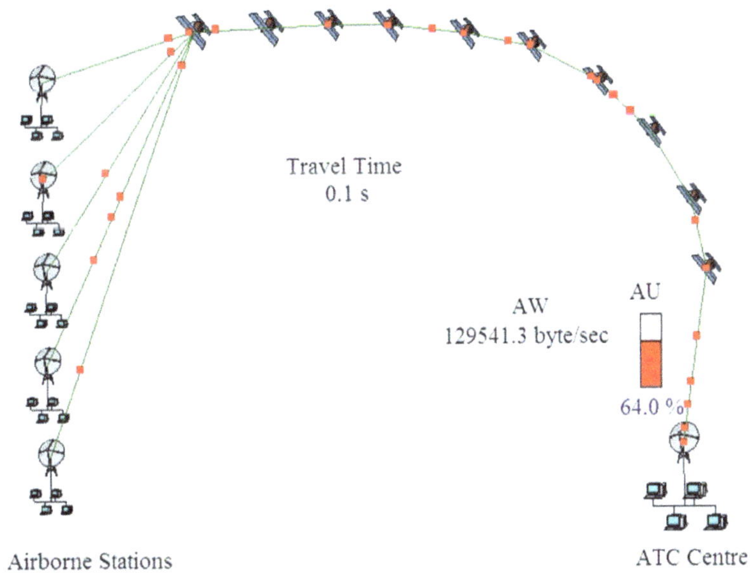

Figure 29. A6S10G1 model (TS = 500-600 Kbits, Uniform PDF; TBT = 0.03 s, Exponential PDF)

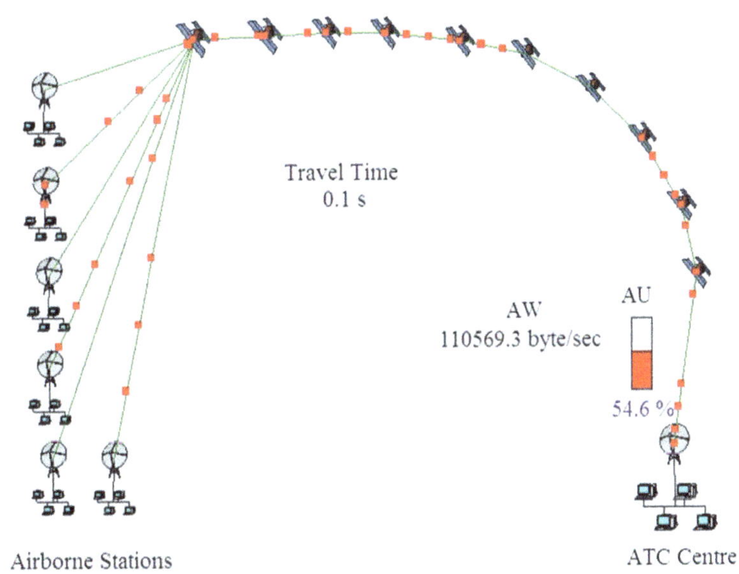

Figure 30. A7S10G1 model (TS = 500-600 Kbits, Uniform PDF; TBT = 0.03 s, Exponential PDF)

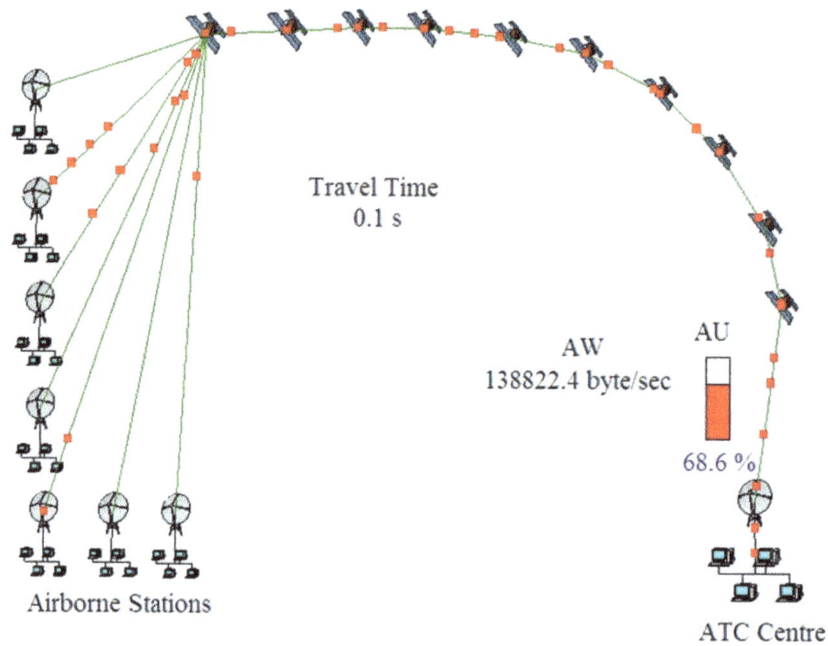

be connected simultaneously in order to reach link saturation. Such kind of calculation is the subject for further analysis.

For comparison, calculations were made with other traffic parameters. In Figures 15-23 the satellite was one, and the number of aircraft varied from three to eleven. The travel time was the same in all cases, and the average utilization of the downlink grew with the increase in aircraft number.

In Figures 24-30 satellites were ten, and the number of aircraft increased from one to seven. In this case, the transactions size was the same, but time between transactions and the distribution law are taken another. As in the previous case, TT parameter was the same for all models, and AU parameter of the downlink increased with the number of aircraft.

In the case of ten satellites, the increase in the average utilization of the downlink occurred faster, which may be due to another distribution law for TBT parameters.

Influence of Parameters Describing Data Traffic

Different distribution laws can describe transaction sizes and time between transactions. In Table 4 the most common Probability Density Functions (PDF) arc given.

Transaction Size Impact

TS parameters of transmitted messages can be distributcd according to different PDF: Constant, Uniform, Exponential, Normal, and LogNormal. The effect of transaction size on AW, AU and TT parameters in reference model (Figure 1) is shown in Table 5, where the marginal value of the link indicators are depicted by bold font (for these values, the model view is shown in Figure 31). From the presented data follows that TS parameter influences on all three parameters – AW, AU and TT.

Table 4. Probability Density Functions

Distribution law	PDF	Probability moment
Normal	$\frac{1}{\sqrt{2\pi\sigma}} exp\left(-\frac{(x-a)^2}{2\sigma^2}\right)$	$m_1 = a$, $\mu_2 = \sigma^2$, $\mu_3 = 0$, $\mu_4 = 3\sigma^4$
Uniform	$\frac{1}{b-a}$	$m_1 = \frac{a+b}{2}$, $\mu_2 = \frac{(b-a)^2}{12}$ $\mu_3 = 0$, $\mu_4 = \frac{1}{80}(b-a)^4$
Exponential	$\lambda e^{-\lambda x}$	$m_1 = 1/\lambda$, $m_2 = 2/\lambda^2$, $\mu_2 = 1/\lambda^2$, $\mu_3 = 2/\lambda^3$, $\mu_4 = 9/\lambda^4$
LogNormal	$\frac{1}{x\sqrt{2\pi\sigma}} exp\left(-\frac{(lnx-a)^2}{2\sigma^2}\right)$	$m_1 = exp(a + 0{,}5\sigma^2)$, $\mu_2 = exp(2a + \sigma^2) *$ $* (exp(\sigma^2) - 1)$
Gamma	$\frac{1}{\beta^a\Gamma(\alpha)} x^{a-1} e^{-x/\beta}$	$m_1 = \alpha\beta$, $m_2 = \alpha(\alpha+1)\beta^2$, $\mu_2 = \alpha\beta^2$, $\mu_3 = 2\alpha\beta^3$, $\mu_4 = 3(\alpha+2)\alpha\beta^4$

Table 5. Transaction Size influence on channel indicators

TS (bytes)	1	100	500	1000	1500	110000
Constant						
AW (Kbit/s)	0	1,6	7,8	15,7	23,6	**1575**
AU (%)	0	0,1	0,5	1	1,5	**100**
TT (ms)	22,6	24,6	32,5	33,9	35,1	**284**
Uniform						
AW (Kbit/s)	0	1,6	7,9	15,8	23,6	**1575**
AU (%)	0	0,1	0,5	1	1,5	**100**
TT (ms)	22,7	24,7	32,6	33,8	35,1	**284**
Normal						
AW (Kbit/s)	1,6	4	8,5	16,5	24,1	**1580**
AU (%)	0,1	0,3	0,5	1	1,5	**100**
TT (ms)	24,7	27,8	32,8	35,7	37,6	**290**
Exponential						
AW (Kbit/s)	1,7	5,3	7,5	16,1	24,2	**1580**
AU (%)	0,1	0,3	0,5	1	1,5	**100**
TT (ms)	24,8	29	32	26,5	39,8	**295**
LogNormal						
AW (Kbit/s)	1	7,8	18	22,8	73,3	**1590**
AU (%)	0,1	0,5	1,1	1,4	4,6	**100**
TT (ms)	24	32,6	37,7	39,1	40,1	**300**

(TBT = 0.5 s, Constant PDF)

Figure 31. Communication channel saturation

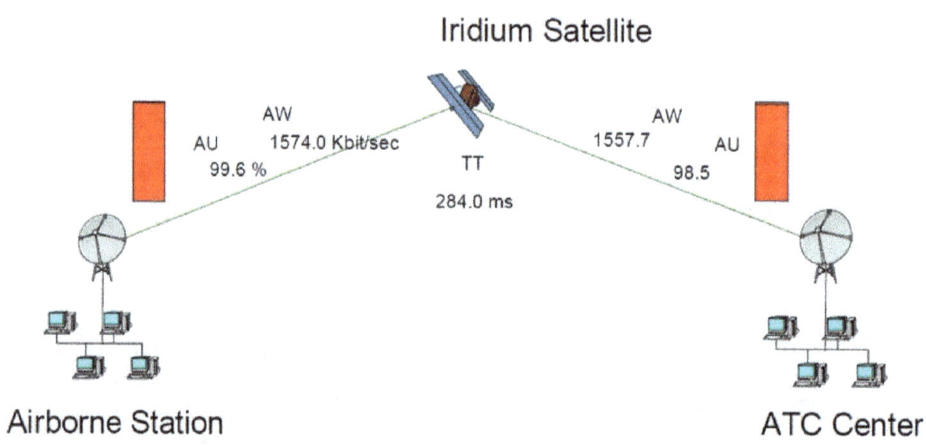

Table 6. TBT impact on traffic parameters

TBT	1 μs	1 ms	10 ms	100 ms	1s	10s
Constant, Exponential, Uniform, Normal						
AW (Kbit/s)	1578	884	88	8,8	0,9	0,1
AU (%)	99,8	56	5,5	0,6	0,1	0
TT (ms)	511,4	24,9	24,9	24,9	24,9	24,9
LogNormal						
AW (Kbit/s)	0,9	0,9	0,9	0,8	0,2	0
AU (%)	0,1	0,1	0,1	0	0	0
TT (ms)	24,7	24,7	24,7	24,7	24,7	0

(TS = 112 bytes, Constant PDF)

Time Between Transactions Impact

TBT parameter means the frequency of data packets transmission and as TS parameter, it can be distributed according to the same laws.

Modifying the value of TBT parameter distributed by Constant law (as it is set by default) it was observed that AW and AU parameters grow with decreasing of TBT parameter. The same results were obtained analyzing Uniform, Exponential and Normal laws. More over TBT parameter influence

Figure 32. Model settings

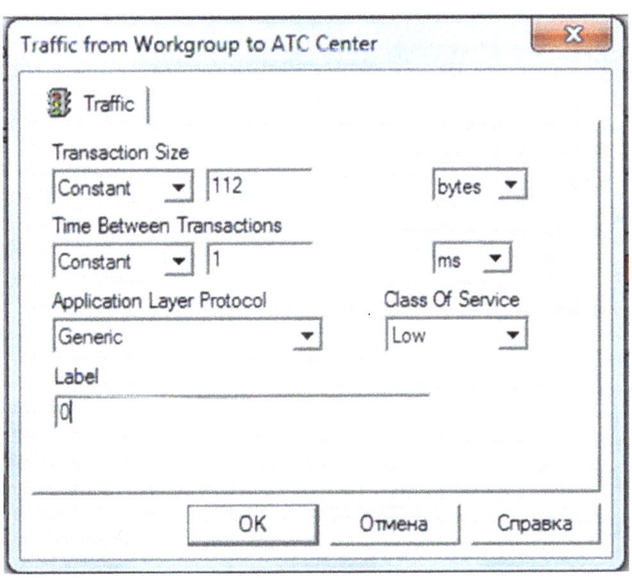

Figure 33. Model read-outs for TBT = 1 ms, Constant PDF

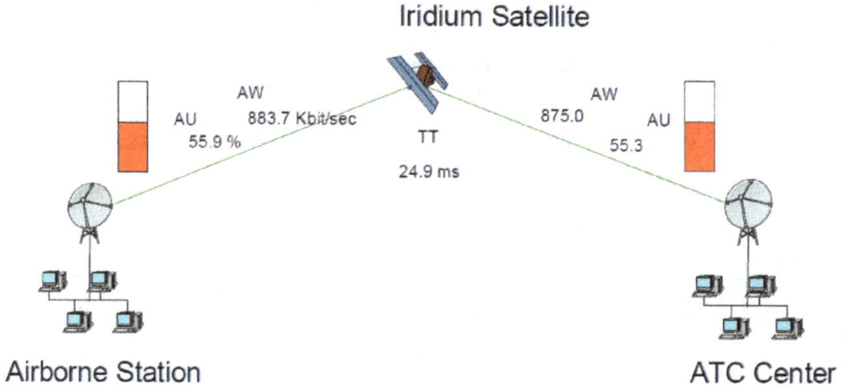

AW and AU parameters at both A-S and S-G links in a same way. That is why results only for A-S link is described in Table 6.

TBT parameter has a significant influence on the link performance. Reducing of TBT parameter leads to decreasing of potential number of aircraft that can be connected to the channel. For example, setting the value of TBT equals to 1 ms for Constant, Uniform, Exponential and Normal distribution laws the capacity of channel is rapidly fall. AU parameter acquires 56%

Figure 34. Model settings

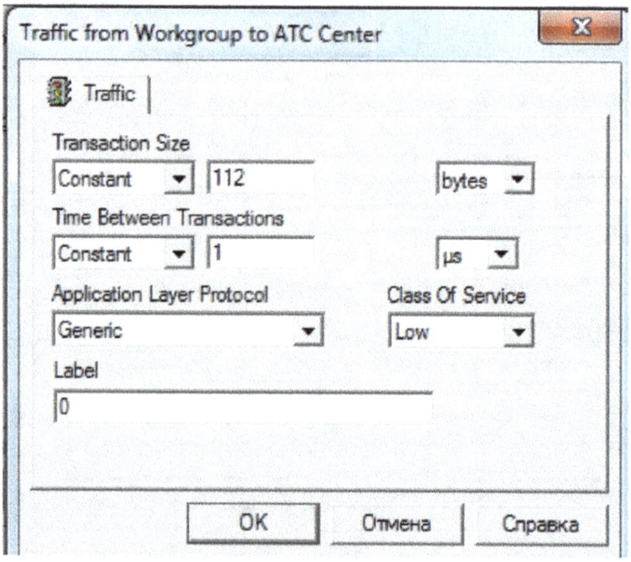

(Figures 32, 33) meaning that no more than two aircraft can be connected to the channel with such value of TBT parameter.

The complete saturation of channel is observed when TBT parameter equals to 1 μs (Figures 34, 35).

Investigation of Link Characteristics Influence

The next important group of parameters for investigation belongs to link characteristics of the channel. The analysis of communication link bandwidth and a BER of the link is described below.

Figure 35. Model read-outs for TBT = 1 μs, Constant PDF

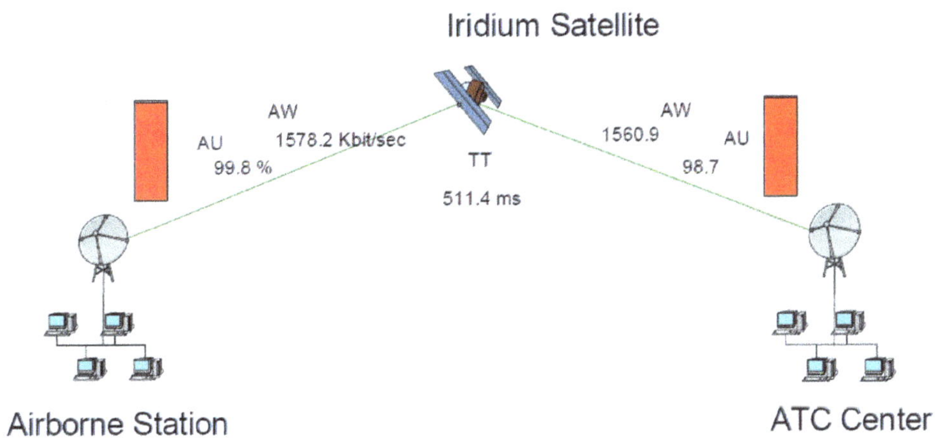

Figure 36. T1 frame format

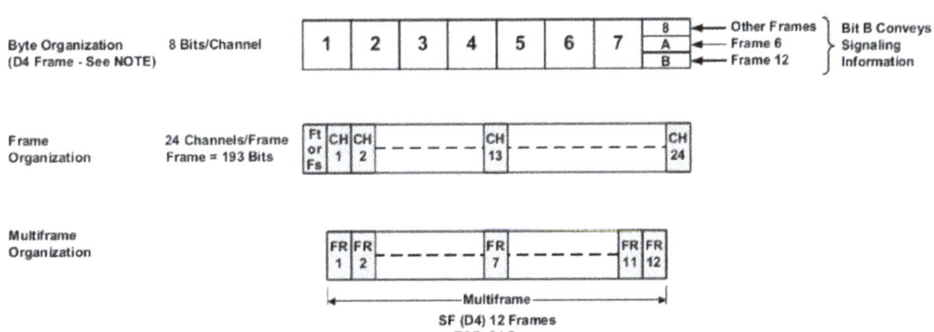

Communication Link Bandwidth Impact

Bandwidth is a characteristic of the communication channel, indicating how much data can be transmitted through this channel per unit of time. Bandwidths usually is expressed in bits per second or multiples of it (bit/s, Kbit/s, Mbit/s, Gbit/s, etc.).

The nominal rate of T1 line operation is 1.544 Mbps. The data that are transferred via T1 line is frame-organized. T1 frame consists of 193 bits (Figure 36). The 193 bits consist of 24 timeslots of eight bits each. Framing and supervision information is carried by an additional timeslot. As a result, the data rate supported by each payload timeslot is 64 kbps. The data rate of the framing slot is 8 kbps.

The nominal rate of E1 line operation is 2.048 Mbps. The data that are transferred via E1 line is frame-organized too (Figure 37). Each E1 frame contains of 256 bits. These 256 bits are organized in 32 timeslots, eight bits in each, which carry the data payload. The data rate supported by each timeslot is 64 kbps, because repetition rate of frame is 8,000 per second. Only 31 timeslots are available for users as "0" timeslot is permanently reserved.

The correspondence between T- and E-carrier is given in Table 7.

Using the capabilities of NetCracker software the bandwidth of each link can be changed. Results of analysis for A-S link are summarized in Table 8.

Figure 37. E1 Frame Format

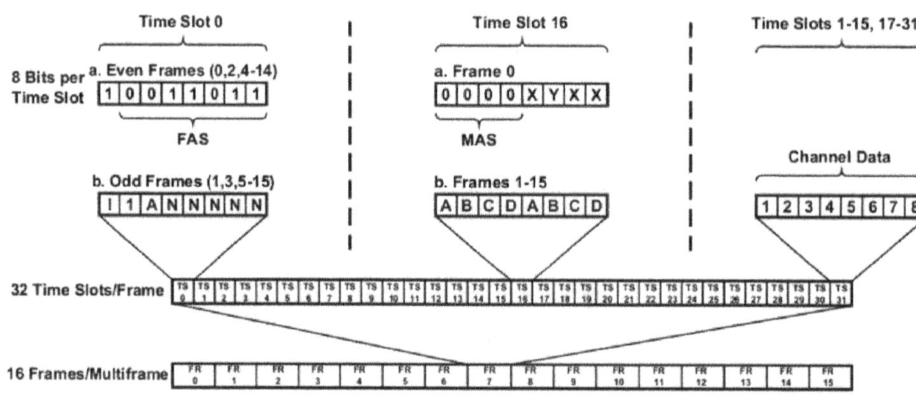

Notes

I	International Bit	ABCD	ABCD Signaling Bits
N	National Bits (S_{a4} through S_{a8})	X	Extra Bit
A	Alarm Indication Signal (Loss of Frame Alignment - Red Alarm)	Y	Loss of Multiframe Alignment
FAS	Frame Alignment Signal, occupies alternate (but not necessarily even) frames	MAS	Multiframe Alignment Signal

Table 7. T-carrier and E-carrier correspondence

T- and E-Carrier Systems	North American	European
Level zero	64 kbit/s (DS0)	64 kbit/s
First level	1.544 Mbit/s (DS1) (24 user channels) (T1)	2.048 Mbit/s (32 user channels) (E1)
Second level	6.312 Mbit/s (DS2) (96 Ch.) (T2)	8.448 Mbit/s (128 Ch.) (E2)
Third level	44.736 Mbit/s (DS3) (672 Ch.) (T3)	34.368 Mbit/s (512 Ch.) (E3)
Fourth level	274.176 Mbit/s (DS4) (4032 Ch.)	139.264 Mbit/s (2048 Ch.) (E4)
Fifth level	400.352 Mbit/s (DS5) (5760 Ch.)	565.148 Mbit/s (8192 Ch.) (E5)

With the increase of data transfer rate AW and TT parameters keep their initial values despite the increasing of link bandwidth. However, AU parameter of the link is decreasing. A response of each link to bandwidth increasing is the same.

Communication Link Latency Impact

Link latency is one of the dominant parameter that characterizes the system performance overall.

The latency of systems developed for aviation purposes should be extremely small in order to make possible the aircraft control in real-time operation because the consequences of the information and data delays could be fatal. The investigation of a latency influence on communication channel indicators was made in three different ways:

1. Modifying the latency value at ADS-B Transceiver – Satellite Transponder link (A-S link);
2. Modifying the latency value at Satellite Transponder – Ground Transceiver link (S-G link) and

Table 8. A-S link bandwidth impact

Bandwidth (Mbps)	1,544 T1	2,048 E1	4	6	8	10	34,368 E3	44,736 T3 T3
AW (Kbit/s)	1,8	1,8	1,8	1,8	1,8	1,8	1,8	1,8
AU (%)	0,1	0,09	0,05	0,03	0,02	0,01	0,0	0,0
TT (ms)	24,9	24,8	24,6	24,5	24,3	24,3	24,3	24,3

3. Modifying the latency value at both links.

Results are presented in Table 9.

A latency does not influence on AW and AU parameters neither of A-S link nor of S-G link. However, the TT parameter is affected and it equals to the sum of initial time of data transmission and a latency value.

Link latency impact for different TS and TBT distribution laws are shown in Table 10 (the latency modification was implemented in A-S link and S-G link simultaneously).

Network Protocol Impact

Network protocols define the conventions and rules of data transmission between two devices. For given communication channel model NetCracker software gives the possibility to set one of the proposed protocols (Figure

Table 9. Link latency impact

Latency (ms)	0	10	100	500	1000	5000	10000
A-S link (latency at S-G link 0 ms)							
AW (Kbit/s)	1,8	1,8	1,8	1,8	1,8	1,8	1,8
AU (%)	0,1	0,1	0,1	0,1	0,1	0,1	0,1
TT (ms)	24,9	34,9	124,9	524,9	1024,9	5024,9	10024,9
S-G link (latency at A-G link 0 ms)							
AW (Kbit/s)	1,7	1,7	1,7	1,7	1,7	1,7	1,7
AU (%)	0,1	0,1	0,1	0,1	0,1	0,1	0,1
TT (ms)	24,9	34,9	124,9	524,9	1024,9	5024,9	10024,9
Both links							
TT (ms)	24,9	44,9	224,9	1024,9	2024,9	10024,9	20024,9

Table 10. Link latency impact for different distribution laws

Latency (ms)	0	10	100	500	1000	5000	10000
Constant, Exponential, Uniform, Normal							
TT (ms)	24,9	44,9	224,9	1024,9	2024,9	10024,9	20024,9
LogNormal							
TT (ms)	31,9	51,9	231,9	1031,9	2031,9	10031,9	20031,9

38): TCP/IP, IPX/SPX, DECNet Phase IV, AppleTalk Phase II, XNS, SNA, and X.25.

After analysis of all these protocols it can be said that difference in their influence on investigated indicators (AW, AU and TT) is insignificant, so the system could be regarded as the one that is not affected by type of protocol.

Conclusions

This section is devoted to the study of the communication channel "Aircraft-Satellite-Ground Station" using NetCracker Professional 4.1 software. A reference model was created for studies with real parameters of the ADS-B and Iridium systems. The dependencies of the traffic parameters on the characteristics and architecture of the communication channel were obtained and described. The response of the system to changes was analyzed using parameters such as Average Workload, Average Utilization, and Travel Time.

1. The number of satellites does not affect Average Workload and Average Utilization. The only parameter affected by the number of satellites in

Figure 38. Network protocols in NetCracker Professional

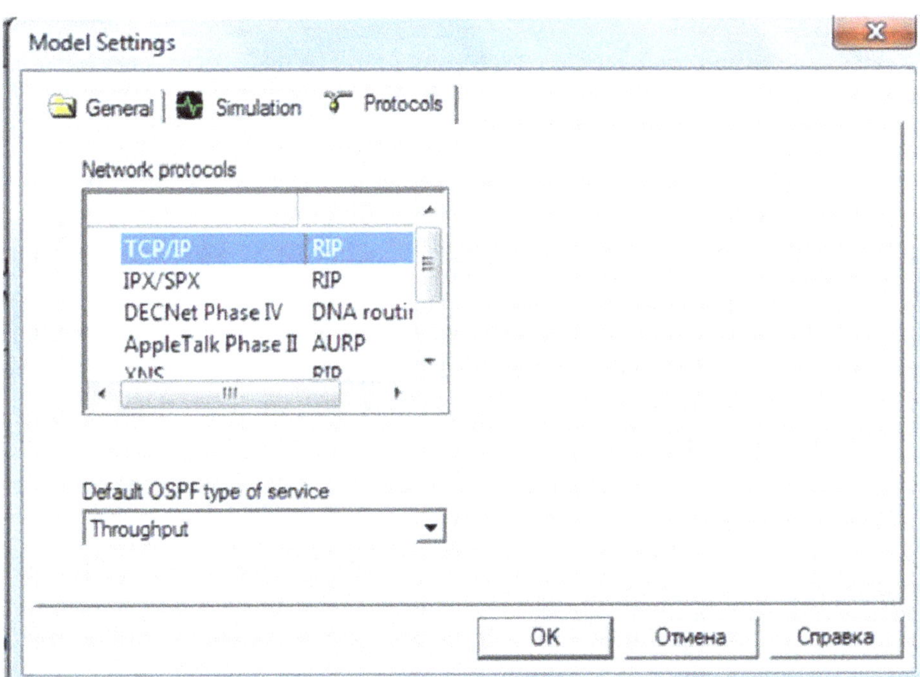

the channel is Travel Time. Its value increases with the complexity of the communication link architecture.

2. Average Workload and Average Utilization increases with the increase in the number of aircraft connected to the channel.
3. Transaction Size affects Average Workload, Average Utilization, and Travel Time. The nature of the effect depends on the probability distribution law.
4. Increasing of Time Between Transactions leads to an increase in the potential number of aircraft that can be connected to the channel.
5. Increasing the Bandwidth of the channel does not affect the value of Average Workload, but Average Utilization of the channel decreases. Travel Time of the message via communication channel is also reduced.
6. The latency in the channel affects Travel Time of a message. This influence depends on the law of traffic distribution.
7. The network protocol practically does not affect the performance of the communication channel.

INVESTIGATION OF ADS-B MESSAGES TRANSMISSION THROUGH SATELLITE TELECOMMUNICATION CHANNEL IRIDIUM USING NETCRACKER PROFESSIONAL

Process of continuous aviation developing leads to necessity of conversion from ground-based CNS systems to satellite-based systems in order to increase the possible operational areas mainly in the Poles regions and regions with reduced radar coverage (EUROCONTROL, 2013). That is why the new programs of ADS-B surveillance systems development using capabilities of satellite communication are produced in the scope of SESAR (Europe) and NEXTGEN (USA).

Communication channels consisting of on-board ADS-B equipment and Iridium satellites could help to resolve the bottlenecks by allowing the relocation of aviation traffic from the regions with the high density to the regions that cannot be used due to its remote location or inability of ground surveillance equipment. For that reason, the deep analysis of traffic parameters for “Aircrafts-Satellites-Ground Stations” channel is urgently needed.

Nowadays aviation is faced the problem of fast information exchange with high quality. The difficulties appear when insufficient number of radars or the lack of radar visibility cause “gaps” in aircraft tracking. This problem needs creation of modern data transmission systems. One of the best ways for their further development is the use of satellite technologies. This leads to a simplification of the equipment and reducing of the cost of installation and maintenance. They also provide coverage even in those places where the ground-based systems could not do this. During last 20 years technologies have stepped forward and today there is a set of satellite communication systems that are able to provide information exchange in general aviation: INMARSAT, COSPAS/SARSAT, Iridium, Globalstar and Thuraya. Each of them has its benefits, but in terms of global data transferring Iridium NEXT has the absolute advantage, that is 81 satellites and coverage of the whole Earth surface. With the help of satellite technologies and airborne ADS-B equipment air traffic management around the Earth will be able to track any aircraft at any point of the globe.

The aim of this study is: 1) to create new models of communication channel "Aircraft-Satellites-Ground Stations", 2) to consider and analyze the dependencies of Average Utilization in downlink and message Travel Time on the number of aircraft and satellites; 3) to investigate satellite links with different architecture; 4) to study effect of communication channel "saturation" during simultaneous data transmission from many planes via different number of satellites (Kharchenko, Wang Bo, Grekhov, & Kovalenko, 2014).

Figure 39. Model A1S5G1 with intersatellite communication link
(TS = 500-1500 Kbytes, Uniform PDF; TBT = 0.1 s, Exponential PDF)

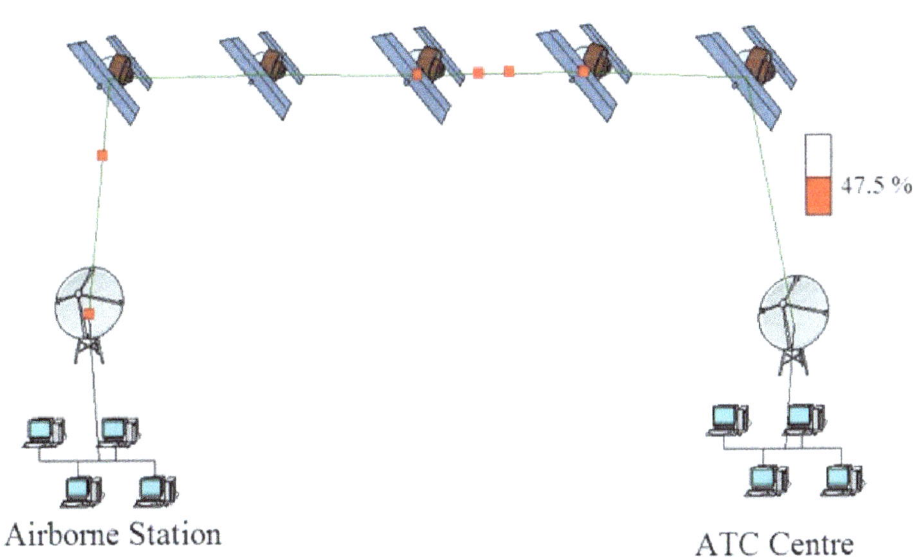

Structure of Models

For modeling of ADS-B messages transmission via satellites in real conditions, new models were developed with data traffic parameters different from those considered in the previous sections.

Figure 39 shows the model ***A1S5G1*** with intersatellite link: one airborne station, which is the aircraft ADS-B system, five Iridium satellites and ATC center. The following parameters of the model were chosen for Iridium satellites and the ADS-B system: Transaction Size – 500-1500 Kbytes with Uniform distribution law; Time Between Transactions – 0.1 s with Exponential distribution law; Packet Latency – 0.02 s with Constant distribution law; Packet Fail Chance – 0.01; aircraft/ground workstations and antennas data rate – 100 Kbyte/s (medium), links bit rates – T1 (1.544 Mbit/s). The same parameters were used in most models, except those for which changes of traffic parameters were defined separately. The same model was shown in Figure 12 but with other traffic parameters (TS = 112 bytes, Constant PDF, TBT = 0.5 s, Constant PDF).

Figure 40 shows the model *A1SGSGSG1* with bent-pipe architecture: one airborne station, three Iridium satellites, two WLAN stations and terrestrial ATC center.

Figure 40. Model A1SGSGSG1 with satellite bent-pipe communication link
(TS = 500-1500 Kbytes, Uniform PDF; TBT = 0.1 s, Exponential PDF)

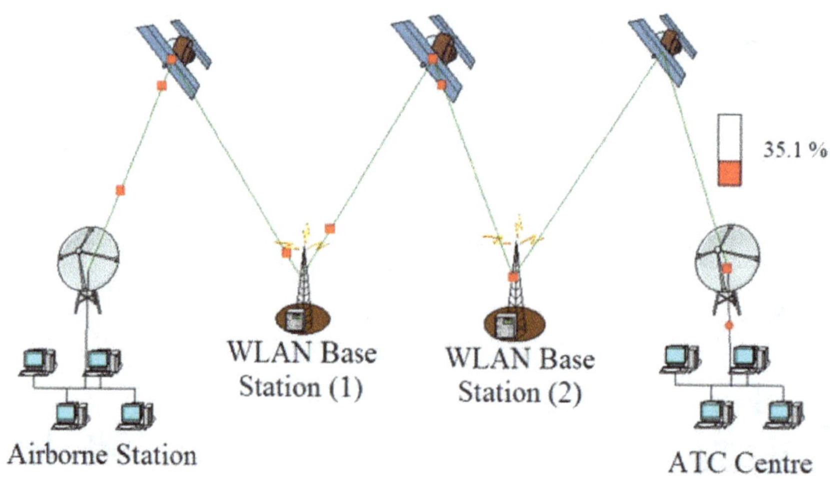

Figure 41 shows the model *A1SCSCSG1* with bent-pipe architecture and clouds: one airborne station, two WAN clouds and terrestial ATC center

In Figure 42, the model *A3S10G1* with intersatellite link is shown: three airborne stations, ten satellites and terrestrial ATC center. The same model

Figure 41. Model A1SCSCSG1 with clouds and satellite bent-pipe communication link
(TS = 500-1500 Kbytes, Uniform PDF; TBT = 0.1 s, Exponential PDF)

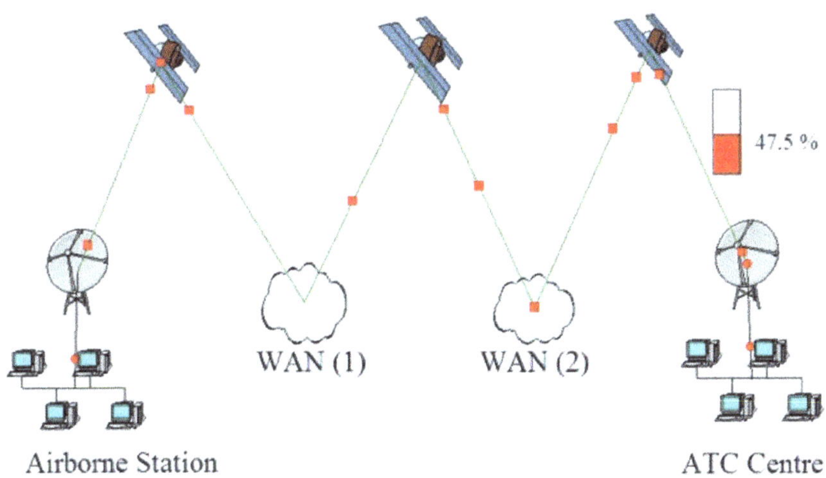

Figure 42. Model A3S10G1 with intersatellite communication link
(TS = 500-1500 Kbytes, Uniform PDF; TBT = 0.1 s, Exponential PDF)

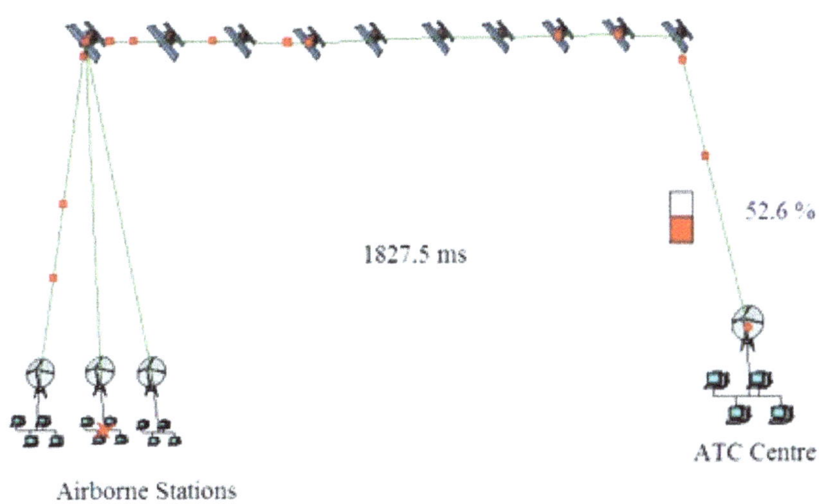

Figure 43. Model A13S4G1 with intersatellite communication link
(TS = 500-1500 Kbytes, Uniform PDF; TBT = 0.1 s, Exponential PDF)

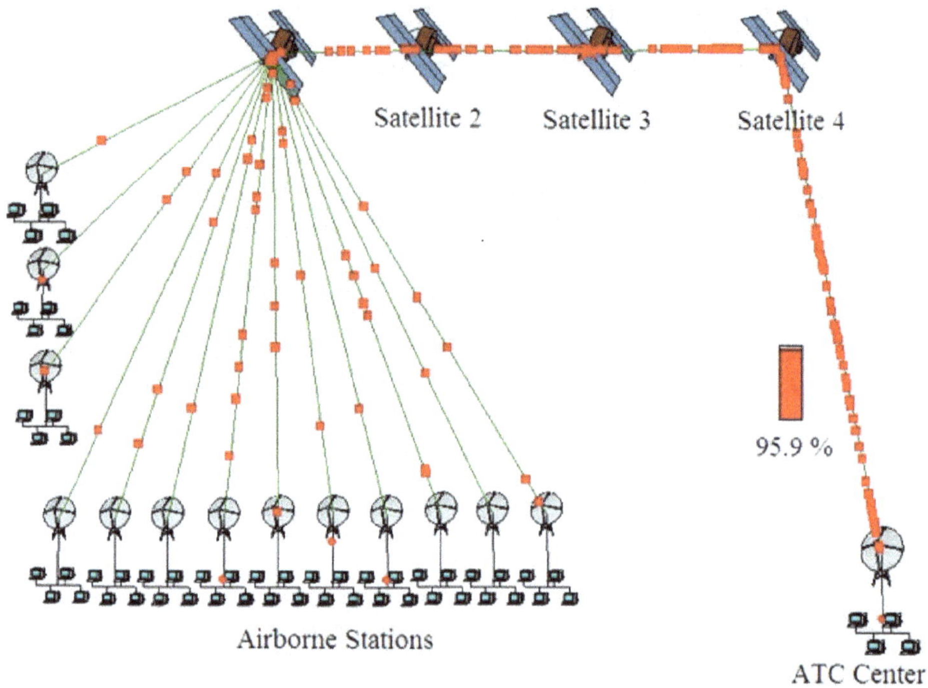

was shown in Figure 26 but with other traffic parameters (TS = 500-600 Kbits, Uniform PDF; TBT = 0.03 s, Exponential PDF).

In Figure 43, the model A13S4G1 with intersatellite links is shown: thirteen airborne stations, four satellites and terrestrial ATC center. A satellite altitude is 780 km, a distance between satellites – 1000 km, a frequency band is 1616 MHz, a bit rate in intersatellite links – T1 (1,544 Mbit/s), initial Bit Error

Table 11. Travel Time and Average downlink Utilization for different link architectures

Model	Message Travel Time (ms)	Average Utilization (%)
A1S3G1 (intersatellite link)	1400	48
A1SGSG1 (WLAN bent-pipe link)	1355	35
A1SCSG1 (WAN bent-pipe link)	1397	47
A1S5G1 (intersatellite link)	1489	48
A1SGSGSG1 (WLAN bent-pipe link)	1402	35
A1SCSCSG1 (WAN bent-pipe link)	1494	47

Rate and Latency equal 0.0, Packet and Circuit Switching are available for the links ADS-B System - Iridium Satellite Transponder and Iridium Satellite Transponder - Ground Transceiver.

Simulation of Data Transmission

Especially important to investigate the dependence of the data Travel Time and downlink Average Utilization on channels architecture. As seen from Table 11 values for Travel Times and Average Utilization of downlink are of the same order for considered models. Increasing the length of the channel (compare models ***A1S3G1*** and ***A1S5G1***, ***A1SGSG1*** and ***A1SGSGSG1***, ***A1SCSG1*** and ***A1SCSCSG1***) results in a slight increase in Travel Times and does not alter the Average Utilization of downlink. The fastest channel is WLAN bent-pipe link, and the slowest - WAN bent-pipe link.

Dependencies of message Travel Time on different number of satellites for several aircraft (Figure 44) was investigated with the help of models ***A(1-3) S(1-10)G1*** (one of which is shown in Figure 42).

♦ ***–A1S(1-10)G1***, ■ ***–A2S(1-10)G1***, ▲***–A3S(1-10)G1***

Figure 44. Dependencies of message Travel Time on the number of satellites

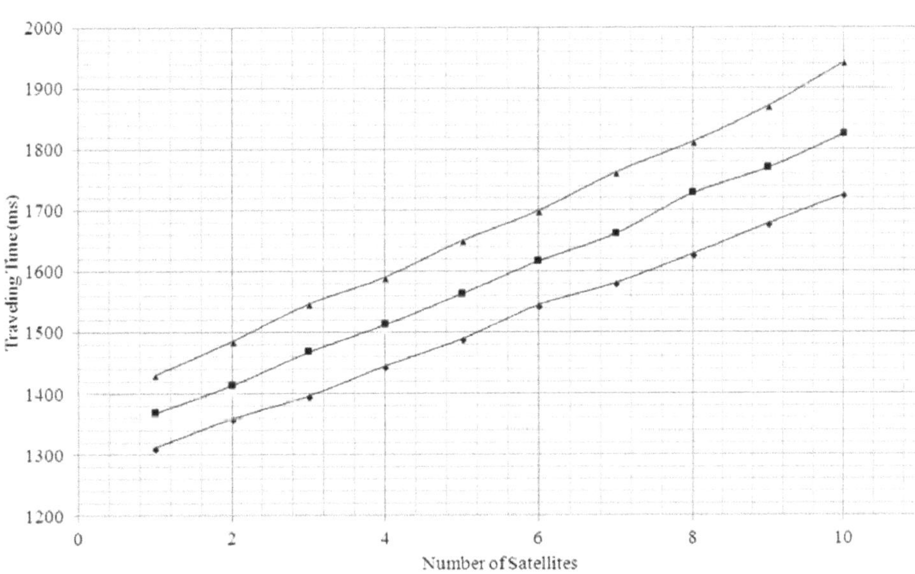

In this case, bit rates of all links were T1. The range of Travel Times variation is rather small (1300-1940 ms) and indicates the possibility of satellite real-time data transmission for air traffic management purposes in real-time. Since here the model estimations take place, it is more correct to speak about the relative changes of travelling times during data transmission due to increasing the number of satellites and aircraft.

When the number of planes simultaneously transmitting data via a single satellite channel to the terrestrial centers becomes very large, a channel will be "saturated" and a data channel will not be able to operate. Modeling showed (Figure 43) that the channel can be "saturated" by 13 aircraft. This means that the communication channel with the given parameters *(TS = 500-1500 Kbytes, Uniform PDF; TBT = 0,1 s, Exponential PDF; Packet Latency – 0,02 s with Constant PDF; Packet Fail Chance – 0,01; aircraft/ground workstations and antennas data rate – 100 Kbyte/s (medium), links bit rates – 1,544 Mbit/s)* is able to serve no more than 13 planes simultaneously. Certainly, for air traffic control purposes it is not enough. However, it should be remembered that this result is an estimation and is valid only for selected software environment, designed model and selected traffic parameters. Moreover, properties of

Figure 45. Dependencies of downlink Average Utilization on its bit rate

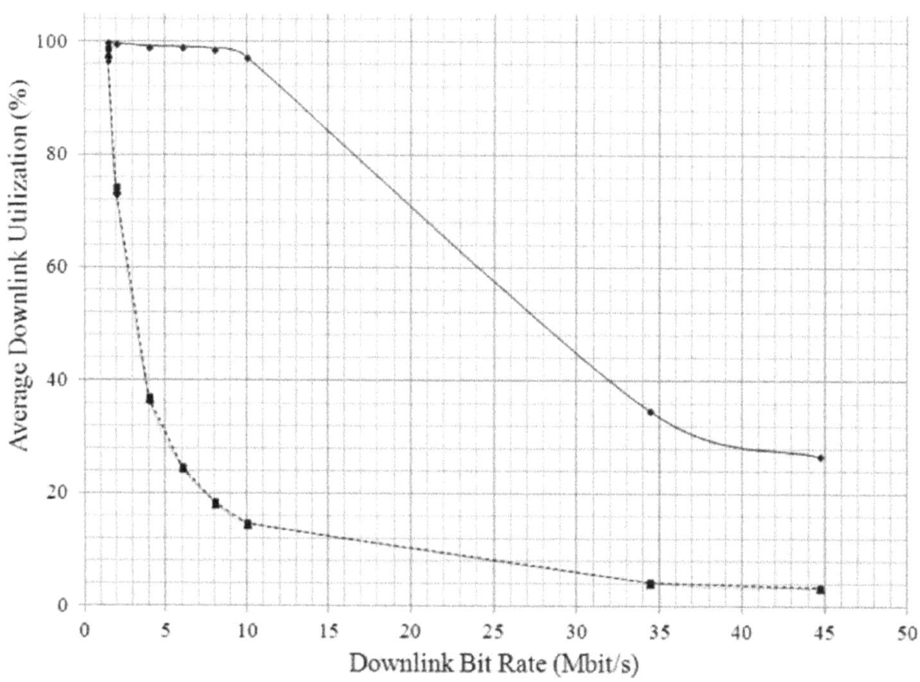

"servicer engine" in the NetCracker software are not determined in details and are given only in the form of a fixed delay of service and absolute speed limit for receiving of requests. However, these assessments can be useful for further research.

Increasing the number of satellites and downlink data bit rate leads to an interesting effect (Figure 45). There is a significant difference between whether the data are transmitted through a single satellite (the model ***A13S1G1***) to the terrestrial ATC center or via several satellites. Increasing the number of satellites (models ***A13S2G1***, ***A13S3G1***, ***A13S4G1***) gives practically the same dependencies of average downlink utilization on the downlink bit rate. Nevertheless, in case of using satellite constellation the average downlink utilization is much less for the same downlink bit rates.

♦ ***–A13S1G1***, ■ ***–A13S2G1***, ▲***–A13S3G1***, ● ***–A13S4G1***

CONCLUSION

1. For modelling of ADS-B messages transmition with the help of low-orbit satellite constellation Iridium new original models of "Aircraft–Satellites–Ground Stations" link were built using NetCracker Professional 4.1 software.
2. Influence of aircraft and satellites number on average downlinklink utilization and message travelling time was studied.
3. Telecommunication channels with intersatellite link and bent-pipe architecture were analized.
4. The effect of communication channel "saturation" during simultaneous data transmission via satellite communication channel from many planes was investigated.

ESTIMATION OF DATA TRAFFIC OVERLOAD FOR SATELLITE COMMUNICATIONS

The way aircraft are tracked is radically changing today. The current radar-based approach is replaced by satellite tracking systems. Aviation telecommunications is dynamically developing due to evolution of communication characteristics and signal processing principles in a link between a satellite and aircraft.

The new system would let air traffic controllers track aircraft using satellite ADS-B network, which is more accurate than today's radar technology. ADS-B promises a ten-fold increase in the accuracy of satellite signals that will let air traffic controllers reduce separation standards between aircraft, significantly increasing the number of aircraft that can be safely managed (Doc. 9750, 1998).

To meet an increasing aviation demand for multimedia services and electronic connectivity across the world, satellite networks will play an indispensable role in the deployment of global aviation networks (Kota, Goyal, Goyal, & Jain 2001).

Satellite networks play an important role for data delivery. They are very effective for data transmitting over large geographic locations, and for reaching remote locations lacking in communication infrastructure. For that reason, the deep analysis of traffic parameters for "Aircraft-Satellites-Ground Station" channel is urgently needed. It is important to develop models of satellite channels for data transmitting and to research ways of channel parameters correction in critical situations. Nevertheless, issues related to the satellite channel parameters estimation leading to data traffic overload still is not investigated in detail.

VHF Data Exchange System (VDES) is a technological concept developed by the International Association of Lighthouse Authorities Committee and widely discussed at International Telecommunications Union, International Maritime Organization and other institutions. VDES was originally developed to address emerging indications of overload for VDL in Automated Information System and simultaneously enabling a wider seamless data exchange for the maritime community (Electronic Communications Committee, 2013).

The advent of unmanned aerial vehicles and other network-centric capabilities on land and sea led to an ever-expanding use of satellite communications. Effective communication on the battlefield is always an integral part of conflict and is decisive in a battle. That means assured communications in remote areas of the world. As technology advances, so does the amount data that is created. Unmanned aerial vehicles alone produce huge quantities of data when they go up in the air and begin sending back crucial operational data that can create data overload for military satellite communications (Turnbull, 2013).

However, present day satellites are limited in their ability to provide high data rate communication services due to the limited availability, and high cost of satellite resources such as power, energy, and frequency bands. Moreover, present communication satellites were designed almost exclusively for supporting stream traffic such as voice, video or bulk data transfers, and are not efficient for the transmission of "bursty" data traffic such as Internet traffic. Therefore, it is necessary to shift from traditional circuit switched technology, used for voice communication, to packet switched technology, used in data networks (Modiano, 2004).

Today there is a set of satellite systems that are able to provide information exchange in aviation: INMARSAT, COSPAS/SARSAT, Iridium, Globalstar and Thuraya. Each of them has its benefits, but in terms of global data transferring Iridium has the absolute advantage, that is 66 satellites and coverage of the whole Earth surface (Iridium, 2009). .

The aim of the study in this section is: 1) to study effect of communication channel overloading during simultaneous data transmission from many aircraft via different number of satellites; 2) to estimate the parameters of aeronautical satellite communication channel (Wang Bo, Kharchenko, Grekhov, & Ali, 2016).

Figure 46. "Aircraft-Satellites-Ground Station" channels

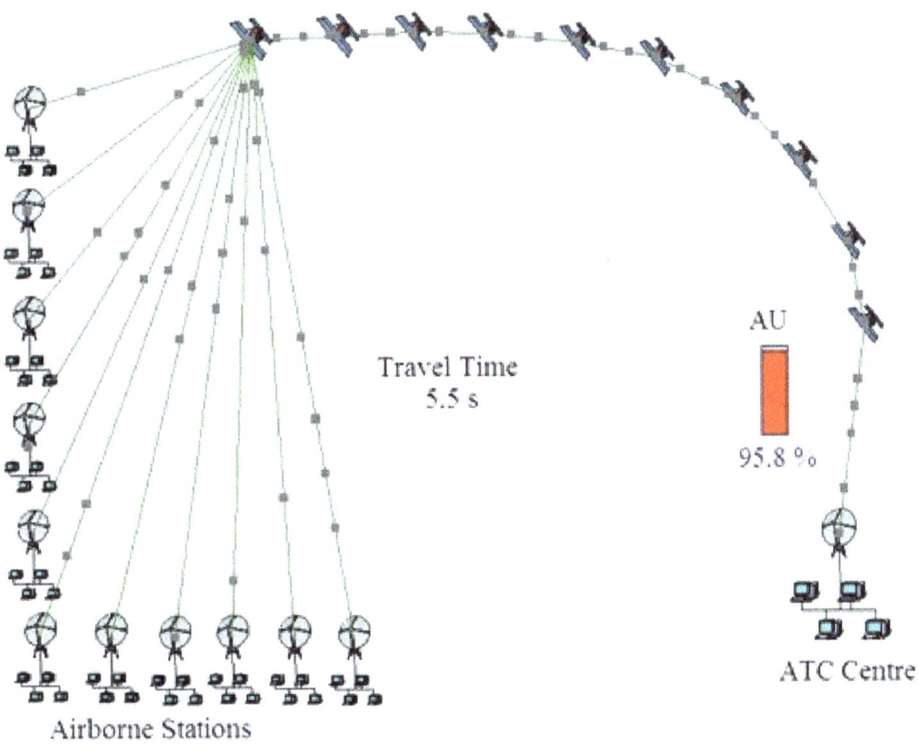

Models for Channel Overloading Study

The model ***A11S10G1*** (Figure 46) with intersatellite links is shown: eleven airborne stations, ten Iridium satellites and terrestrial ATC center. In addition, in this section models ***A11S1G1*** and ***A1S1G1*** are used, but with different traffic parameters not considered earlier.

Models contain packet switching circuits with bandwidth T1=1,544 Mbit/s. Packet Latency, Packet Fail Chance, and a BER assumed to be zero except in special cases. Different types of traffic profiles were investigated, but results in this section are given only for InterLAN traffic.

There are local network nodes, which are located geographically at a distance of more than 12 500 km (orbital space stations and centers). Despite these distances, similar networks are still referred to as local.

Figure 47. Number of aircraft for which Average Utilization of Downlink is 100%

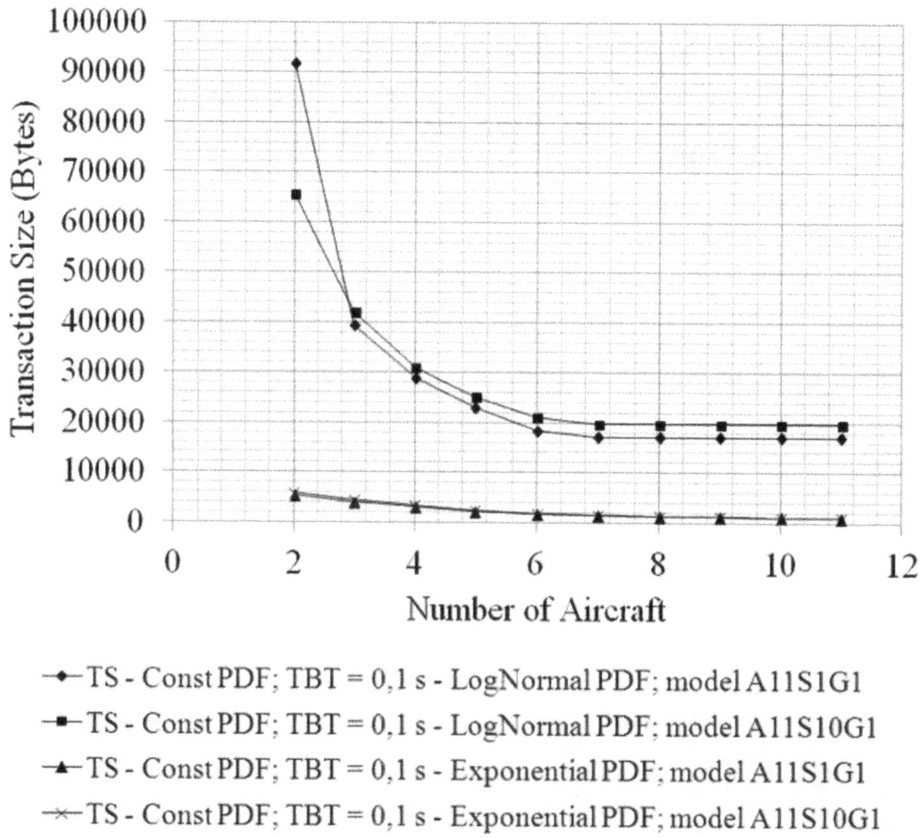

The following Probability Density Functions (PDF) for TS and TBT parameters were considered: Constant, Uniform, Exponential, Normal, and Lognormal.

Aeronautical Satellite Communication Channel Simulation

It is important to understand how much and what kind of messages will be able to pass with a large number of aircraft, simultaneously using satellite link. Figure 47 shows at what value of the transaction size channel congestion occurs when the number of aircraft is increasing. In this case, Constant PDF for TS parameter and Lognormal and Exponential PDF for TBT parameter were taken. It can be seen that the results are radically different: the size of the transaction, which can be transmitted without congestion, is more on the order for Lognormal PDF. Common for considered distribution laws is that by increasing the number of aircraft the size of the maximal transaction is reduced to a finite value. The increase in the number of satellites has little

Figure 48. Dependencies of downlink Average Workload on Transaction Size

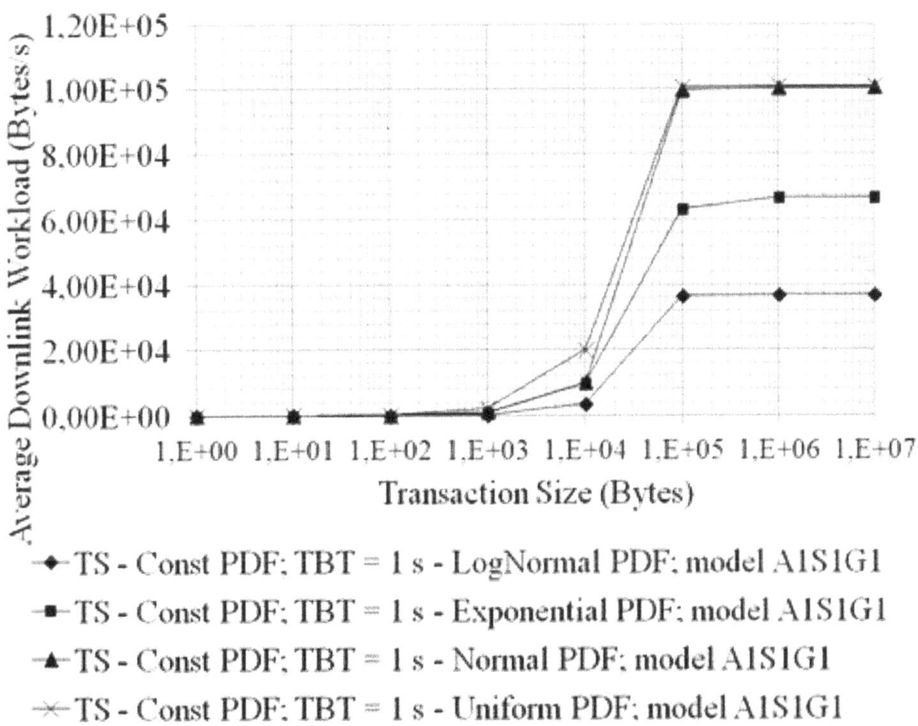

effect, and "saturation" of the channel occurs at slightly larger values of a transaction size.

Figure 48 shows the variation of AW parameter in dependence on TS parameter (for a fixed value of TBT parameter) when changing the type of a PDF for TBT parameter. The lowest value of AW parameter is observed for Lognormal distribution, and the highest value - for Uniform PDF. All types of distributions are characterized by "saturation" (for Lognormal PDF AW ≈ 36846 bytes/s, for Exponential PDF - AW ≈ 67004 bytes/s, for Normal and Uniform PDFs AW ≈ 99650 bytes/s).

In Figure 49 dependencies of AU parameter on a BER are shown (at TBT = 0,2 s and Exponential PDF) for TS = 10^3 bytes and TS = 10^5 bytes with

Figure 49. Dependencies of downlink Average Utilization on BER

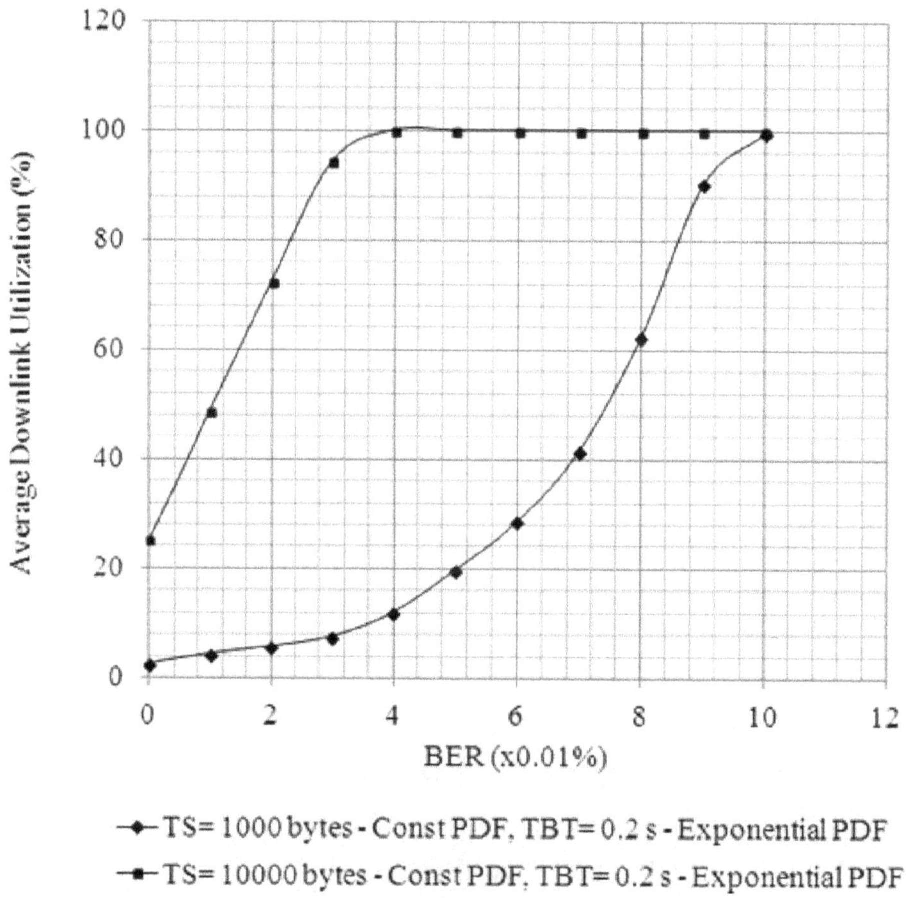

Constant PDF. For BER = 0% an average downlink utilization is 3% and 25% respectively. With increasing a BER AU parameter for TS = 10^3 bytes is increasing too but slowly (data are transmitted with a small number of errors), and for TS = 10^5 bytes increases rapidly up to 100%. For other types of distributions, the nature of change of these curves is similar.

In Figure 50 dependencies of AU parameter on a PFC are given (at TBT = 0,2 s and Exponential PDF) for TS = 10^3 bytes and TS = 10^5 bytes with Constant PDF.

It is seen that for TS = 10^3 bytes AU parameter grows slowly, giving a total AU = 5% at PFC = 0,5. For large transactions (TS = 10^5 bytes) at selected traffic settings AU parameter is constant and equal to 25%. Notice, that with increasing a probability of packet fail chance AU parameter is not increased.

Figure 51 demonstrates dependencies of AU parameter on the types of a distribution law for TS parameter. These are the same values up to the

Figure 50. Dependencies of downlink Average Workload on Packet Fail Chance

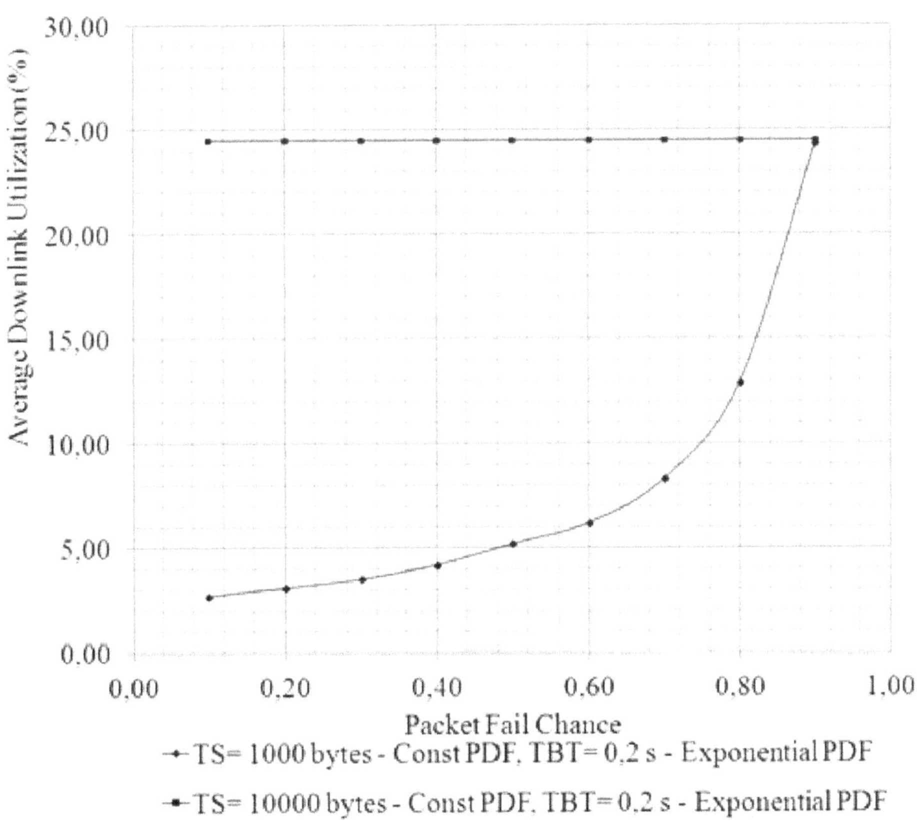

parameter values TS = 10^3 bytes, and a significant difference only for large transactions.

This is because for Constant distribution law all transactions have the same dimensions, and at Uniform law, different-sized transactions within the prescribed limits are sent with equal probability.

Figure 52 shows dependencies of AU parameter on TBT parameter with Exponential PDF for three different TS parameters with Constant PDF. The result was trivial - the more TS parameter, the greater is AU parameter. However, attention is drawn to the character of AU parameter recession with an increasing of TBT parameter, - the larger the transaction size, the slower is decreasing of AU parameter during an increasing of TBT parameter.

A "special case" is given in Figure 53 for a dependence of AU parameter on TBT parameter for ADS-B messages with transaction size 112 bits (in Mode S Extended Squitter). AU parameter reaches 100% at TBT = 0,1 s with Exponential PDF. Then there is a rapid decline of AU parameter with increasing of TBT parameter.

Figure 54 shows dependencies of TT parameter on TS parameter with different distribution laws for a time between transactions. The "Best" is Lognormal distribution, and the "Worst" - the Uniform one.

Figure 51. Dependencies of downlink Average Utilization on Transaction Size

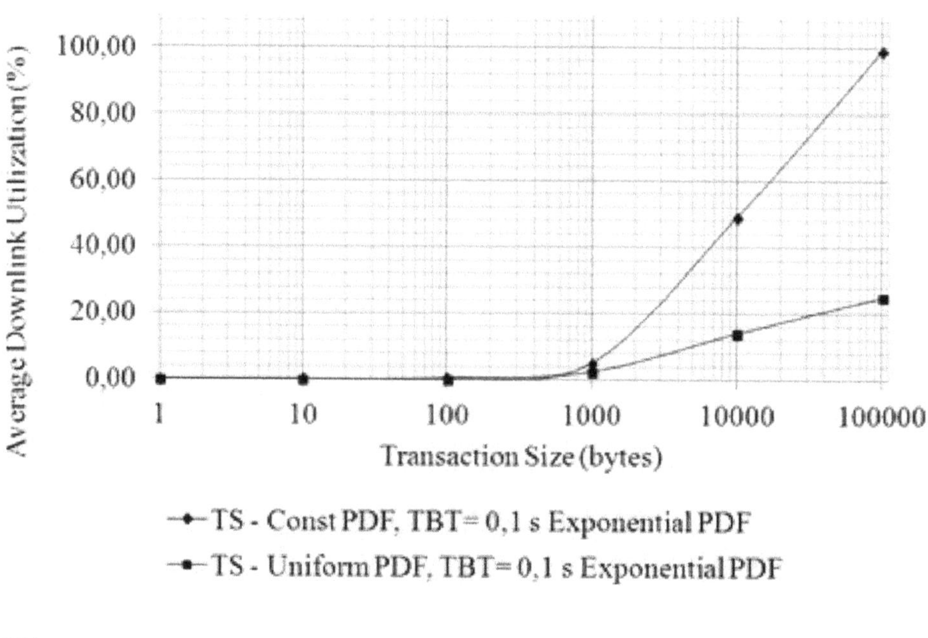

Figure 52. Dependencies of downlink Average Utilization on Time Between Transactions

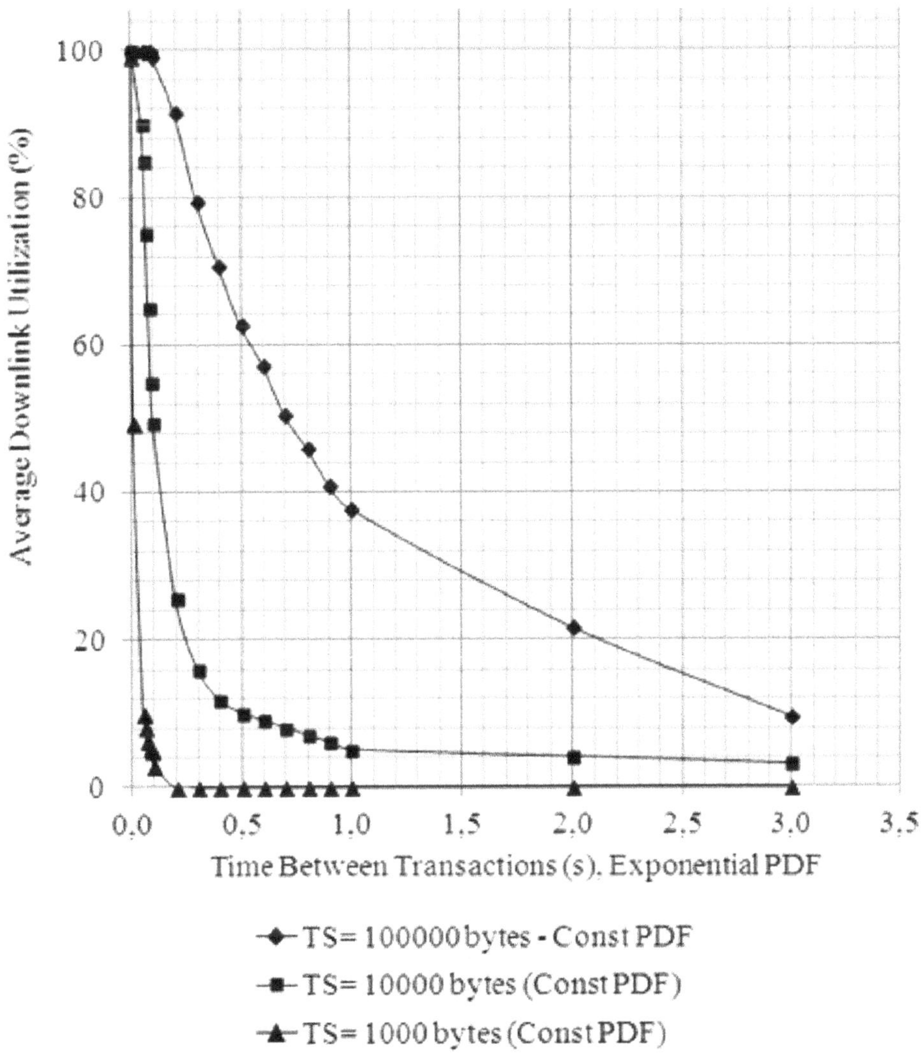

CONCLUSION

1. A number of satellites does not influence on an average workload and an average utilization neither "Aircraft–Satellite" link nor "Satellite–Ground station" link. The only parameter that is affected by a number of satellites in the link is a message Travel Time.

Figure 53. Transmission of ADS-B messages

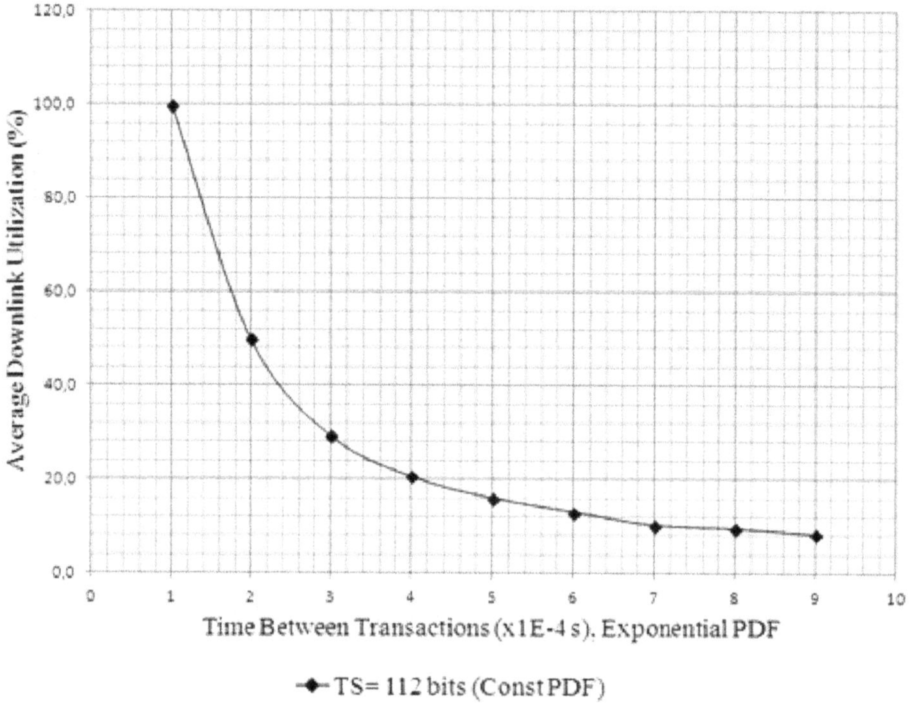

2. Average Utilization and Average Workload of "Satellite-Ground station" link are growing proportionally to the number of aircraft connected with the satellite channel.
3. Average Workload and Average Utilization of each link are not affected by the cable length. The only indicator that is sensitive to cable length changes is Travel Time.
4. Transaction Size affects Average Workload, Average Utilization and Travel Time. The behavior of influence depends on probability distribution law. The system that transmits messages with small transaction size should implement Constant and Uniform distribution laws. For the systems where transaction size is in the range of 100 – 1000 bytes, the best operational characteristics are achieved with Constant, Uniform, Normal and Exponential laws. In case of big size of transactions (more than 1000 bytes) the Lognormal law should be used.

Figure 54. Dependencies of Travel Time on Transaction Size

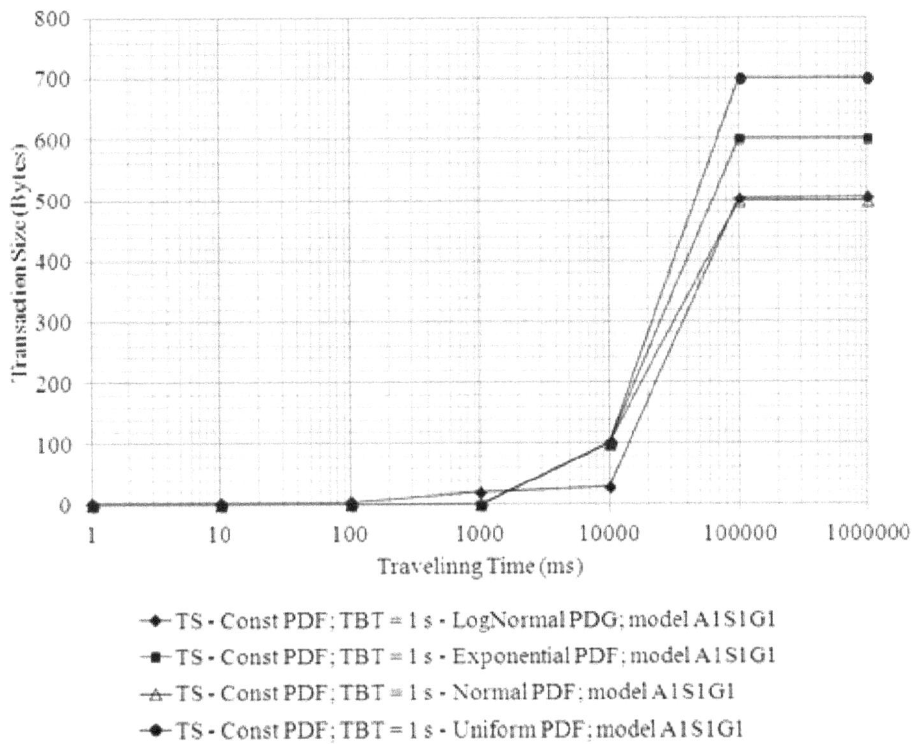

5. Increasing of Time Between Transactions leads to increasing of potential number of aircraft that can be connected to the channel. In the case of Constant, Uniform, Exponential or Normal distribution laws for Time Between Transactions its value should be within 100 ms – 10s range. The most reliable and acceptable result from Average Utilization point of view is obtained by applying Lognormal law to distribution of Time Between Transactions.
6. The increasing of link bandwidth does not influence on Average Workload value. However, Average Utilization of the link where changes are applied is decreasing. Travel Time of message transmission over the communication channel is declining as well.
7. Link Latency directly affects Travel Time. This influence depends on the distribution law of traffic transmitted by the link. The most acceptable

results are obtained for Constant, Exponential, Normal and Uniform distribution laws.

8. Different suitable traffic profiles slightly influence on the performance of communication channel.
9. A BER and Packet Fail Chance are affected by link Average Workload, Average Utilization and messages Travelling Time. In order to get the best operational capabilities of the communication channel the values of a BER and Packet Fail Chance should tend to zero.
10. It was discovered that a BER and Packet Fail Chance values extremely grow when an Average Utilization of the link is increasing. That means that all parameters that cause Average Utilization changes, namely Transaction Size, Time Between Transaction and Link Bandwidth, should be chosen in such a way to prevent the exceeding of Average Utilization the acceptable level.

TRAFFIC LOSS ESTIMATION IN SATELLITE CHANNEL

With the increase in the intensity of flights compliance with security will require a significant increase in the transmitted volumes of information, as well as reliable and adequate mechanisms for data transfer in the air/land and air/air channels. However, existing communication systems will not be able to support the estimated increase in aviation traffic using the existing operational concept (John, Ndujuiba, Okonigene, & Kenechukwu, 2013).

EUROCONTROL assessed the performance of various technologies that could become part of the aviation communications infrastructure of the future. Studies have shown that such an infrastructure should be based on a combination of terrestrial and satellite technologies (Pouzet & Fistas, 2008). Satellite systems will play an indispensable role in the deployment of global aviation networks and meet the growing demand for aeronautical telecommunications around the world (Kota, Goyal, Goyal, & Jain, 2001).

The problems that arise in the operation of an aeronautical satellite communication channel are very important. Even a slight deterioration in the channel parameters affects the data rate or changes the coverage, which immediately affects flight safety and operating costs. It is important to know how to maintain optimal channel parameters. Therefore, it is actual to develop real models for aeronautical satellite communication channels and to study methods for correcting channel parameters in critical situations.

Real aeronautical satellite data networks have a complex composition and algorithms of operation. For their study, simulation models must be used, built, in particular, using the NetCracker software (NetCracker, 2018). This program makes it possible to simulate the transmission and reception systems of aircraft, satellite and ground station data, connect them with communication channels with real properties and simulate the functional characteristics of the network, taking into account data transfer protocols. It is possible to manage a set of factors: the statistical parameters of transaction flows and the time between messages, a type of a traffic, time delay of messages in channels and satellite transponders, the probability of messages loss, the probability of bit errors. In the program the data rate depends on the bandwidth of the communication link, and the data rate is determined by the intensity of messages and packets volume. The Netcracker Professional 4.1 network simulation system allows you to accurately predict network performance.

The effect of data exchange on the network capacity is considered in paper (John & Atayero, 2008). Overloading of the satellite communication channel in the transmission of military data was considered in article (Turnbull, 2013). Satellite networks play an important role in the delivery of data. They are very effective in transferring aviation data over long distances and to remote locations where there is no communication infrastructure (Modiano, 2004).

The purpose of this section is analyzing the dependence of Average Workload of the downlink and the data loss during transmission (Wang Bo, Kharchenko, Grekhov, & Ali, 2015).

Data Loss Coefficient Dependence on Transaction Size

To estimate the quality of transmission via satellite communication channel, Data Loss Coefficient K was introduced:

$$K = \frac{S - R}{S} \quad ,$$

where S is the number of transactions transferred by the aircraft, and R is the number of transactions accepted by the ground station (Figure 1). When the number of transactions transmitted by an airplane and received by a ground station is the same, Data Loss Coefficient $K = 0$. With increasing data traffic, the "channel stops coping" with the load and the ground station ceases to accept transactions ($K = 1$).

It is important to understand what conditions cause a minimal loss of information in the transmission link. Dependencies of Data Loss Coefficient *K* on Transaction Size for three different Times Between Transactions (1 μs, 1ms and 1 s) and for different types of distribution laws were obtained.

In the case when Time Between Transactions equals to 1 μs, the full load of the channel is achieved from the very beginning for all types of distribution laws except Lognormal PDF. That is why the relationship between a Data Loss Coefficient *K* and TS parameter is given in Figure 55 only for Lognormal PDF. From the data obtained, it follows that messages can be transmitted up to 10 Kbytes without data loss while using this law.

In the case when Time Between Transactions is equal to 1 ms, for Constant, Uniform, Exponential, and Normal PDF a channel overload occurs at Transaction Size 100 bytes (Figure 56), and for Lognormal PDF at 10 Kbytes.

Figure 55. The dependence of Data Loss Coefficient on Transaction Size (Constant PDF) for TBT = 1 μs (Lognormal PDF)

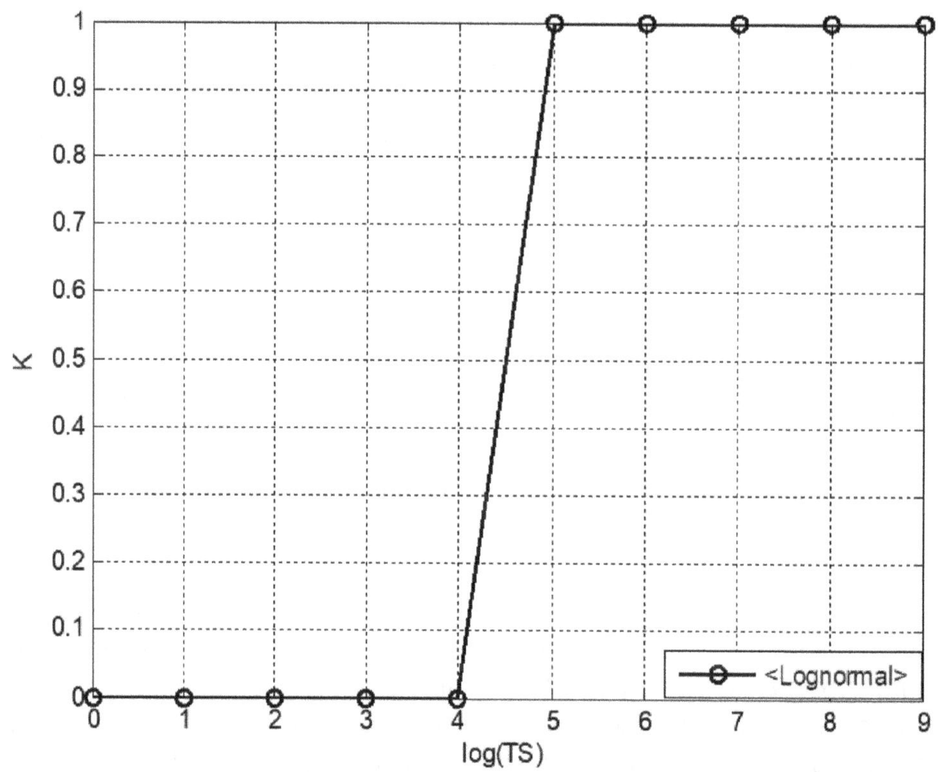

Figure 56. The dependence of Data Loss Coefficient on Transaction Size (Constant PDF) for TBT = 1 ms

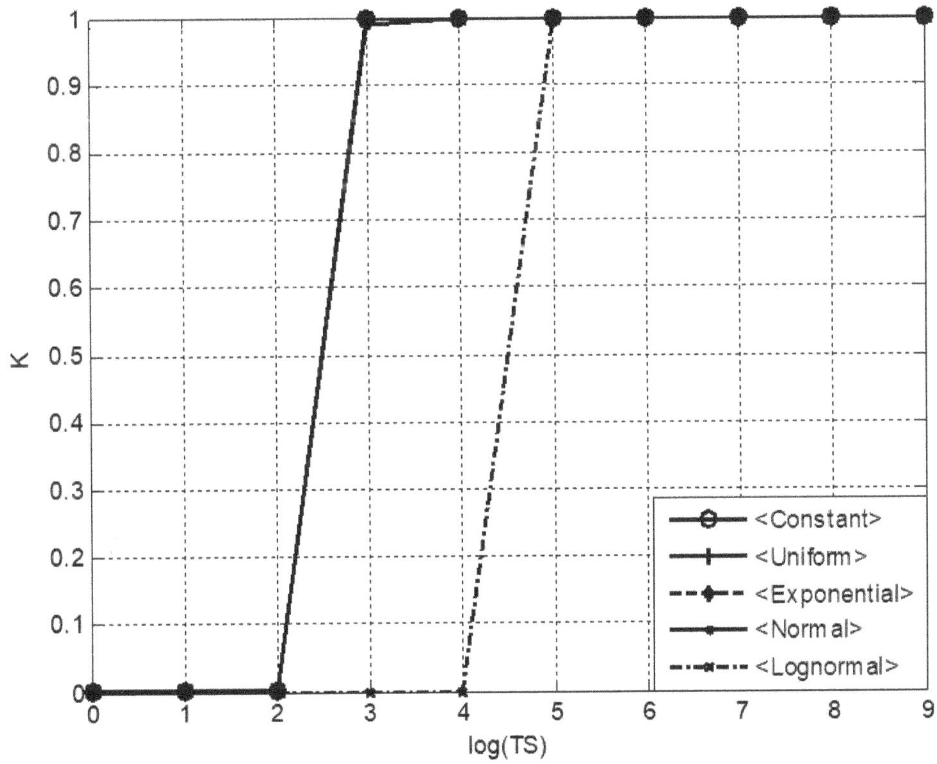

In the case when Time Between Transactions is equal to 1 s, for Constant, Uniform, Exponential, Normal, and Lognormal PDF a channel overload occurs at Transaction Size 10 Kbytes (Figure 57). The channel becomes closed when sending messages of 100 Kbytes and greater.

Data Loss Coefficient Dependence on BER

For safe and reliable operation of the data transmission system, it is necessary to investigate the influence of all factors that can cause a drop in level of integrity and accuracy. The main causes of deterioration in the quality of data transmission are signals interference, distortion and attenuation. They lead to the fact that some bits of the sent message will be received with errors. A BER indicates the number of incorrectly received bits to a total number of bits sent.

Figure 57. The dependence of Data Loss Coefficient on Transaction Size (Constant PDF) for TBT = 1 s

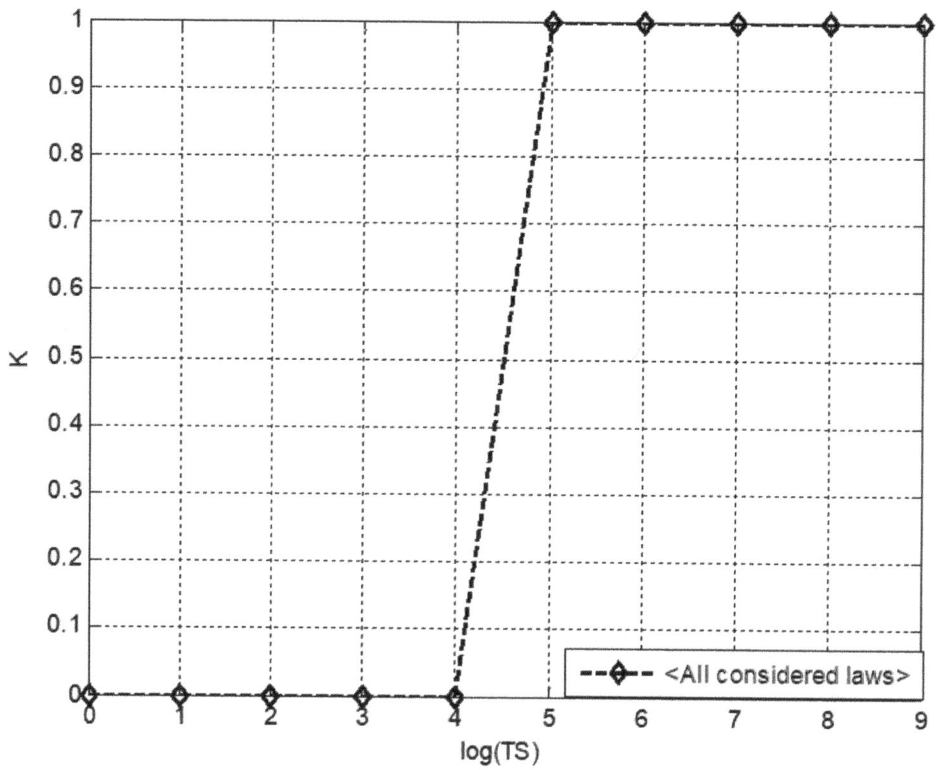

Data Loss Coefficient for considered model was investigated (Figures 58-61) on the influence of a BER (1%, 10% and 50%). From the obtained results follows that a BER significantly affects both the rate of data loss, and the maximum size of messages that can be sent through the data channel at a particular TBT parameter and its distribution laws.

Comparison of the data for BER = 0% with the results for BER = 1% shows (Figure 58) that for TBT = 1 μs with Lognormal PDF data loss is 42% (in this case log(TS) = 2) when data are sent by a 100-byte packet. At BER = 10% data loss is 100% for 100-byte packet.

Dependencies of Data Loss Coefficient on a BER for TBT= 1 ms and TBT = 1 s are shown in Figures 59-61. The increase in the time between transactions to TBT = 1 s (Figure 61) allows transfer of transactions up to 10^5 bytes using all the considered distribution laws in the case of BER = 0%, and at BER = 1% - up to 10^3 bytes. As in all previous cases, the Lognormal distribution

Figure 58. Dependencies of Data Loss Coefficient on BER for TBT= 1 μs, Lognormal PDF (TS in bytes, Constant PDF)

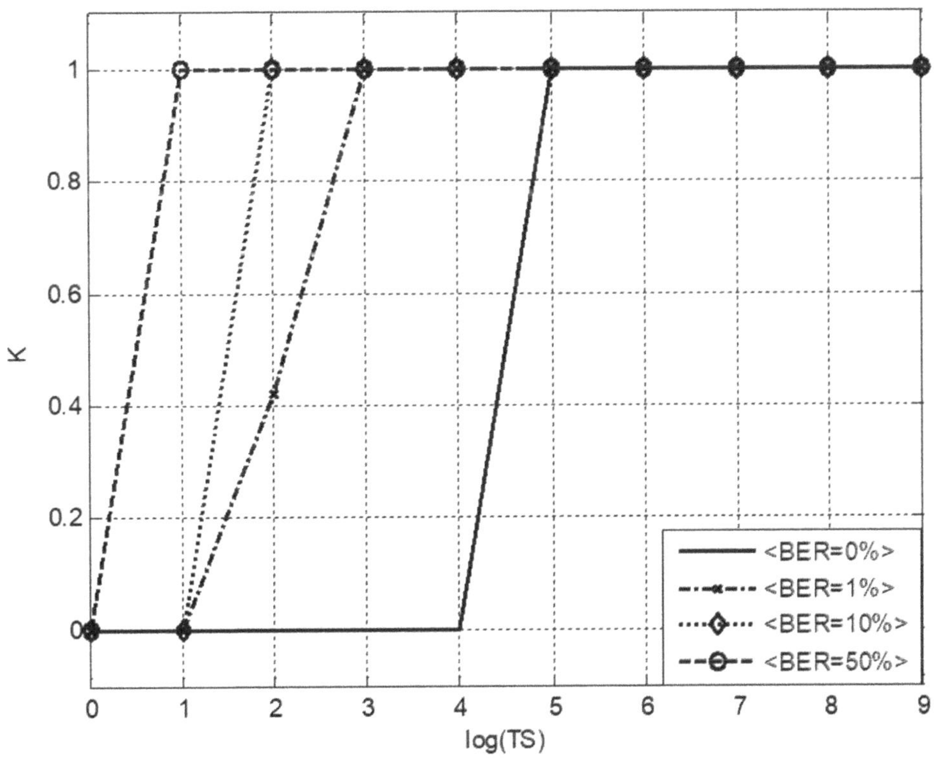

proves to be preferable from the point of view of the least information loss during transmission.

Data Loss Coefficient Dependence on Channel Latency

Delays in the communication line are one of the most important parameters that indicate the performance and effectiveness of the system. Delays can be caused by numerous reasons, but the most significant are:

1. Long path between the source of information and the destination;
2. Atmospheric conditions (rains, clouds, etc.);
3. Complex network architecture.

Regarding the reliability and safety of aviation traffic, it is advisable to investigate the direct causes of communication channel delays and,

Figure 59. Dependencies of Data Loss Coefficient on BER for TBT = 1 ms, Constant, Uniform, Exponential and Normal PDF (TS in bytes, Constant PDF)

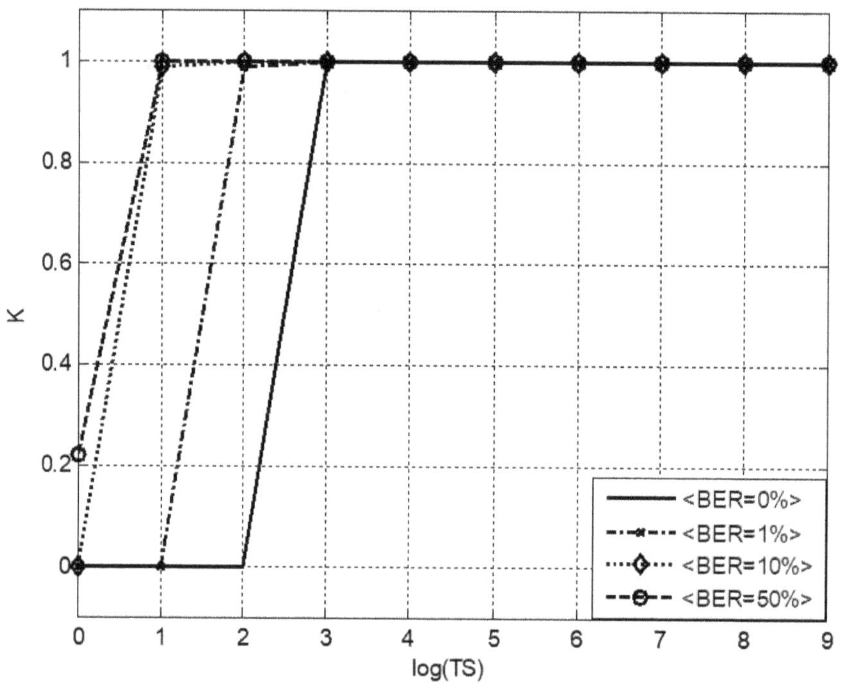

consequently, to understand how to eliminate these delays, since large delays in the data exchange system can cause irreversible disasters.

The analysis of the latency effect on data transmission over the satellite communication link is carried out here. During the study, it was found that the latency in the communication channel does not affect either Average Workload or Average Utilization rate. However, it was found that delays significantly affect Data Loss Coefficient. This means that, due to latency, the transmitted data of the ADS-B system cannot be immediately received by a ground station or other aircraft, which can negatively affect the efficiency of the communication channel. The level of data loss depends on the delay time and is presented in Table 12.

Figure 60. Dependencies of Data Loss Coefficient on BER for TBT= 1 ms, Lognormal PDF (TS in bytes, Constant PDF)

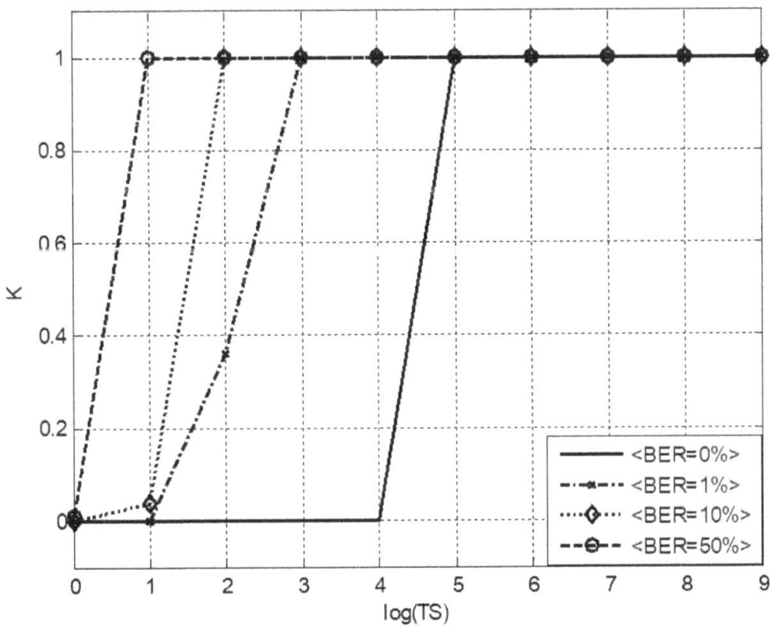

Figure 61. Dependencies of Data Loss Coefficient on BER for TBT= 1 s for all considered distributions (TS in bytes, Constant PDF)

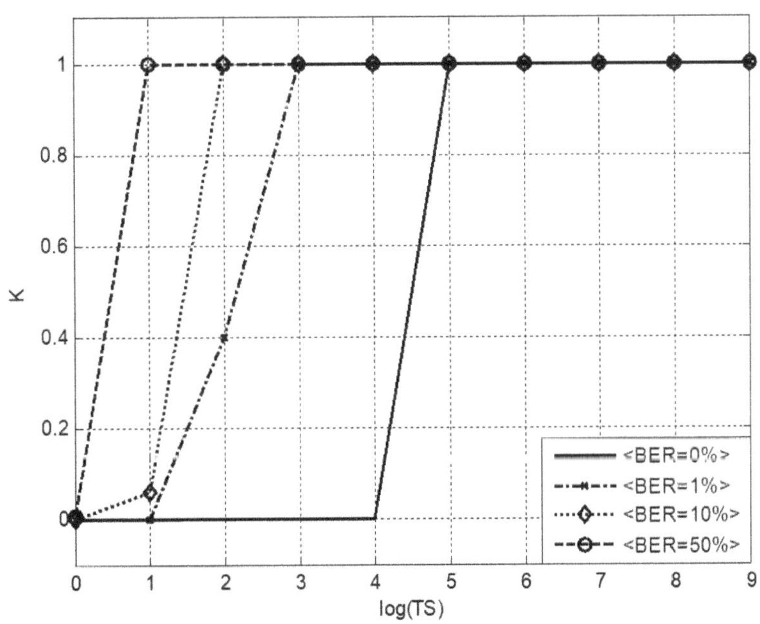

Table 12. Dependence of Data Loss Coefficient on Latency

Latency (ms)	0	10	100	1000	10000	100000	1000000
Data Loss Coefficient (%)	0	0	3	14	58	63	96

Figure 62. Comparison of T1 and E3 bandwidths impact on Data Loss Coefficient for Constant, Uniform, Exponential and Normal PDF of TBT (TBT = 1 s, TS in bytes, Constant PDF, BER = 0%)

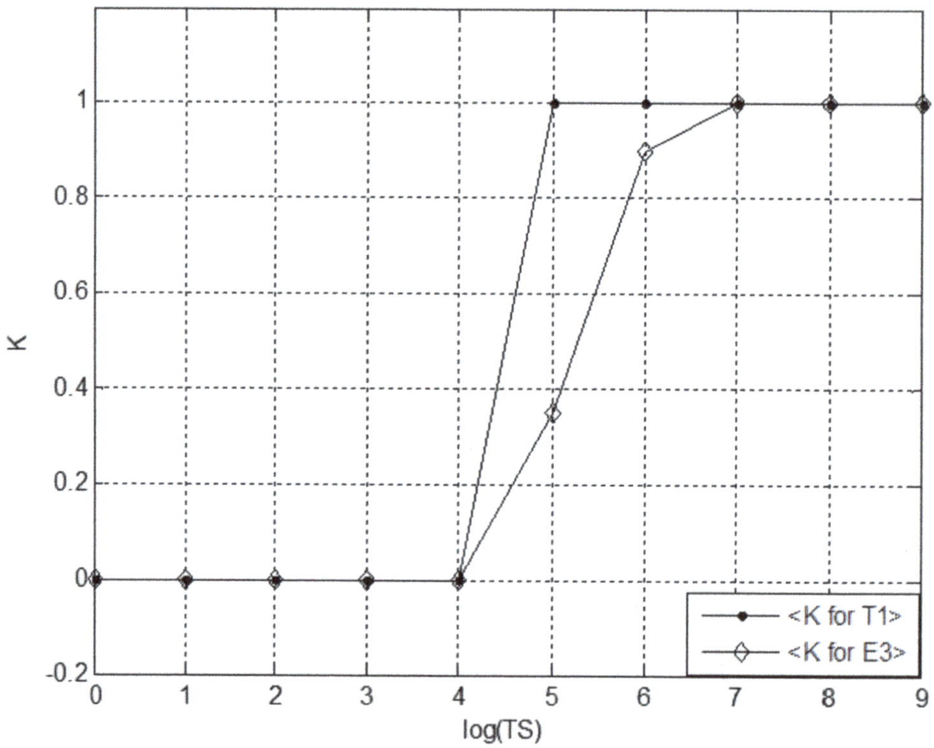

Data Loss Coefficient Dependence on Bandwidth

The highest possible data rate in any channel determines its throughput and the above data is calculated for the data transfer rate T1 corresponding to the primary level of the American standard of the Plesiochronous Digital Hierarchy (PDH). In the NetCracker Professional 4.1 software it is possible to simulate networks with T-carriers (T1 = 1.544 Mbit/s, T2 = 6.312 Mbit/s,

T3 = 44.735 Mbit/s) and E-carriers (E1 = 2.048 Mbit/s, E2 = 8.448 Mbit/s, E3 = 34.368 Mbit/s) corresponding to the primary level of the European standard of the PDH hierarchy.

During the investigation, it was found that bandwidths T1 and E1 gives almost the same results, T3 and E3 affects the Data Loss Coefficient nearly equally also. That is why results for the lowest and for the fastest transmission rates, i.e. T1 and E3, are presented in Figures 62, 63. The data transfer rate was changed in the "Up/Down" channels simultaneously.

In Figure 62, the difference between bandwidths T1 and E3 is shown for Constant, Uniform, Exponential and Normal PDF of TBT parameter.

Figure 63 shows the difference between bandwidths T1 and E3 for Lognormal PDF of TBT parameter.

An increase in the data transfer rate significantly affects the quality of information exchange. The higher the data transfer rate, the higher the maximum transaction size threshold: transactions up to 10^7 bytes with

Figure 63. Comparison of T1 and E3 bandwidths impact on Data Loss Coefficient for Lognormal PDF of TBT (TBT = 1 s, TS in bytes, Constant PDF, BER = 0%)

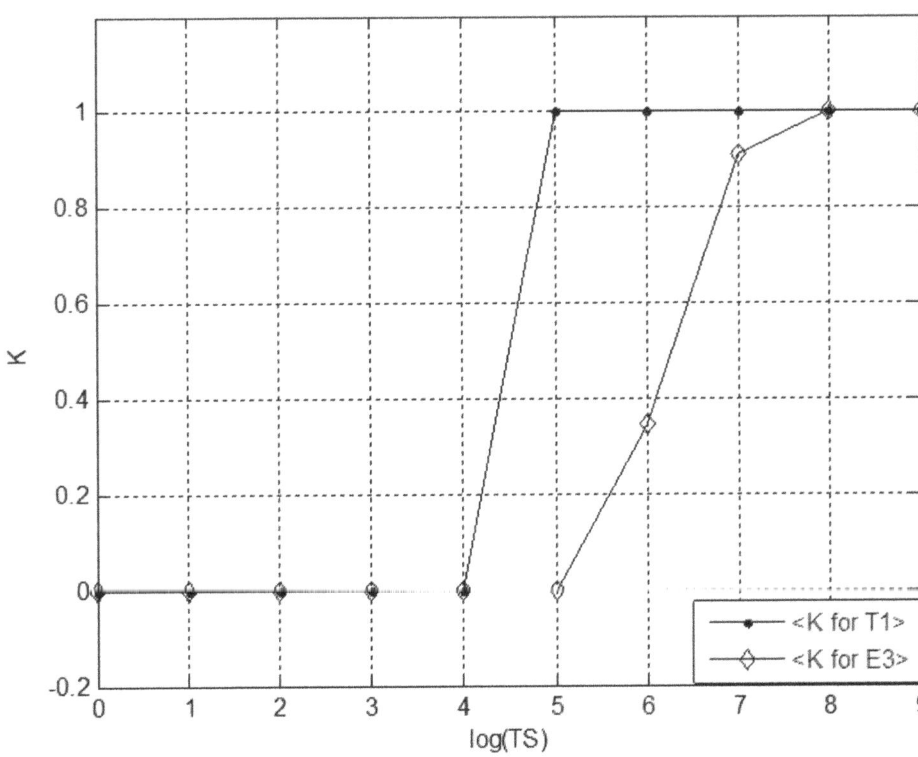

Constant, Uniform, Exponential and Normal PDF are possible, and up to 10^8 bytes for Lognormal PDF.

CONCLUSION

Data Loss Coefficient depends on the same parameters as channel Average Workload. The most useful dependencies were obtained when taking into account the level of bit errors. It is important to consider the fact that the process of aviation messages transmitting through the satellite communication channel is extremely sensitive to the magnitude of a BER. This means that to reduce a BER and the amount of delay in the transmission line it is necessary to eliminate the causes of their occurrence (noises, outdated equipment, imperfect software, etc.).

The average time for transactions travel through the channel is also an important indicator, since the transmission of aviation messages must occur in real time with minimal delays for the timely use. It was obtained that the travel time does not exceed 40 ms for messages up to 10^3 bytes, which agrees well with the known data for the low-orbit satellite constellation Iridium.

REFERENCES

Akimaru, H., & Kawashima, K. (1999). *Teletraffic – theory and applications* (2nd ed.). Springer-Verlag London. doi:10.1007/978-1-4471-0871-9

Bo, W., Kharchenko, V. P., Grekhov, A. M., & Ali, I. (2015). Method for Estimation of Traffic Loss in Aviation Satellite Communication Channel. *Bulletin of Engineering Academy of Ukraine*, *4*, 51–55.

Bo, W., Kharchenko, V. P., Grekhov, A. M., & Ali, I. (2016). Estimation of Data Traffic Overload for Satellite Communications. *Proceedings of the National Aviation University*, *66*(1), 11–17. doi:10.18372/2306-1472.66.9865

Doc. 9750. (1998). *Global Air Navigation Plan for CNS/ATM Systems*. ICAO. Retrieved from http://www.icao.int/publications/Documents/9750_2ed_en.pdf

Electronic Communications Committee. (2013). Information paper on VHF data exchange system. *3rd Meeting CPG PTC*. Retrieved from cept.org/documents/cpg-pt-c/13547/cpg-tc

EUROCONTROL. (2013). *ADS-B & WAM Implementation in Europe*. Retrieved from http://www.eurocontrol.int/publications/ads-b-wam-implementation-europe

Final report on research work No. 721-DB11. (2013). "Development of the methodology for the modernization of ADS-B surveillance systems on base of low-orbit satellite systems", number of state registration 0111U000542.

Flood, J. E. (1998). *Telecommunications switching, traffic and networks*. New York: Prentice-Hall.

Iridium Adds ADS-B to its Iridium NEXT Constellation. (2012). Retrieved from https://www.aviationtoday.com/2012/06/20/iridium-adds-ads-b-to-its-iridium-next-constellation/

Iridium Satellite Reports Record 2008 Results. (2009). Retrieved from http://investor.iridium.com/news-releases/news-release-details/iridium-satellite-reports-record-2008-results

John, S. N., & Atayero, A. A. (2008). Simulation of the effect of data exchange mode analysis on network throughput. *European Journal of Scientific Research*, *24*(2), 244–252.

John, S. N., Ndujuiba, C., Okonigene, R., & Kenechukwu, N. (2013). *Simulation and monitoring of a university network for bandwidth efficiency utilization*. Retrieved from http://worldcomp-proceedings.com/proc/p2013/MSV2312.pdf

Kennedy, I. G. (2003). *Traffic simulation*. School of Electrical and Information Engineering, University of the Witwatersrand.

Kharchenko, V. P., Barabanov, Y. M., Grekhov, A. M., Ivanenko, M. S., & Lobanov, R. I. (2012). Modeling of ADS-B messages transmission through satellite telecommunication channel Iridium using Netcracker Professional 4.1. *Proceedings of the National Aviation University*, *50*(1), 81–86.

Kharchenko, V. P., Bo, W., Grekhov, A. M., & Kovalenko, M. A. (2014). Investigation of ADS-B messages traffic via satellite communication channel. *Proceedings of the National Aviation University*, *61*(4), 7–14. doi:10.18372/2306-1472.61.7580

Kota, S., Goyal, M., Goyal, R., & Jain, R. (2001). Multimedia satellite networks and TCP/IP traffic transport. *International Journal of Computers and Applications*, *23*(2), 115–128. doi:10.1080/1206212X.2001.11441640

Manual for ICAO Aeronautical Mobile Satellite (Route) Service Part 2-Iridium Draft V4.0. (2007).

Manual on Detailed Technical Specifications for the Aeronautical Telecommunication Network (ATN). Doc. 9880-AN/466. (2015). Retrieved from https://www.icao.int/safety/acp/repository/doc%20 9880-partiiafinal%20-.pdf

Minimum Aviation System Performance Standards for Automatic Dependent Surveillance-Broadcast (ADS-B). RTCA, Inc. (2002). DO-242A.

Modiano, E. (2004). Satellite data networks. *Journal of Aerospace Computing, Information, and Communication*, *1*(10), 395–398. doi:10.2514/1.12800

NetCracker Proffesional 4.1. (2018). Retrieved from https://www.netcracker.com

Penttinen, A. (1999). *Introduction to teletraffic theory*. Helsinki University of Technology.

Pouzet, J., & Fistas, N. (2008). Air Traffic Management (ATM) communications and satellites: An overview of EUROCONTROL's activities. *Space Communication Magazine, 21*, 103-108. Retrieved from https://www.icao.int/safety/acp/Inactive%20working%20groups%20library/ACP-WG-M-Iridium-8/IRD-SWG08-IP05%20-%20AMS(R)S%20Manual%20Part%20II%20v4.0.pdf

Turnbull, G. (2013). *Data overload: satellite communications on the battlefield*. Retrieved from http://www.airforce-technology.com/features/feature-data-overload-satellite-communications-battlefield

Chapter 5
Antenna for ADS-B Signals

ABSTRACT

This chapter deals with calculations of the microstrip antenna and the linear phased array that were based on the method of moments using Antenna Magus software. The analysis of the characteristics of these antennas at different emitter numbers and different types of the amplitude distribution was carried out. Calculations of electric field intensity and directional patterns for collinear antennas were provided. The method of moments in the framework of two program complexes was used. Comparison has shown high level of results coincidence. The sample of the antenna is described, which is used in operating system for reception of ADS-B signals from airborne transponders.

INTRODUCTION

The direction of radiation and the shape of the corresponding radiation pattern in phased array antenna is controlled by a change in the amplitude-phase distribution of currents or excitation fields on the radiating elements. The radiating element is a component of the antenna array with a given relative excitation. The required directional pattern is formed due to the interference of electromagnetic waves radiated into space by its radiating elements. For this, the necessary relative amplitudes and initial phases of the alternating currents or excitation fields of each radiating element is provided. The amplitude-phase distribution is not fixed, it can be controlled during operation. Due to this, it is possible to move the beam (the main lobe of the radiation pattern) of the antenna array in a certain sector of space. These and some other properties

DOI: 10.4018/978-1-5225-8214-4.ch005

of phased arrays, as well as the ability to use modern means of automation and computer technology, have made them promising and widely used in radio communications. Phased array antenna containing a large number of controllable elements are part of various ground (stationary and mobile) aviation and space radio engineering systems. Intensive developments are underway in the direction of the further development of the theory and technology of phased arrays and the extension of their field of application.

An antenna array of N radiating elements allows an increase in the directional effect factor by approximately N times, and to narrow the beam for increasing noise immunity. An increase of electric field intensity in the antenna array is possible in comparison with an aperture antenna equipped with a single irradiator. An important advantage of the phased array is the ability to scan quickly the space by "swinging" the beam of the radiation pattern by electric methods. Such a phased array is an antenna with electric beam scanning. Functional capabilities of the phased array are expanded by using the active transceiver module. There are a number of design and technological advantages in comparison with other classes of antennas.

The excitation of radiators is performed either by feeder lines or by freely propagating waves (in the so-called quasioptical phased array). The feeder paths of excitation, along with the phase shifters, sometimes contain complex electrical devices (so-called diagrammering circuits). That ensure the excitation of all radiators from several inputs, which allows creating in the space simultaneously scanning rays. Quasi-optical phased arrays are basically of two types: through-pass (lens), in which the phase shifters and main radiators are excited (by means of auxiliary radiators) by waves propagating from the common illuminator, and reflective - the main and auxiliary radiators are combined, and reflector outputs are installed at the outputs of the phase shifters. Multibeam quasioptical phased arrays contain several irradiators, each with its own beam in space. Sometimes focusing devices (mirrors, lenses) are used. The above-mentioned phased arrays are sometimes called passive.

The most powerful are active phased arrays, in which each transmitter or module is connected to a phase-controlled transmitter or receiver. Phase control in active phased arrays can be performed in intermediate frequency paths or in excitation circuits of coherent transmitters, receiver heterodynes, etc. Thus, in active phase shifters, phase shifters can operate in wave bands different from the frequency range of the antenna. Losses in a phase shifter in some cases do not directly affect the level of the main signal. Transmitting active phased arrays allow the addition in space of powers from coherent electromagnetic waves generated by individual transmitters. Joint processing of signals received by individual elements allows obtaining more complete information about the sources of radiation.

Because of the direct interaction of the emitters, the characteristics of the phased array vary with the beam swing. To combat the harmful effects of the mutual influence of radiators in the phased array, sometimes special methods are used for compensation the mutual connection between the elements.

Forms, sizes and designs of modern phased arrays are very diverse. Their diversity is determined both by the type of radiators used and by the nature of their location. In a phased array with a fast wide-angle beam oscillation, low-directional radiators are usually used: symmetrical and asymmetrical vibrators, often with one or more reflectors.

Depending on the required form of antenna pattern and the necessary spatial sector, different relative positions of the elements are used: along the line (straight or arc), on the surface (for example, flat - in the so-called flat phased arrays, cylindrical, spherical), or in a given volume (volumetric phased arrays). Sometimes the shape of the radiating surface of a phased array is determined by the configuration of the object on which the phased array is mounted.

In widespread flat phased arrays, the beam can scan from the direction of the normal to the opening (as in an in-phase antenna) to the direction along the opening (as in the antenna of a traveling wave). The gain of a flat phased array decreases when the beam deviates from the normal to the opening. To ensure wide-angle scanning without a noticeable reduction in the directional antenna action, phased arrays with a nonplanar aperture are scanned in these systems by excitation of appropriately oriented emitters and their phasing.

By the nature of the distribution of radiators in the opening, there are distinguished equidistant and non-equidistant phased arrays. In equidistant phased arrays, the distances between neighboring elements are the same throughout the opening. In flat equidistant phased array, radiators are most often located at the nodes of a rectangular lattice or in nodes of a triangular grid. The distances between radiators in equidistant phased arrays are usually chosen sufficiently small (often less than the working wavelength), which makes it possible to generate in the scanning sector antenna pattern with one main lobe and low side lobes. However, in order to form a narrow beam it is necessary to use a large number of elements. In non-equidistant phased arrays, the elements are located at unequal distances from one another. In such a phased array, even at large distances between neighboring radiators, it is possible to avoid the formation of parasitic rays and to obtain a single main beam. This allows, in the case of large openings, to form a very narrow beam with a relatively small number of elements. Nevertheless, such non-equidistant, large-opening phased arrays with a small number of emitters have a higher level of side lobes. In non-equidistant phased arrays with small distances between the emitters at equal powers of the waves emitted by individual elements, it is possible to obtain (with the uneven distribution of the radiation density in the antenna opening) antenna pattern with a lower level of side lobes than in equidistant phased arrays with the same opening and the same number of elements.

The theory and methods of antennas building are based on the theory of elementary electric oscillator radiation, published by H. Hertz in 1889. Elementary electric oscillator is a conductor with a length many times less of the emitted wave λ, a high-frequency current of the same amplitude and phase over its entire length. Its radiation pattern in the plane passing through the axis has the form of an "eighth". In a plane perpendicular axis, the radiation absent, and the diagram has the form of a circle. The elementary oscillator has the gain equal to 1,5. An example of practical implementation of elementary oscillator is the Hertzian dipole. Any antenna can be considered as a set of a large number of elementary oscillators.

The methods of model representation use general methods for calculating the characteristics of antenna arrays. They usually consider a system of half-wave oscillators. The problem of emitting a system of thin half-wave oscillators is analogous to the problem of emitting a single oscillator. The difference is in replacing one oscillator with a system of oscillators, each of which is excited by its third-party source. By doing this, it is possible to establish connections between external sources and the parameters of the antenna array. The currents in the antenna array radiators can be found from the joint solution of integral equations. Such a solution turns out to be more complicated than for a single radiator, and it makes it very difficult to identify the main regularities of the antenna array. That is why antenna theory uses approximate methods in which the general problem of the antenna array calculating is conditionally divided into two problems.

The solution of the internal problem is to determine the amplitude-phase distribution in the antenna array for given external sources, which is necessary for the antenna array excitation. The solution of the external problem is to find the antenna directivity characteristics with the known amplitude-phase distribution of currents (fields) over the antenna array elements. This distribution is considered to be known from the solution of the internal problem and is achieved by appropriate selection of external sources of excitation. The solution of the external problem can be carried out in a general form for various antenna arrays and then to establish directional characteristics. It should be noted that the methods for solving the internal problem are different for different types of antenna array radiators. The radiation field of the antenna array is the result of individual radiators fields interference. Therefore, it is necessary to find a separate field from each radiator at a given point in space, and then the sum of the fields of all radiators, taking into account the amplitude and phase relations, as well as the polarization of the fields.

The calculation of the antenna patterns of such systems is carried out as follows:

Determine the amplitude and phase diagrams of the individual elements radiation. Then the phase center of each radiator must be determined. The emitters must be replaced with point emitters, arranging them in the phase centers of the real radiators in the antenna array. Each point source is assigned a uniform phase and amplitude directivity diagrams of the real radiator. Then the point emitter in the external action will be equivalent to the real radiator.

Determine the amplitudes and phases of the fields created by equivalent point emitters at an arbitrary point in space (each separately). It is necessary to consider the field at a big distance from the observation point to all the radiators. The phase calculation should be carried out taking into account the difference in the distance to each radiator. When determining the difference in distances for simplicity, it is necessary to consider the directions to the observation point as parallel to all the radiators. When calculating the phases, it is necessary to determine the phases with respect to the field phase of any one radiator taken as the initial one.

Determine the amplitude and phase of the entire antenna field by summing the fields of all its constituent radiators, taking into account the amplitude and phase relationships, as well as the polarization of the fields.

Programs based on the method of moments are used for the calculation and analysis of antennas. The most common are NEC2 (Numerical Electromagnetics Code, 1981) and MININEC3 (Mininec History, 1980). Despite the venerable age, these programs consider antennas with sufficient for practice accuracy. However, direct work with them is very difficult, since the input of information and its output is possible only in text mode. MMANA-GAL (MMANA-GAL, 2012) is one of the programs that allows you to comfortably prepare data for calculations in a modified MININEC3 and analyze the result. To create an antenna model and output the results in MMANA, you can use both text and graphic modes. In addition to the preparation and processing of MININEC3 data, MMANA includes many additional features that make life easier for the antenna designer.

For use in MMANA-GAL, MININEC3 was translated into C/C++ and significantly modified (primarily to improve accuracy, speed up calculations, and automatically correct many user errors in the antenna description) and is directly included in the code of the program. Naturally, MMANA-GAL inherited the limitations of MININEC3, the main of which is that the calculation of the input resistance and near-field parameters does not take into account losses in the ground. This leads to an error in the calculations. MININEC3 in these cases gives an error the greater, the more ground parameters differ from the ideal ones. Therefore, if your case falls under these limitations, then calculate your model in a program using NEC2, for example, GAL-ANA. When calculating the radiation pattern, the influence of the real earth parameters in MININEC3 is always taken into account correctly.

Since the calculations are universal for any arrangement of wires, they must be based only on the most general formulas. Actually, they are based on them: the basis of the calculations is the Maxwell equations system. However, for numerical methods it is more convenient to convert this system into the so-called electric-field integral equation (EFIE). In fact, this is the same system of Maxwell equations, but expressed in a more suitable form for calculations. EFIE allows you to calculate the intensity of the emitted field and the dependence on the current distribution in the antenna. Two properties of EFIE make it indispensable for the calculation of antennas:

EFIE allows solving radiation and scattering problems in an unlimited region (the boundary of which is at infinity). In other words, you can calculate the radiating antenna (its field and goes to infinity).

EFIE can be solved by numerical methods, in particular, by the method of moments. As the initial results, EFIE requires the distribution of currents in the antenna. To calculate this distribution, all antenna wires are divided into segments. It is clear that if we split the antenna into n segments, then in calculating the current distribution, a square matrix with side n is formed (for each of the n segments we consider n currents: one own and all induced ones). Therefore, the time of its calculation and the memory necessary for this increase in proportion to the square of the number of segments.

The basic errors of modeling are connected with the partitioning of the antenna into segments. That is, the number of segments and the method of partitioning. The current in each segment is assumed to be linearly varying. If this condition is not met in the antenna, then the calculated current distribution will be incorrect. Consequently, calculated on the basis of this distribution, the antenna field and its characteristics will be incorrect too.

Violation of the above condition may occur, for example, if:

- Length of the segment is more than 0,1 of wavelength; on such a long segment, the linear approximation of the current is already noticeably different from the actually existing sinusoidal distribution; this error is called insufficient segmentation density;
- In those sections of the antenna, where the current passes through zero on the sinusoidal distribution, the actual place within the limits of one segment may not coincide with the calculated on the basis of linear approximation of the current in the segment; therefore, at the ends of the antenna and in the sections of the prospective antinodes, the length of the segments should be reduced (variable segmentation density).

Basic requirements for on-board antennas:

- effective localization of radiation in the service area, creating the maximum uniform power flux density within this zone and a minimum radiation outside of it;
- ensuring the required polarization characteristics and high spatial selectivity;
- small transport dimensions and weight; ability to withstand large acceleration and vibration.

The antennas of the onboard radioelectronic equipment should be placed for reasons of obtaining the necessary radiation patterns for each type of antenna. It is necessary to ensure maximum possible decoupling between antennas working in one frequency range or at close frequencies.

The antennas of earth stations usually do not have such strict limitations as onboard ones. The more the size of the terrestrial antenna, the more it can "overcome" signals attenuation and ensure a greater throughput of the system as a whole.

LINEAR PHASED ARRAY MODELING ON THE BASE OF PLANAR RADIATORS

Modeling of linear antenna arrays for aviation satellite systems is an important problem, since the evolution of an airplane in flight makes it necessary to use weakly directed onboard antennas, which results in significant signal weakening at the input of the onboard receiving system. Significant gain in the energy indicators of the signal at the input of the onboard receiving system in the communication channel "Aircraft-Satellite" can be achieved by using a narrowly oriented directional diagram of antenna arrays, the main lobe of which is oriented in the direction of the satellite. The widespread use of satellite systems for aviation mobile communications, necessitates designing of new phased antenna arrays.

Typically, antenna arrays are systems of weakly directed emitters that are located at a certain distance, usually smaller than the wavelength. Emitters for placement on aircraft widely use planar emitters.

Currently, there are a large number of software complexes that use cluster procedures for simulation of three-dimensional emitting structures such as communication tower, aircraft, communication satellite, etc., and analyze the distribution of electromagnetic fields. These objects have dimensions that can be tens or hundreds of wavelengths. One of the main methods of modeling is Method of Moments (MoM) based on Maxwell equations (Carlson, 2016). The well-known and powerful simulation software for electromagnetic problems include Computer Simulation Technology (CST STUDIO SUITE, 2012), Microwave Office (Antenna Design, 2018), FEKO Suite (FEKO Suite, 2018), Ansoft HFSS (Ansoft HFSS, 2018) and others. As an integral part of these complexes, the Antenna Magus software is widely used to investigate the radiation characteristics of any single antenna or system of emitters in free space, as well as to export models of generated antennas for further calculations for complex objects. Using cluster procedures for parallel computing in workstations, it is possible to calculate the resulting radiation field of a complicated radiating structure whose geometric dimensions can be n*(100 ... 1000) λ.

Antenna Magus sofrware allows you to synthesize a phased array antenna with a given pattern, with a minimum level of side lobess, to find the optimal excitation circuit of the phased array elements, to automate the process of individual antennas designing. For modeling the phased array antenna, within the framework of the obtained Classroom license, the Antenna Magus 3.2.1 software environment was used (Antenna Magus, 2018).

In this section, an equidistant linear phased array antenna with various methods of elements excitation based on a microstrip antenna (MSA) using the Antenna Magus sofrware is investigated. This antenna can be used for placement on an aircraft fuselage and provide a satellite link in the L (1,5-1,6 GHz) band (Kharchenko, Barabanov, Grekhov, & Tereschenko, 2012).

Modeling a Single Emitter Using Antenna Magus Software

Antenna modeling utilized the Antenna Magus 3.2.1 software with MSA element of a phased array antenna (Figure 1).

The design of MSA is modeled by a thin (about tens of microns) flat metal plate placed on the dielectric layer - a substrate of thickness h = (0,003 ... 0,08) λ, which is below the screen. As a substrate, materials with a relative permittivity ε = 2 ... 10 are usually used, but, depending on the application, a wider range of values ε is possible.

Figure 1. Model of microwave emitter: 1) plate of the radiator; 2) a coaxial probe; 3) dielectric lining; 4) screen

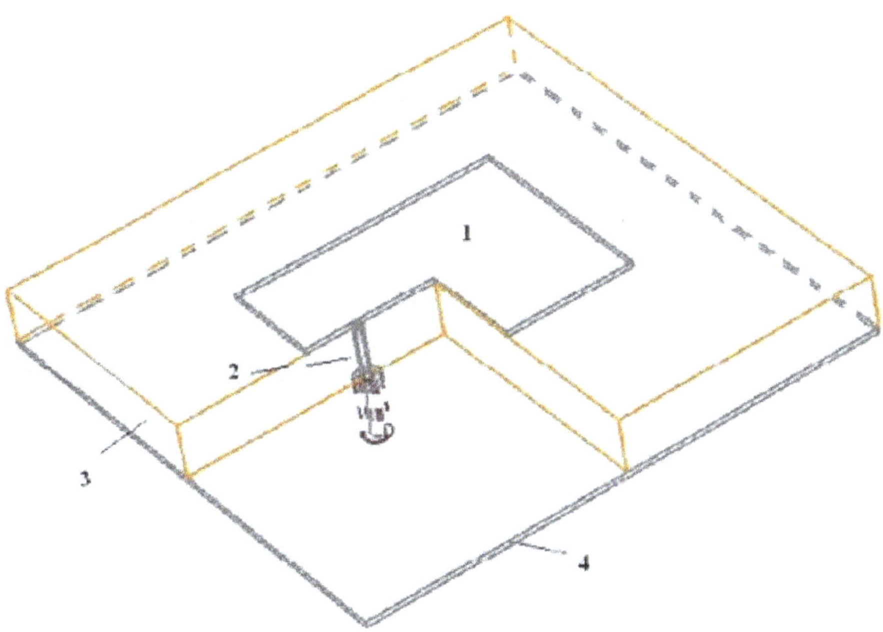

The main requirement for a substrate material is small losses characterized by the tangent of dielectric loss $tan\delta$. MSA plates most often have a rectangular or round shape, but in principle, it is possible arbitrary shape with known resonant frequency.

The choice of the plate shape can be used for significant improvement the matching of the MSA with the feeder line, and to realize the circular polarization of the antenna radiation. The advantages of MSA are especially important for space and airborne radio engineering systems, the requirements for technical and constructive parameters of which are very rigid and contradictory. Small weight and overall dimensions are also relevant for portable equipment.

Calculations of MSA electric characteristics at 1,6 GHz were performed using the Antenna Magus software. The following parameters of the microwave antenna were taken: the input resistance of 50 Ω, the dielectric substrate thickness 2.492 mm, the relative permittivity $\varepsilon_r = 2.08$, the tangent of dielectric loss $tan\delta = 10^{-3}$. The geometric dimensions of the MSA are as follows: the diameter of the radiator plate $D = 74.83$ mm, the displacement of the point of the coaxial probe connection $S_f = 11.0$ mm (Figure 2).

MSA structural parameters allow calculating the following radiation characteristics: the dependence of the input impedance (Figure 3), the reflection coefficient (Figure 4), the static wave coefficient (Figure 5), and the Smith diagram (Figure 6) on the frequency, with a shift relative to the resonant frequency within ± 20 MHz. Antenna directional diagram is presented in Figure 7.

Linear System of Microwave Emitters Simulation Using Antenna Magus Software

In the simulation of a linear equidistant phased array antenna 5, 7 and 11 identical MSAs were taken, the design and characteristics of which were discussed above. Three variants of antenna excitation are accepted: 1) uniform and in-phase distribution of MSA excitation currents along OX-axis, located at a distance of $0{,}5\lambda$ one relative to one; 2) distribution by the Dolf-Chebyshev method (Orfanidis, 2010), in which the amplitude distribution is such that the level of the side lobes is minimal for the given main lobe width, or the main lobe width is minimal for a given level of side lobes; 3) excitation by the law of Villeneuve, which can be represented as modification of the Taylor method for optimization of discrete side lobes.

Figure 2. Antenna Magus interface when calculating MSA parameters

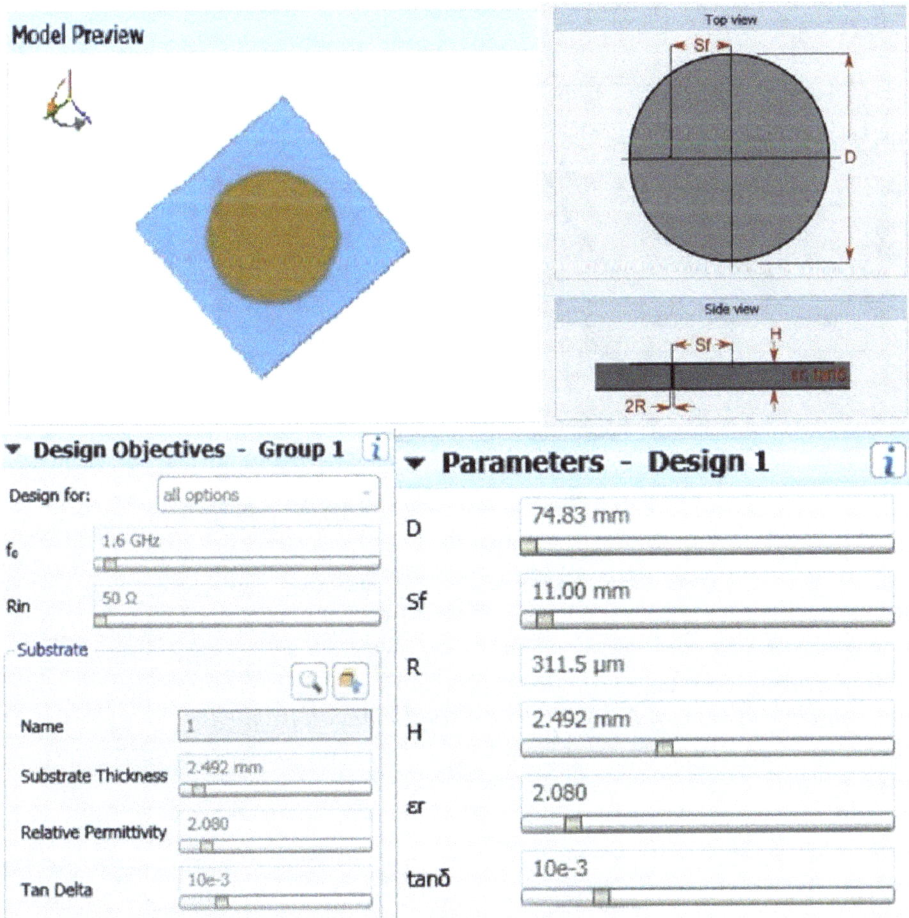

The results of the radiation field simulation for a linear phased array antenna are presented in Table 1, with the level of the main lobe set to 30 dBi for all variants of excitation.

All designed and analyzed MSAs can be used as onboard aircraft antennas for receiving of ADS-B messages.

Figure 3. Dependence of MSA input impedance on frequency

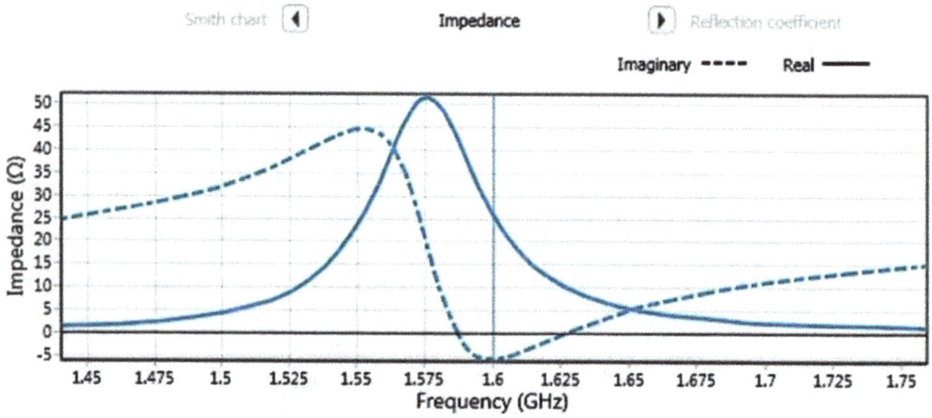

	Design 1
Peak real impedance @ frequency	51.32 Ω @ 1.575 GHz
Real impedance @ zero-crossing frequency	42.34 Ω @ 1.587 GHz 10.05 Ω @ 1.626 GHz
Mean real impedance	11.04 Ω
Mean imaginary impedance	17.98 Ω

Figure 4. Dependence of MSA reflection coefficient on frequency

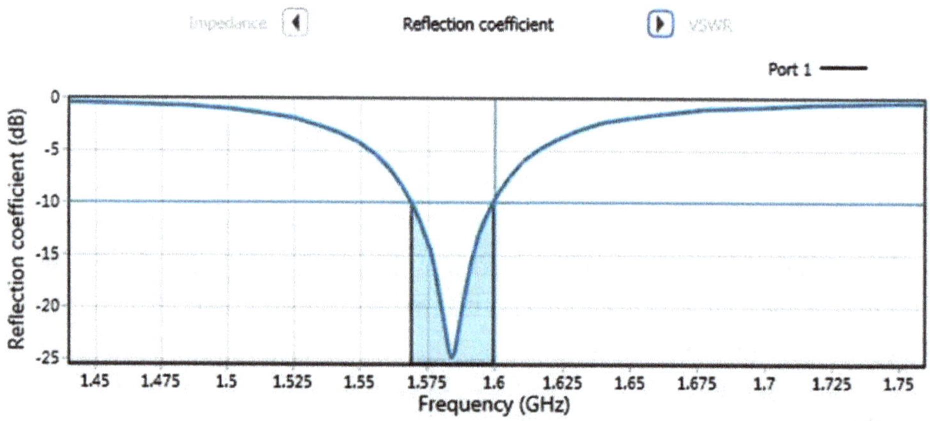

	Design 1
Reference impedance @ port 1	50 Ω
Frequency at which S11 = -10 dB	1.569 GHz 1.599 GHz
Minimum S11 value	-24.83 dB @ 1.584 GHz

Figure 5. Dependence of standing wave coefficient (VSWR) on frequency

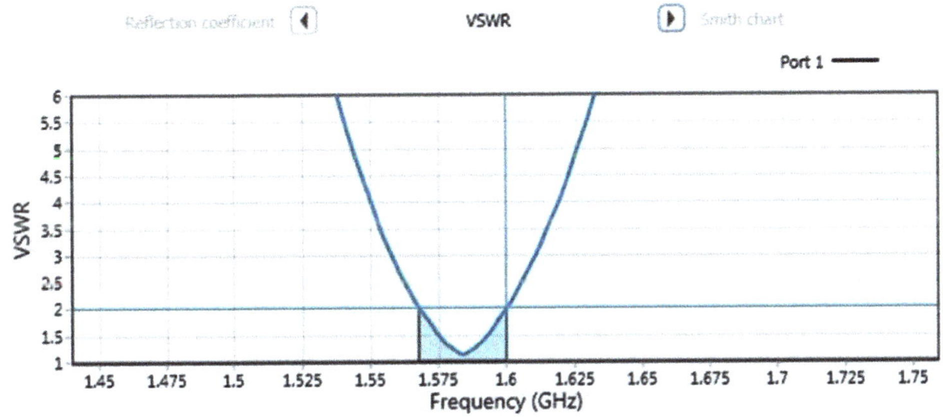

	Design 1
Reference impedance @ port 1	50 Ω
Frequency at which VSWR = 2	1.568 GHz 1.600 GHz
Minimum VSWR value	1.122 @ 1.584 GHz

Figure 6. MSA Smith diagram

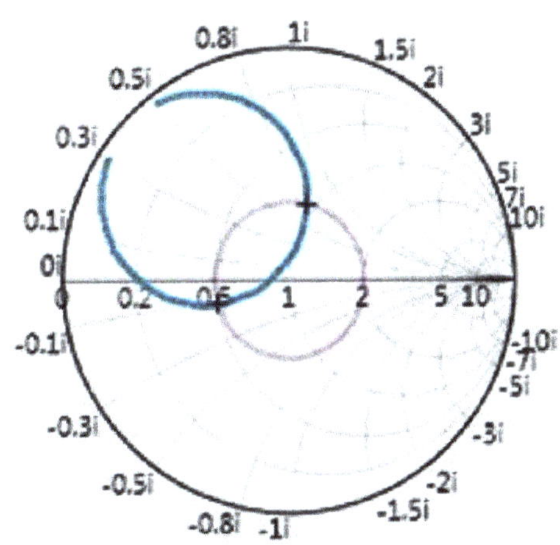

	Design 1
Reference impedance @ port 1	50 Ω
Frequency at which VSWR = 2	1.568 GHz 1.600 GHz
Minimum VSWR value	1.122

Figure 7a. MSA directional diagram in H and E planes with diagram width at half-power level: a) directional diagram in the Cartesian coordinate system

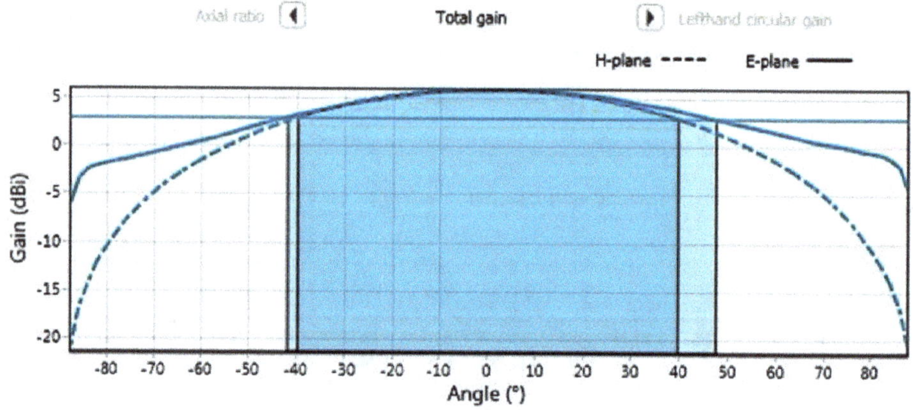

	Design 1
Peak gain @ angle (freq) [φ = 0 °]	5.922 dBi @ θ = 2 ° (1.6 GHz)
Peak gain @ angle (freq) [φ = 90 °]	5.915 dBi @ θ = 0 ° (1.6 GHz)
Main 3dB beamwidth (freq) [φ = 0 °]	89.19 °
Main 3dB beamwidth (freq) [φ = 90 °]	79.13 °

Figure 7b. Directional diagram in the polar coordinate system for linear vertical polarization

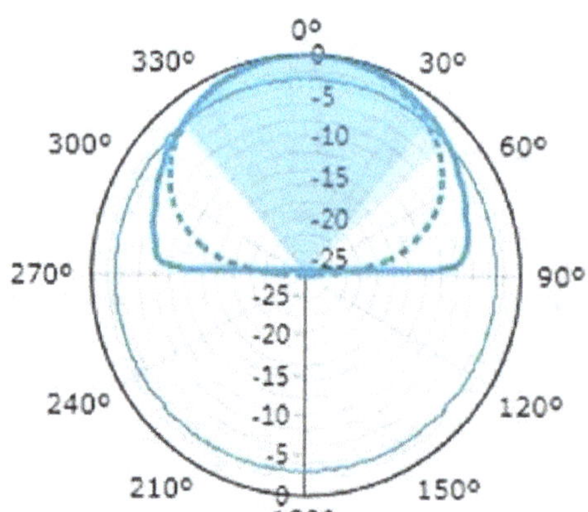

	Design 1
Peak gain @ angle (freq) [φ = 0 °]	5.922 dBi @ θ = 2 ° (1.6 GHz)
Peak gain @ angle (freq) [φ = 90 °]	5.915 dBi @ θ = 0 ° (1.6 GHz)
Main 3dB beamwidth (freq) [φ = 0 °]	89.19 °
Main 3dB beamwidth (freq) [φ = 90 °]	79.13 °

Figure 7c. Directional diagram with the ratio of radiation along the axes in the polar coordinate system, for the linear, right and left-circular polarization

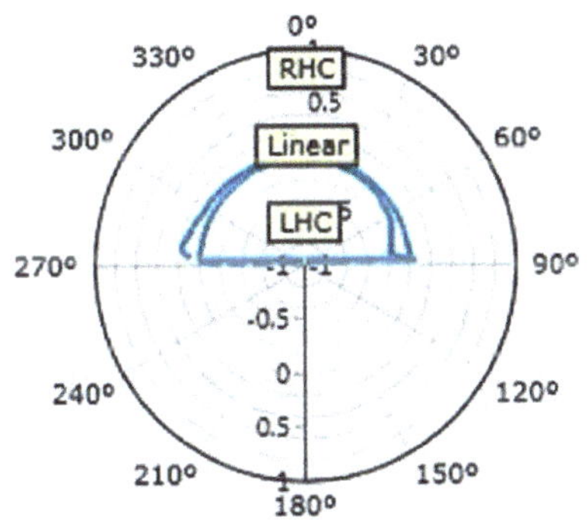

	Design 1
Maximum axial ratio @ angle [φ=0 °]	0 @ θ=-88 °
Minimum axial ratio @ angle [φ=0 °]	0 @ θ=-88 °
Maximum axial ratio @ angle [φ=90 °]	179.1e-3 @ θ=-82 °
Minimum axial ratio @ angle [φ=90 °]	-179.1e-3 @ θ=82 °

Figure 7d. Three-dimensional directional diagram for all components of the electromagnetic field

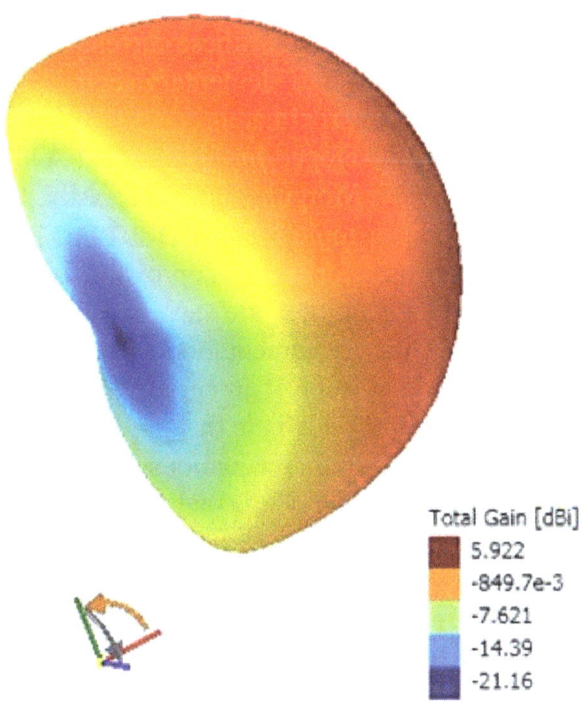

INVESTIGATION OF COLLINEAR ANTENNA PARAMETERS FOR ADS-B SIGNALS RECEIVING

ADS-B is currently being introduced in Europe, the United States and other countries. The CASCADE EUROCONTROL Programme (EUROCONTROL CASCADE Programme, 2013) coordinates the implementation of ADS-B systems in Europe. The system receiving ADS-B signals can be used as a virtual radar to create a real-time picture of air traffic and consists of four components: an antenna, receiver decoder and computer software. To create antenna sample, it is necessary to analyze antenna performance for a system that receives ADS-B signals (Kharchenko, Barabanov, Grekhov, & Tereschenko, 2013).

The purpose of a study in this section is calculation of the electromagnetic field for the collinear antenna, which is used in the pilot system for ADS-B signals receiving at the frequency of 1090 MHz in onboard aircraft transponder with Mode S.

For calculations of the electromagnetic field intensity and the corresponding antenna pattern, numerical methods of electrodynamics are used, in particular the method of moments (Carlson, 2016), in which the metal elements of the antenna are replaced by equivalent surface electric currents and create an equivalent network model of the object. Then the problem of calculation the electromagnetic field, which is created by these currents, must be solved. For this, the metal surface is divided into elementary segments and the electric current within the segment is approximated.

In the software complex developed in the National Aviation University (Ukraine) (Barabanov, Ivanov, Morgun, & Chernyavsky, 2007; Report on research work of NAU on the topic N° 006-DB01-03, 2003), piecewise-constant basic functions are used for current approximation and the boundary conditions for an electromagnetic field are superimposed along the current metal wire. As a result of the implementation of boundary conditions in discrete points, we obtain a system of integral equations with respect to coefficients for basic functions, which are the amplitude of currents within the elementary segment. The system of equations in the software (Report on research work of NAU on the topic N° 006-DB01-03, 2003) is solved with respect to interrelated currents.

The accuracy of the method of moments is higher, the smaller the size of the elementary segment is. It is believed that in order to obtain an acceptable accuracy, the size of a segment should not exceed $\lambda/10$, where λ is the wavelength in free space. The number of integral equations is equal to the number of elementary segments N, which increases with increasing the size of the object or with increasing the frequency. Therefore, solving the problem of scattering radio waves at an object using the method of moments requires solving the system of integral equations in a matrix form of large dimension.

Calculations of Antenna Pattern

The vector of the electric field intensity, which is excited by the clectrodynamic object, is generally determined by the vector sum of projections, which needs to be determined:

$$\bar{E}_{\Sigma} = \bar{1}_r E_{r0\Sigma} + \bar{1}_{\theta} E_{\theta\Sigma} + \bar{1}_{\varphi} E_{\varphi\Sigma} \cdot$$

For the numerical calculation using the method of moments, the collinear antenna is divided into N segments, which length z is much smaller than the wavelength. The complex amplitudes of the vector components in the far radiation zone of any segment are determined by the following relations:

$$E_{\theta} = -i\frac{Ik^2}{\omega\varepsilon} G\left(z, z'\right)\left\{r_2\left[sin\theta_2 cos\theta\cos\left(\varphi_2 - \varphi\right) - cos\theta_2 sin\theta_2\right] - r_1\left[sin\theta_1 cos\theta\cos\left(\varphi_1 - \varphi\right) + cos\theta_1 sin\theta\right]\right\},$$

$$E_{\varphi} = -i\frac{Ik^2}{\omega\varepsilon} G\left(z, z'\right)\left\{r_2 sin\theta_2 \sin\left(\varphi_2 - \varphi\right) - r_1 sin\theta_1 \sin\left(\varphi_1 - \varphi\right)\right\},$$

$$E_{\varphi} \cong 0.$$

The indices 1 and 2 refer to the coordinates of the points $\left[r, \theta, \phi\right]$, that determine the beginning and end of antenna elementary segment with the complex amplitude of the current I;

$$G\left(z, z'\right) = \frac{e^{-ikrr}}{4\pi r},\ r = \sqrt{\left(z - z'\right)^2 + 0^2},\ \omega - 2\pi f,\ k = \frac{2\pi}{\lambda},$$

where a is the radius of the antenna rod, z is the coordinate of the observation point on the surface, z' is the coordinate of the point on the axis of the grid element with length *L*, in which the partial source of the linear current is located

$$I(z') = \int_L j(z)dz\,.$$

The procedure for determining the currents in the elements of the model consists in the numerical solution of the Poklington equation:

$$\int_{-\frac{L}{2}}^{\frac{L}{2}} I(z')\left[\frac{\partial^2 G(z,z')}{\partial z'^2} + k^2 G(z,z')\right]dz' = -i\omega\varepsilon E_z^i(z)\,.$$

When solving this equation in accordance with the method of moments, unknown distributions of currents $I(z')$ along the z axis of any fragment of a model are decomposed in the system of selected basic functions $J_n(z')$:

$$I(z') = \sum_{n=1}^{N} I_n J_n(z'),$$

where unknown constant coefficients I_n must be determined. After that, the integral equation of Poklington takes the following form:

$$\sum_{n=1}^{N} I_n \int_{-\frac{L}{2}}^{\frac{L}{2}} j_n(z')\left[\frac{\partial^2 G(z,z')}{\partial z'^2} + k^2 G(z,z')\right]dz' = -\,i\omega\varepsilon E_z^i(z).$$

Taking into account the weight functions W_m of the mutual influence of the *m*-th element on the *n*-th element the Poklington equation takes the form:

$$\sum_{n=1}^{N} I_n \int_{L_n}\int_{L_m} W_m(z) j_n(z')\left[\frac{\partial^2 G(z,z')}{\partial z''^2} + k^2 G(z,z')\,dz'\right]dz = -\,i\omega\varepsilon \int_{L_m} W_m(z) E_z^i(z)dz.$$

Table 1. Antenna patterns in the planes

Type of excitation and amplitude distribution	Plane	Phased array antenna diagrams in free space		
		N=5	*N*=7	*N*=10
Uniform and in-phase distribution 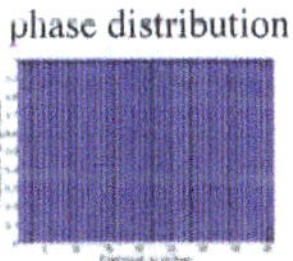	XZ	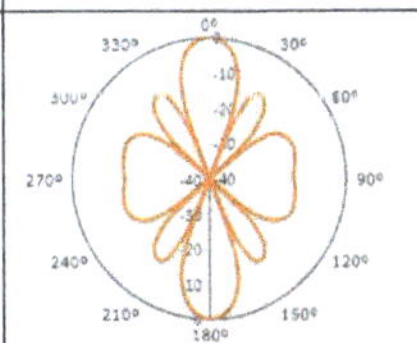	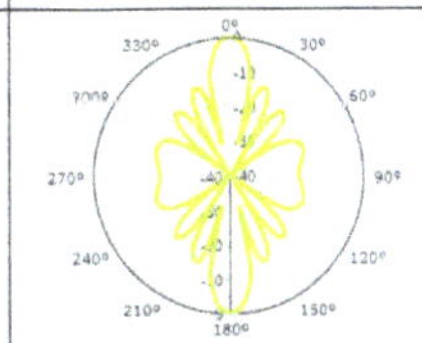	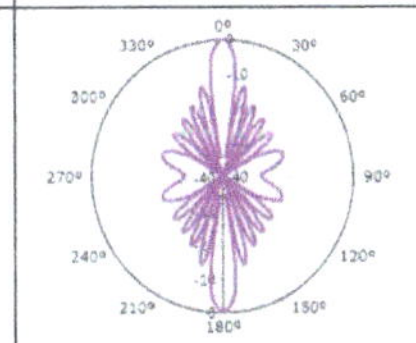
	XY	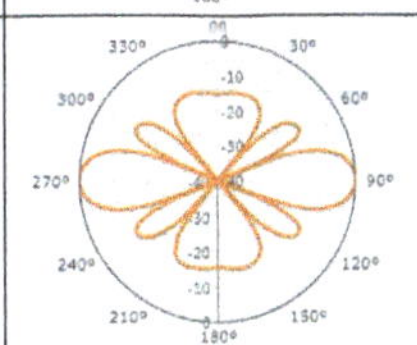	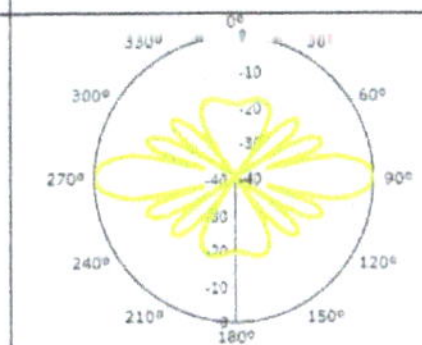	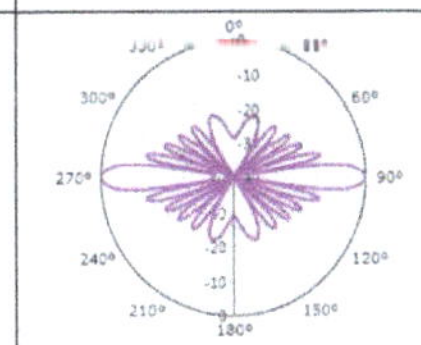
Dolf-Chebyshev distribution 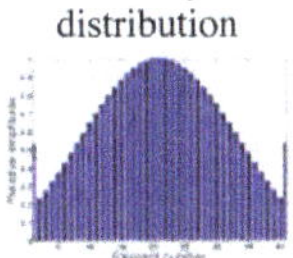	XZ	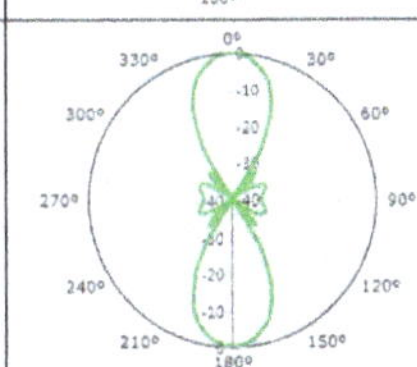	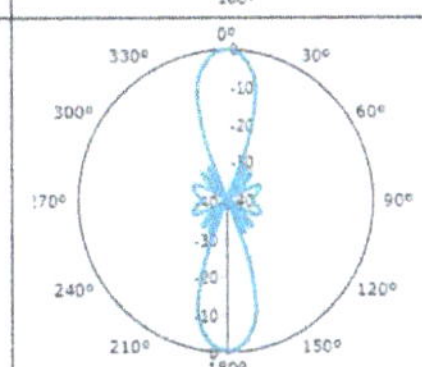	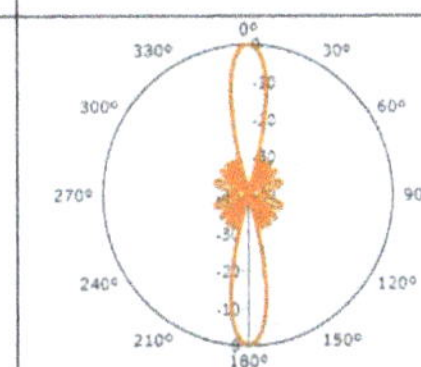
	XY	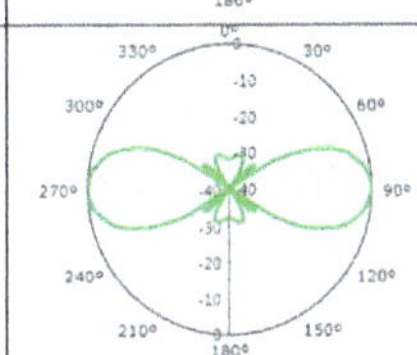	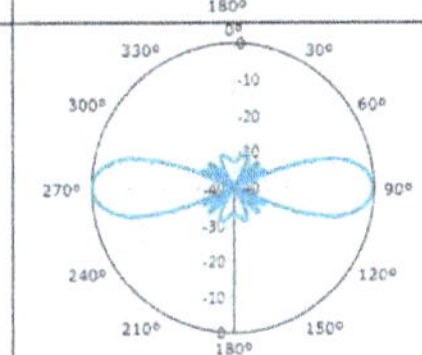	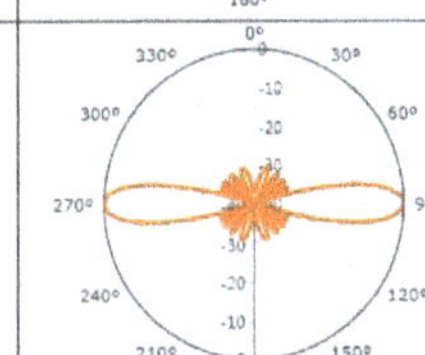
Villeneuve distribution 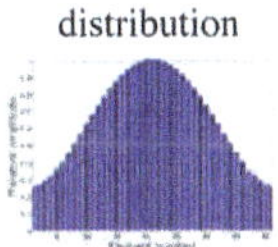	XZ	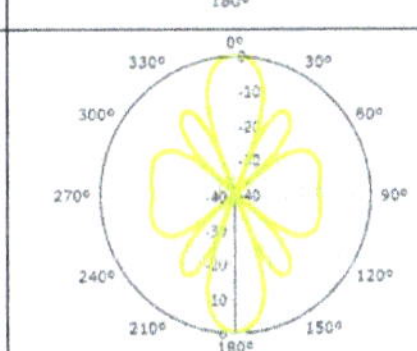	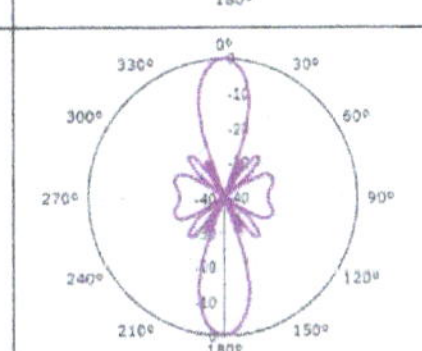	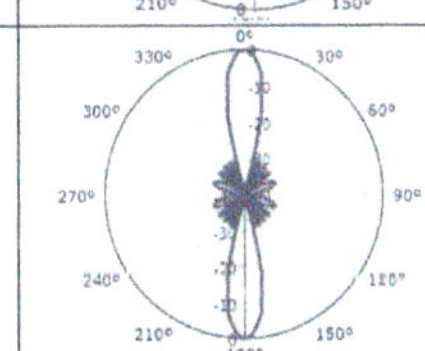
	XY	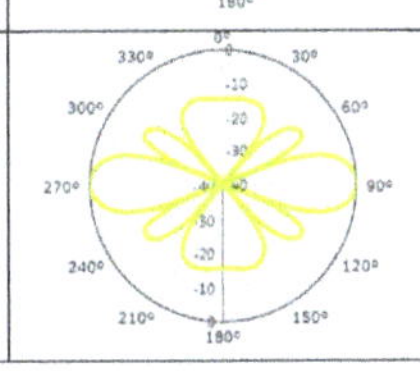	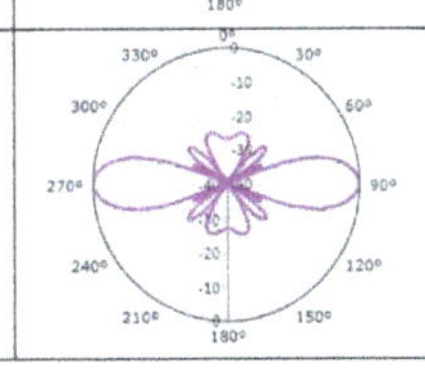	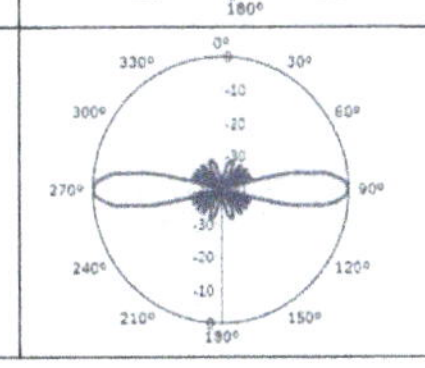

Figure 8. The three-dimensional antenna pattern

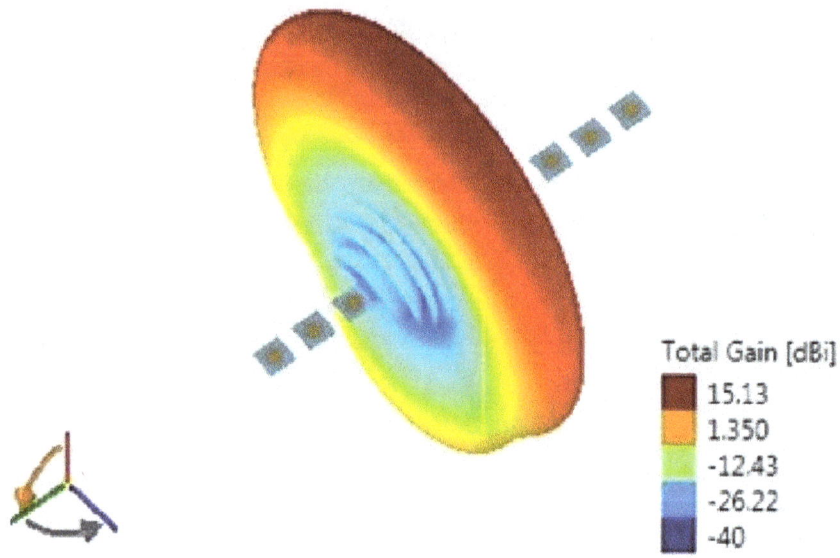

According to the Galerkin method, the weight functions W_m are chosen the same as the basic functions $J_m(z')$:

$$W_m = J_m(z').$$

The integral part of the Poklington equation has a resistance dimension, denoted by Z_{mn}, and is called a generalized impedance. The right side of the Poklington equation is the generalized excitation voltage U_m. The obtained system of equations in a matrix form has the following form:

$$[Z]\cdot[I] = [U],$$

where [*Z*] is the matrix of generalized impedances, [*I*] is the matrix column of unknown coefficients of currents expansion in the object model, [*U*] is the matrix-column of excitation sources. The solution of the system of equations in the matrix form is carried out using special programs after its transposition:

$$[I] = [Z]^{-1}[U],$$

or transformation into an equivalent system of equations:

$$I_1 = Y_{11}U_1 + Y_{12}U_1 + \ldots + Y_{1n}U_n,$$

$$I_2 = Y_{21}U_1 + Y_{22}U_1 + \ldots + Y_{2n}U_n,$$

$$I_n = Y_{n1}U_1 + Y_{n2}U_1 + \ldots + Y_{nn}U_n.$$

where I_p is the amplitude and phase of current flowing through the clamps of the element *p* of the antenna model, Y_{pq} - the mutual complex conductivity between *p* and *q* elements, U_p - the amplitude and phase of the voltage on the clamps of the element *p*. The system of equations in the software (Report on research work of NAU on the topic N° 006-DB01-03, 2003) was solved with respect to interrelated currents. Their values were used to determine the modules of complex projections of electric field intensity and their normalized values.

These ratios form the basis of the calculation part in the software (Report on research work of NAU on the topic N° 006-DB01-03, 2003) for calculating the intensity of the radiation field and the antenna pattern by the method of moments.

The collinear antenna for reception of ADS-B signals receives signals of onboard aircraft transponders at the frequency of 1090 MHz of the secondary radar with Mode S. An antenna of the ground part of the system for reception of ADS-B signals should provide in a horizontal plane a circular pattern with vertical polarization of the field. The range of the system's operation on the response channel of the secondary radar with Mode S must provide a maximum service area of the air traffic control center not less than 350 km for aircraft flights in the upper airspace. Therefore, in order to provide a high gain in the vertical plane, we selected a collinear antenna in the form of three in-phase half-wave wire segments, between which two phase-shift laps of length $\lambda/2$ are installed to provide in-phase excitation of vibrators.

The radiation field of the collinear antenna was calculated using software (Barabanov, Ivanov, Morgun, & Chernyavsky, 2007; Report on research work of NAU on the topic N° 006-DB01-03, 2003) and (MMANA-GAL, 2018), and the results were comparable.

Initial data for calculating the radiation field were identical. The case of antenna arrangement above the surface of the earth with the following characteristics is considered: the relative permittivity $\varepsilon = 9{,}0$ and the specific conductivity of the underlying surface $\gamma = 0{,}01$ (Ohm • m)$^{-1}$. In both codes, the method of moments with the splitting of antenna conductors into elementary segments was used. Software (Barabanov, Ivanov, Morgun, & Chernyavsky, 2007; Report on research work of NAU on the topic N° 006-DB01-03, 2003) allows you to perform calculations with a constant number of segments. In the MMANA-GAL sofrware (MMANA-GAL, 2018), you can select the variable density of segments from 40 to 400 along the antenna. To compare the results, the number of segments was assumed to be 400. The antenna is set at $\lambda/10$ above the surface of the earth.

Figure 9a shows the selected antenna configuration. As a phase-shift element of the antenna, a twisted wire of length $\lambda/2$ was taken, which was approximated by an octagon (Figure 9b).

Figure 10 shows the program interface (Report on research work of NAU on the topic N° 006-DB01-03, 2003), which was used to enter the parameters of the calculations and the geometry of the antenna.

Results of calculations of the collinear antenna pattern for the underlying surface with parameters $\varepsilon = 9{,}0$ and $\gamma = 0{,}01$ (Ohm · m)$^{-1}$ are given in Table 2.

In software (Report on research work of NAU on the topic N° 006-DB01-03, 2003) and (MMANA-GAL, 2018) for which the results are compared,

Figure 9a. Antenna configuration in a three-dimensional space with the distribution of excitation currents (a)

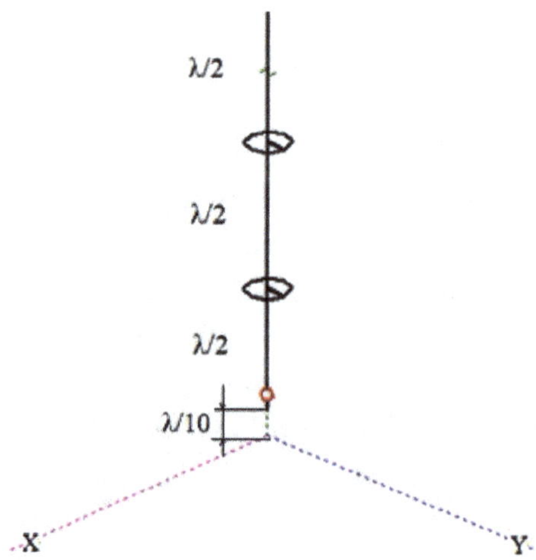

Figure 9b. The projection of the phase-shift element of the antenna on the XOY plane (b)

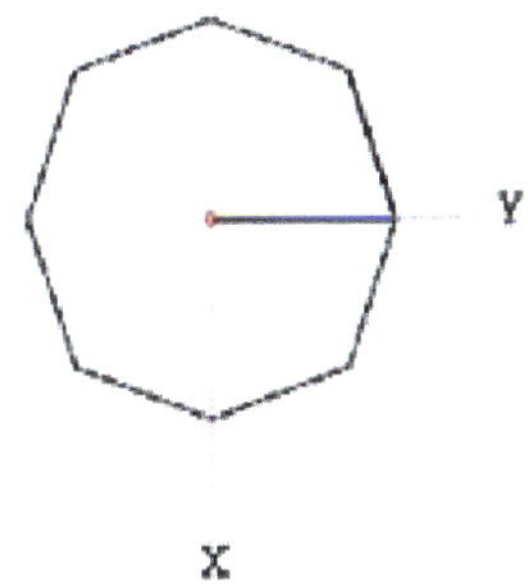

Figure 10a. Program interface: (a) - panel for data entry

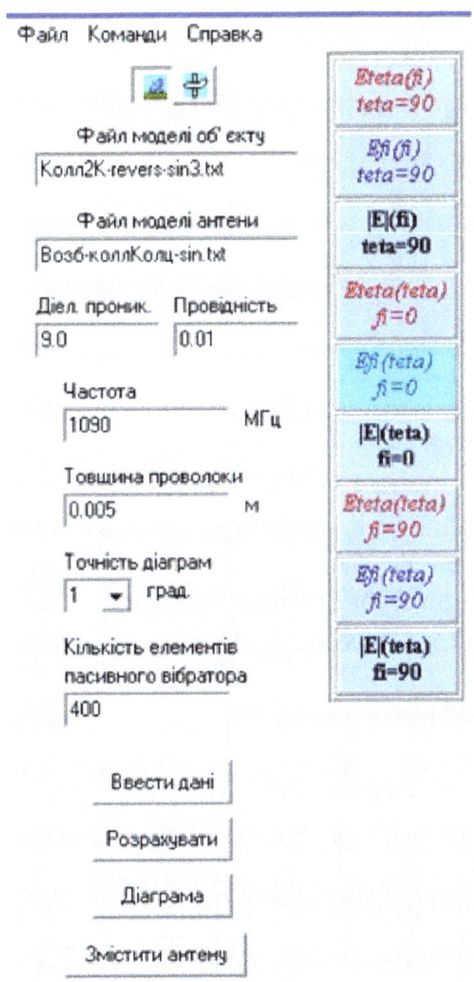

Figure 10b. Antenna geometry

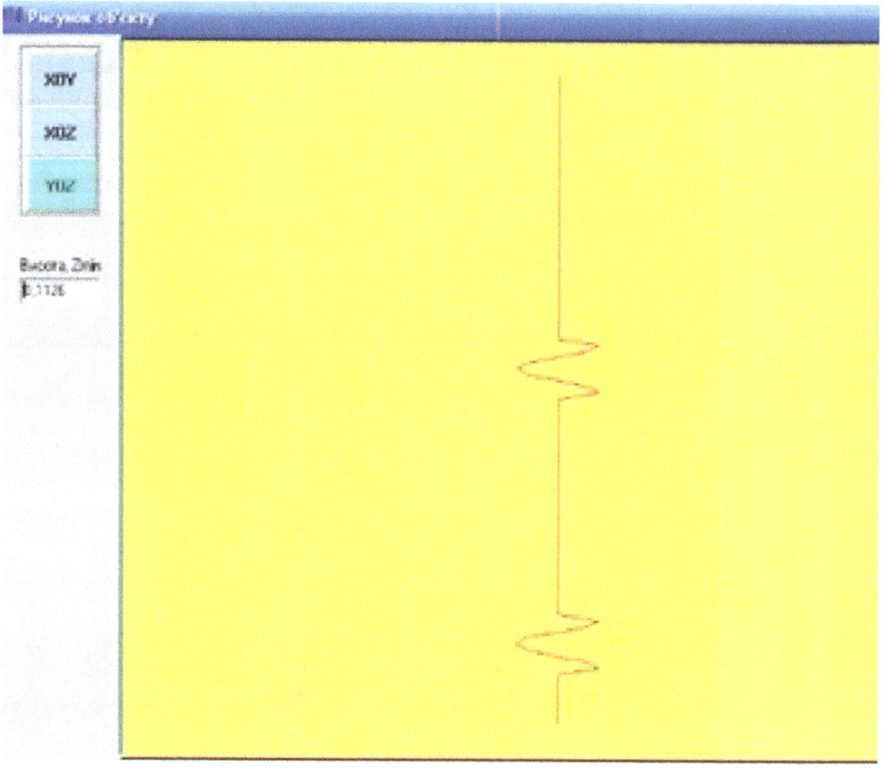

different methods are used to process the output data in field calculations. In the software (Report on research work of NAU on the topic N° 006-DB01-03, 2003), two methods are used for the electric field intensity and the antenna pattern normalization, which are derived in relative units: 1) the method of local rationing within one plane of the antenna pattern, and 2) the method of global normalization of the antenna pattern for all individual calculations. In this case, the voltage maximum of the electric field intensity is 1,0 in relative units.

The software (MMANA-GAL, 2018) uses the method of logarithmic global normalization of the electric field intensity module. With the help of the perfect graphical interface in the software (MMANA-GAL, 2018) the character of antanna pattern change was analyzed in the transition from the ideal to the real underlying surface. It can be seen that the antenna pattern is significantly improved by reducing "failures" (Figures 11, 12).

Table 2. Antenna patterns

	Software (Report on research work of NAU on the topic № 006-DB01-03, 2003)	Software MMANA – GAL (MMANA-GAL, 2018)
Electric field in the vertical plane XOZ $E_\theta(\theta)\|_{\varphi=0^\circ}$	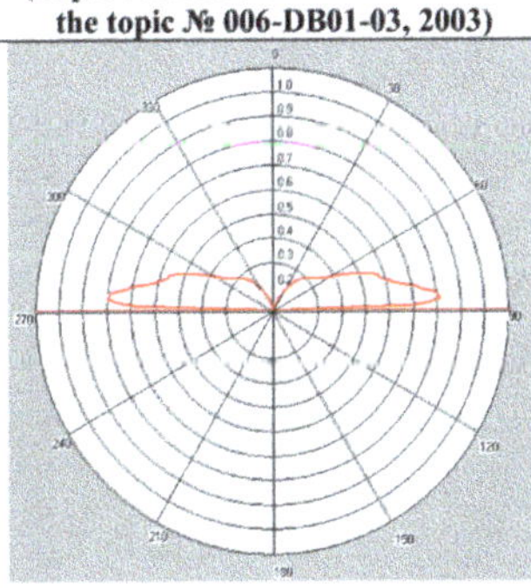	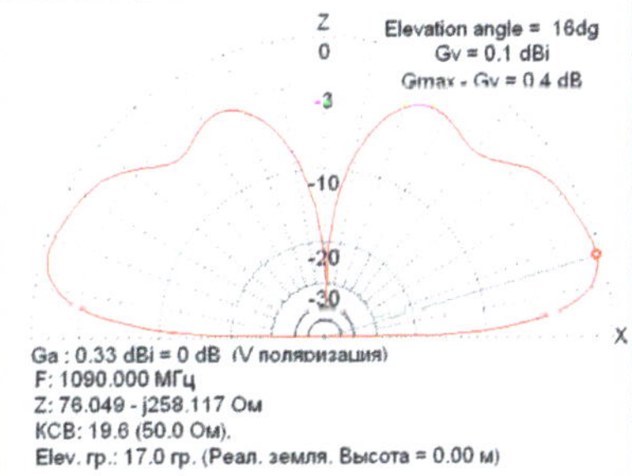
Electric field in the vertical plane YOZ $E_\theta(\theta)\|_{\varphi=90^\circ}$	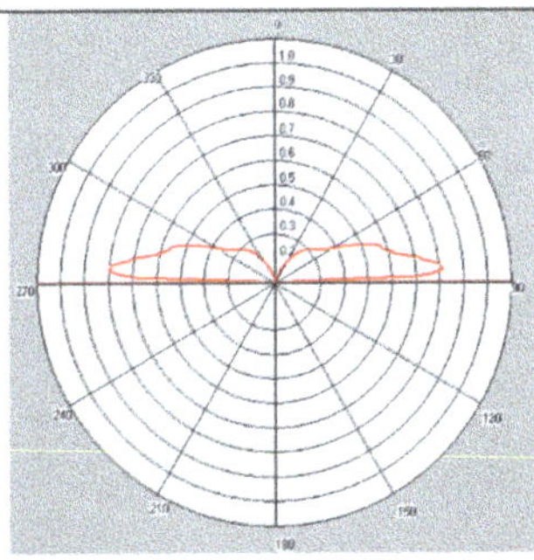	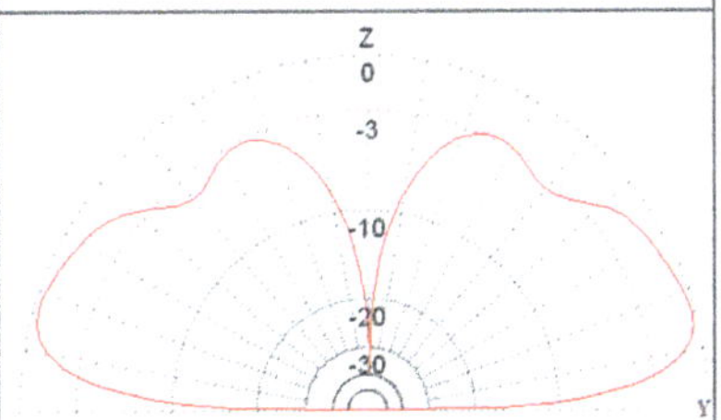
Electric field in the horizontal plane XOY $E_\theta(\varphi)\|_{\theta=90^\circ}$	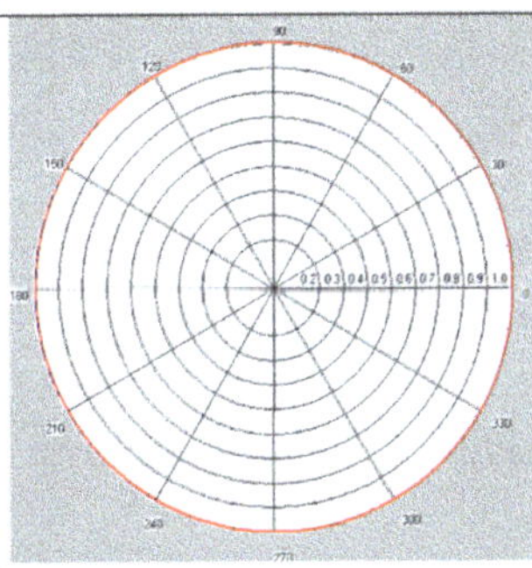	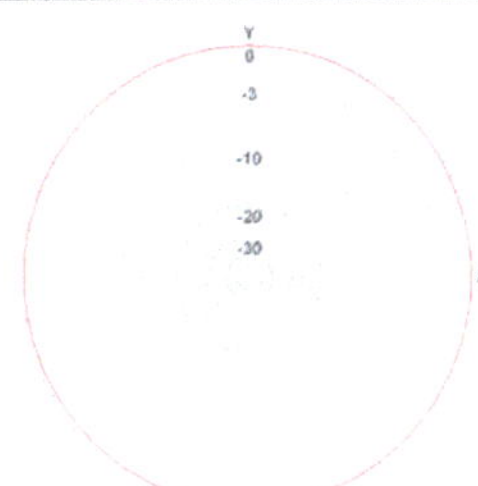

Experimental Antenna Sample

On the basis of performed calculations a collinear antenna was manufactured. Between the half-wave elements, phase-shift elements (horizontal cable turns of $\lambda/2$) were included, which perform currents phase shift of 180 degrees. They exclude the formation of currents in the opposite direction on the elements of the collinear antenna. The synchronous power of these antennas depends

Figure 11. 3D antenna pattern of the collinear antenna for an ideal underlying surface

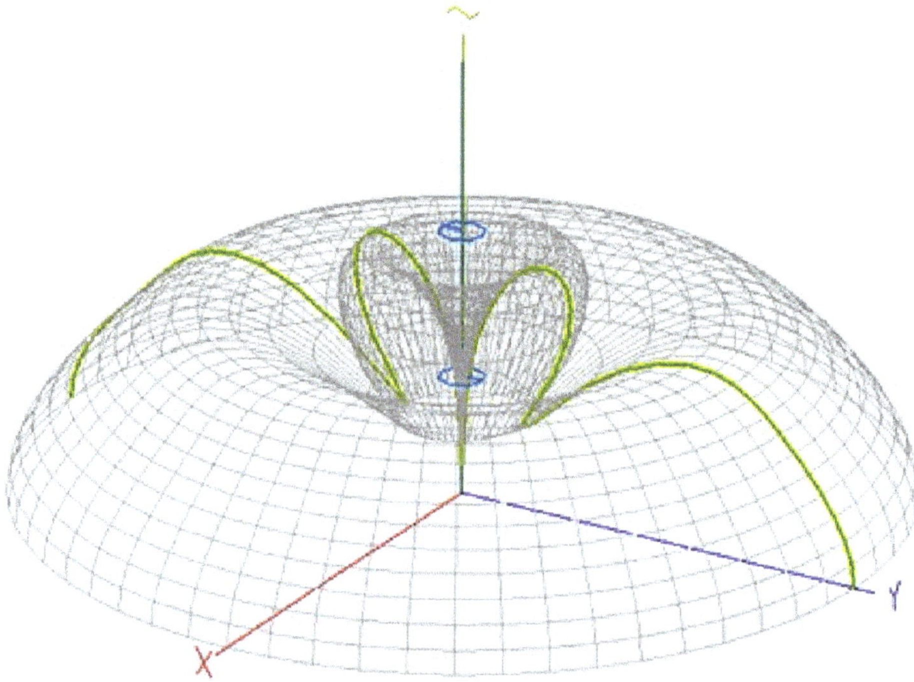

on the length of the radiating elements and the distance between them, so this antenna is narrowband.

The created collinear antenna has a good harmonization at a frequency of 1090 MHz, Ga = 0,33 dBi gain in a vertical plane at an angle of 16°, and an input impedance Z = (76,049-j258,117) Ohm. It allows receiving signals from onboard transponders via Mode S channel for the operation of the ground ADS-B system.

Observation Results

The following experiments were carried out using the created system for reception of ADS-B signals:

- An observation of air traffic and results comparison (number of aircraft and information about them) with the data on the site Flightradar24.com;

Figure 12. 3D antenna pattern of the collinear antenna for the underlying surface with parameters ε = 9,0 and γ = 0,01 (Ohm · m)$^{-1}$

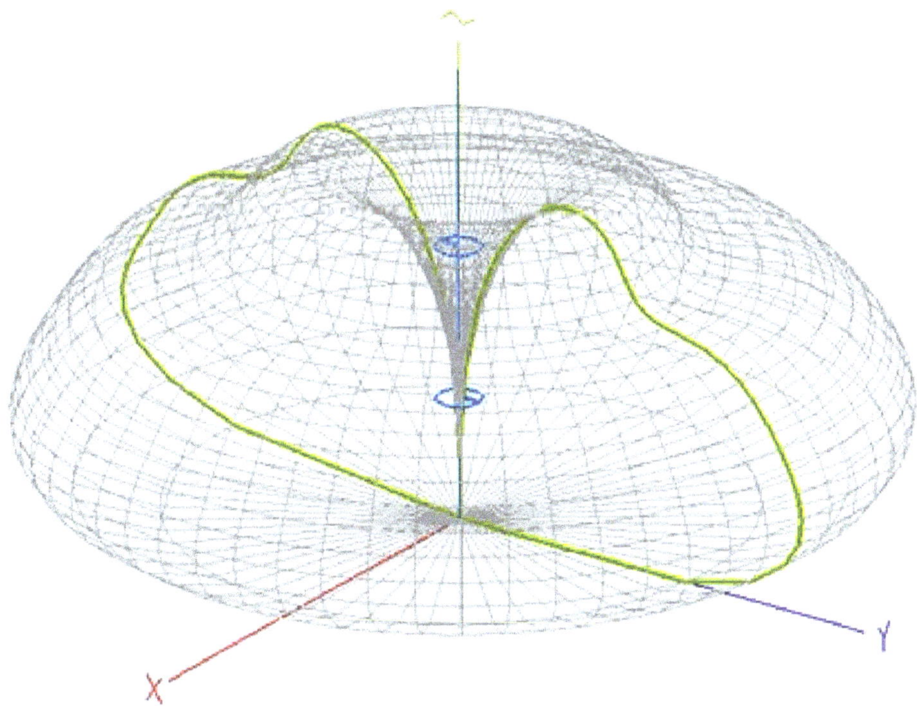

- Finding system possibilities for determining the maximum detection distance.

The results of observantions are shown in Figure 13. The distance from the system's antenna to the Boryspil airport is 47 km, and to the Kyiv Zhulyany airport is 23 km. With such distances to the airports, there are obviously related features of the distances and observed heights of planes. Airplanes or landed (depart), or at high altitudes traveled past Kiev. Nevertheless, it was possible to observe planes at a distance of 160 km and heights up to 12000 m. It turned out that the antenna has a small "blind funnel" and allows you to track aircraft virtually "over yourself". As can be seen from Figure 13, the antenna created for reception of ADS-B signals allowed to detect a plane and observe its motion at considerable distances from the receiver.

Figure 13. Observation results

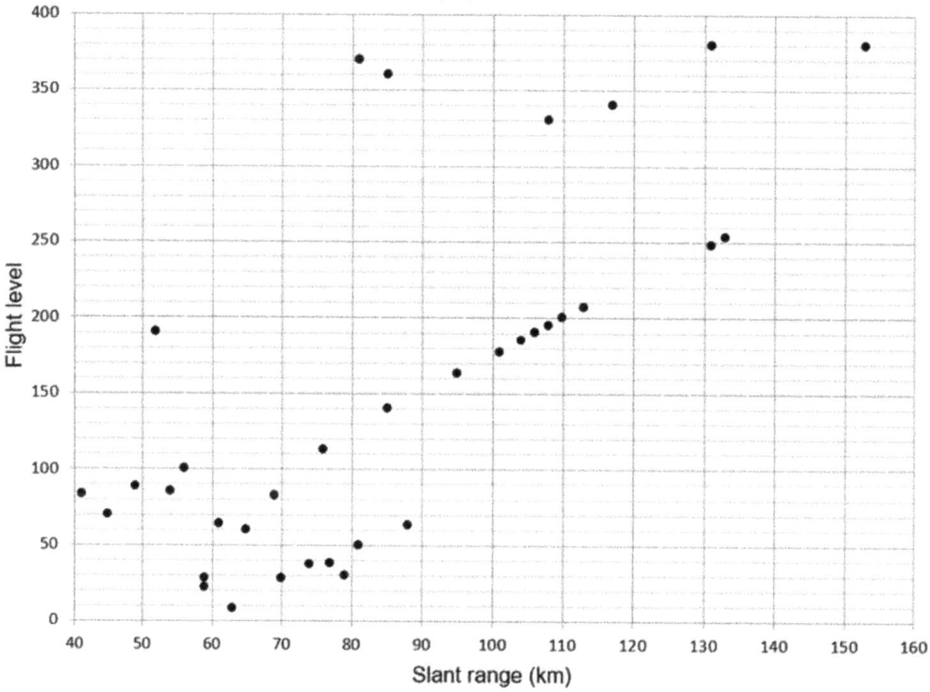

CONCLUSION

1. Calculations of antenna pattern for collinear emitters are carried out using method of moments.
2. Software (Report on research work of NAU on the topic N° 006-DB01-03, 2003) and (MMANA-GAL, 2018) for calculating the electric field intensity used in this study yield close results. They allow to provide a qualitative analysis of the directional patterns for the vertical field component.
3. The created antenna sample and its field tests showed a sufficiently high gain and a significant range of action.

REFERENCES

AnsoftH. F. S. S. (2018). Retrieved from http://ansoft-hfss.software.informer.com

Antenna design. (2018). *Microwave Office*. Retrieved from http://www.awrcorp.com/products/ni-awr-design-environment/microwave-office

Antenna Magus. (2012). Retrieved from http://www.antennamagus.com

Barabanov, Yu. M., Ivanov, V. O., Morgun, O. A., & Chernyavsky, I. I. (2007). Field of the antenna mounted on the body of aircraft. *Electronics and Control Systems*, *3*(13), 88–95.

Carlson, M. (2016). *Method of Moments for Maxwell's equations based on higher-order interpolatory representation of geometry and currents* (Master's thesis EX066). Chalmers University of Technology, Gothenburg, Sweden.

CST Studio Suite. (2012). *System assembly and modeling*. Retrieved from https://www.cst.com/-/media/cst/solutions/articles/journal-article/cst-studio-suite-2012-system-assembly-and-modeling/attachments/2011-12_mwj_cst2012.ashx

EUROCONTROL CASCADE Programme. (2013). Retrieved from http://www.eurocontrol.int/sites/default/files/publication/files/2013-ads-b-wam-leaflet.pdf

FECO Suite. (2018). Retrieved from https://www.rfglobalnet.com/doc/feko-suite-0001

Kharchenko, V. P., Barabanov, Y. M., Grekhov, A. M., & Tereschenko, D. I. (2012). Linear Phased Array Modeling Using Planar Irradiators. *Bulletin of Engineering Academy of Ukraine*, *2*, 35–38.

Kharchenko, V. P., Barabanov, Y. M., Grekhov, A. M., & Tereschenko, D. I. (2013). Investigation of Collinear Antenna Parameters for ADS-B Signals Receiving Using Numerical Calculations. *Proceedings of the National Aviation University*, *55*(2), 13–20.

Mininec History. (1980). Retrieved from http://wb0dgf.com/mininec.htm

MMANA-GAL. (2012). Retrieved from http://hamsoft.ca/pages/mmana-gal.php

Numerical Electromagnetics Code. (1981). *Method of Moments*. Retrieved from http://www.nec2.org/

Orfanidis, S. J. (2010). *Electromagnetic waves and antennas*. Rutgers University.

Report on research work of NAU on the topic N° 006-DB01-03. (2003).

Chapter 6
Air Traffic Monitoring Using ADS-B System

ABSTRACT

This chapter describes the system created for ADS-B messages receiving. For this an antenna, ADS-B signals receiver, a decoder was made and software was installed. This system was allocated at the Department of Air Navigation Systems in the National Aviation University and was used for students training and investigations. Original software for modeling of real-time TCAS operation was developed using MATLAB. The experimental model with data exchange between onboard systems via Wi-Fi network was created. This model was used for modeling of aircraft approaching. Such model can be used as a base for creation a collision avoidance system of RPAS.

INTRODUCTION

Advantages of ADS-B

ADS-B is a new surveillance technology and is designed to modernize aviation to next generation (NextGen and Single European Sky) transport systems. The ADS-B allows ATC to control aircraft with greater accuracy and over a much larger portion of the earth's surface than ever before. For NextGen and SESAR, ADS-B technology is one of the most important in terms of converting ATC from the use of radar surveillance to the satellite global positioning system GPS. ADS-B uses a combination of satellites, transmitters and receivers to

DOI: 10.4018/978-1-5225-8214-4.ch006

provide flight crews and ground control personnel with specific information on the location and speed of aircraft in the area. Navigation satellites give the exact time. Unlike conventional radar, ADS-B operates at low altitudes and on land, can be used to monitor traffic on taxiways and airport runways. ADS-B is also effective in remote areas where there is no radar coverage or where radar coverage is limited.

Advantages of ADS-B are as follows (Richards, O'Brien, & Miller, 2010):

- ADS-B makes it possible to maintain or improve existing safety standards while increasing the efficiency and power of the system.
- ADS-B significantly improves situational awareness of the crews, since
- they know where the other planes are.
- ADS-B provides an overall picture of real-time monitoring and the ability to exchange information quickly if necessary, if other aircraft deviate from their prescribed flight paths.
- ADS-B offers more accurate and more frequently updated generalized information about traffic. All participants in the flights have a common operational picture.
- ADS-B provides more accurate and timely information than radiolocation; provides more frequent updates than radar.
- ADS-B significantly improves the detection of possible conflicts and their resolution compared to any other system.
- ADS-B can provide a significant increase in the number of flights that can be serviced by the ATC. More aircraft can simultaneously occupy this airspace while changing the separation standards.
- ADS-B significantly reduces the separation standards while maintaining safety standards, improves the accuracy and integrity of reports.
- Increases runway capacity with improved arrival accuracy to the metering fix.
- Helps maintain runway approaches using a cockpit display of traffic information in marginal visual weather conditions.
- Enhances visibility of all airplanes in the area to allow more airplanes to use the same runway.
- Allows 5 nmi of separation in non-radar airspace compared to current procedural separation, from 5 to 3 nmi in radar airspace.
- ADS-B improves flight efficiency, and also increases throughput.
- This helps the ATC understand the actual separation between airplanes and allows controllers to avoid inefficient vectoring. This helps aircraft

to move to optimum operational altitudes in remote areas, which reduces impact on the environment.

- Aircraft can fly closer to each other, because controllers have more accurate data that is updated more often.
- The amount of fuel consumed decreases, because planes fly more effective way.
- Existing digital communications allow ADS-B to be quickly deployed at relatively low cost.
- Effective surveillance of all air and land transport is available, even at the airport on taxiways, airstrips and in airspace where the radar is inefficient or inaccessible.
- General aviation aircraft can use ADS-B datalinks to obtain flight information.
- Airlines can reduce the cost of a passenger-kilometer by flying on more direct routes at more efficient altitudes and speeds without continuous rises and descents.
- Engine emissions and airplane noise are reduced.

With the continuous aviation development amount of flights through the Poles and regions with reduced radar coverage grows annually due to its economic efficiency. So aviation met the challenge that consists mainly in conversion from ground-based CNS systems that are not available at any point of the globe and expensive at service to satellite-based systems that are cheap at service and cover 100% of the Earth. This conversion will be beneficial for either airspace users due to decreased air navigation service cost and air navigation service providers due to low cost of equipment maintenance and constant awareness of air traffic situation in any point over the globe.

From that point of view the communication channel consisting of on-board ADS-B equipment and Iridium satellite could help to resolve the bottlenecks by allowing the relocation of aviation traffic from the regions with the high density to the regions that can not be used nowadays due to its remote situation or inability of ground surveillance equipment installation.

ADS-B System

ADS-B is defined by ICAO as: "ADS-B is a surveillance application transmitting parameters, such as position, track and ground speed, via a broadcast mode data link, and at specified intervals, for utilisation by any air and/or ground users requiring it" (ICAO. Doc 9694-AN/955, 1999).

ADS-B acronym means:

- **Automatic:** Periodically transmits information without pilot or operator interference;
- **Dependent:** Position and speed data are received from Flight Management System (FMS) or Global Positioning System (GPS);
- **Surveillance:** Defines the location of aircraft, vehicles and obstacles;
- **Broadcast:** Transmitted data are available for anyone, who has the appropriate receiving equipment.

ADS-B relies on the regular and frequent transmission of position reports via a broadcast data link. The position reports are sent periodically by the aircraft with no intervention from the ground function. Position reports may be received by any recipient in range of the transmitting aircraft. Recipients may be communications receivers ('data acquisition units') on other aircraft, ground vehicles or at fixed ground or satellite sites. If received by a data acquisition unit, the position report will be processed with other surveillance data and may be forwarded to a controller/pilot display (EUROCONTROL, 2004).

Three technologies for ADS-B data transmission are available:

1. 1090ES - 1090 MHz Extended Squitter (international standard);
2. VDL Mode 4 - Very high frequency Data Link Mode 4 (standard for 'Southern Ring´ and Northern Europe);
3. Universal Access Transceiver - UAT (standard for USA).

ADS-B system provides two functions: *ADS-B OUT* (Figure 1) and *ADS-B IN* (Figure 2).

ADS-B Out periodically broadcasts information about aircraft through an onboard transmitter. The information contains data about identification, altitude, current position and velocity. ADS-B Out provides air traffic controllers with real-time position information that is in times more precise than the information obtained from current radar-based systems. Obtaining more accurate information, air traffic controller is able to control and separate aircraft with improved timing and precision.

ADS-B In function refers to an ability of an aircraft that is equipped by this function to receive and display ADS-B Out information of another aircraft and ADS-B In services provided by ground systems, such as TIS-B (Traffic Information Service-Broadcast), FIS-B (Flight Information Service-

Figure 1. ADS-B OUT

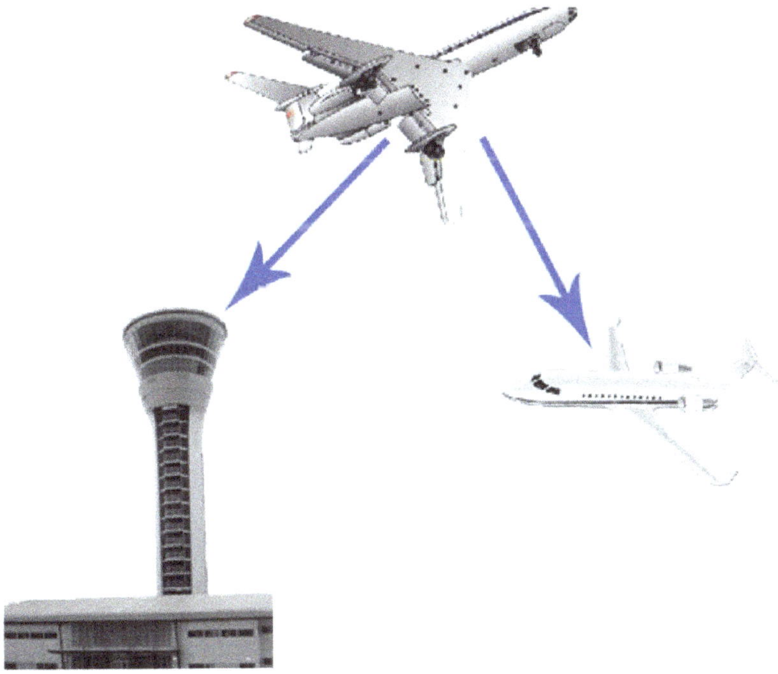

Figure 2. ADS-B IN

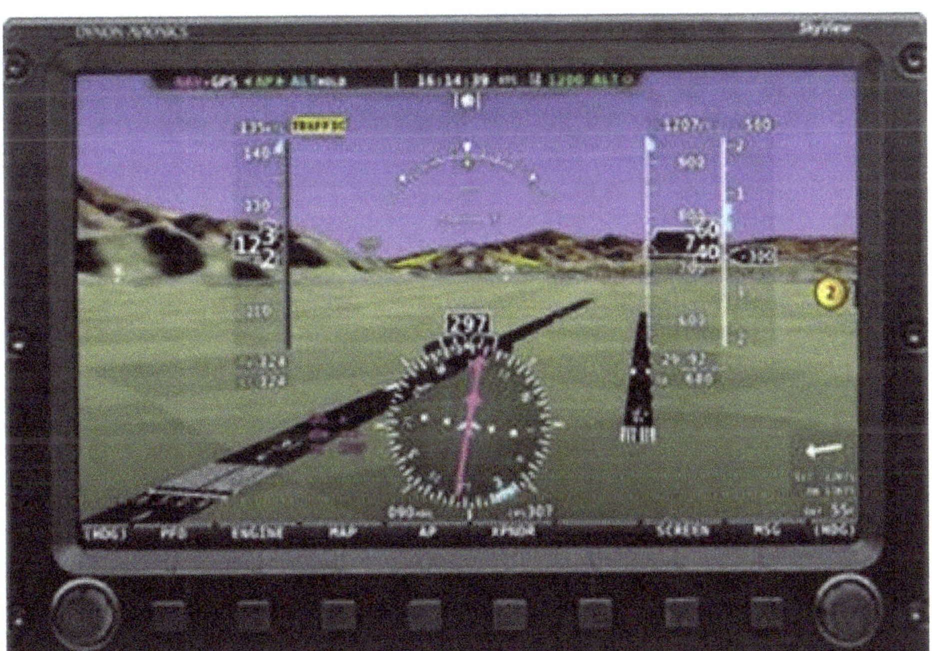

Broadcast) and ADS-R (Automatic Dependent Surveillance-Rebroadcast). Such displaying of the information in the cockpit greatly improves situational awareness of the pilots in comparison to an aircraft not equipped with TCAS (Traffic Alert and Collision Avoidance System) and ACAS (Airborne Collision Avoidance System).

ADS-B Message

ADS-B message is an information packet broadcasted by an aircraft which is equipped with ADS-B system. ADS-B message includes a predefined set of aircraft surveillance parameters. Multiple messages can be sent to transmit all required data, such multiple messages are merged by receiving device into one complete message. Linked to the ADS-B message are two functions *ADS-B Out* and *ADS-B In*.

Via ADS-B messages the following surveillance parameters can be delivered:

1. ICAO 24-bit address;
2. Aircraft identification;
3. Code of Mode-A;
4. Emergency status;
5. SPI -Special Position Indication;
6. Emitter category of ADS-B;
7. Geodetic horizontal position (World Geodetic System (WGS84) latitude and longitude), both for airborne aircraft and those, which are on the ground;
8. Version number of ADS-B;
9. Pressure altitude;
10. Quality indicator of geodetic horizontal position;
11. Geometric altitude by WGS84;
12. GVA -Geometric Vertical Accuracy;
13. Quality indicator of velocity;
14. Velocity over ground, both for airborne aircraft and on those, which are on the ground;
15. Antenna offset of PS);
16. Aircraft width and length that are coded;
17. Selected Altitude of Mode Control Panel / Flight Control Unit (MCP/FCU) equipment;

18. Vertical rate: barometric vertical rate when the aircraft is required and capable to transmit this data item via the Mode S protocol, or Global Navigation Satellite System (GNSS) vertical rate;
19. In case when TCAS II is installed on the board - Active resolution advisories of Airborne Collision Avoidance System (ACAS);
20. Settings of barometric pressure.

Message Format

The Extended Squitter is 112 bits long and contains 56 bits of ADS information. The added message information is a 56 bit field inserted between the 24 bit Aircraft address and the Parity information.

An Extended Squitter has the following format (Figure 3):

1. **Preamble:** this is special bit sequence to allow the receiver to identify and synchronise with a received message. Total radiated pulses – 2 μs;
2. Data Block: 112 bits encoded of 112 μs containing the ADS-B position reports. Total radiated pulses - 56μs;
3. **Downlink Format (DF Field):** This is used to indicate the type of message being transmitted. It is set to 17 for Extended Squitter messages;

Figure 3. 1090 MHz ADS-B Message Format

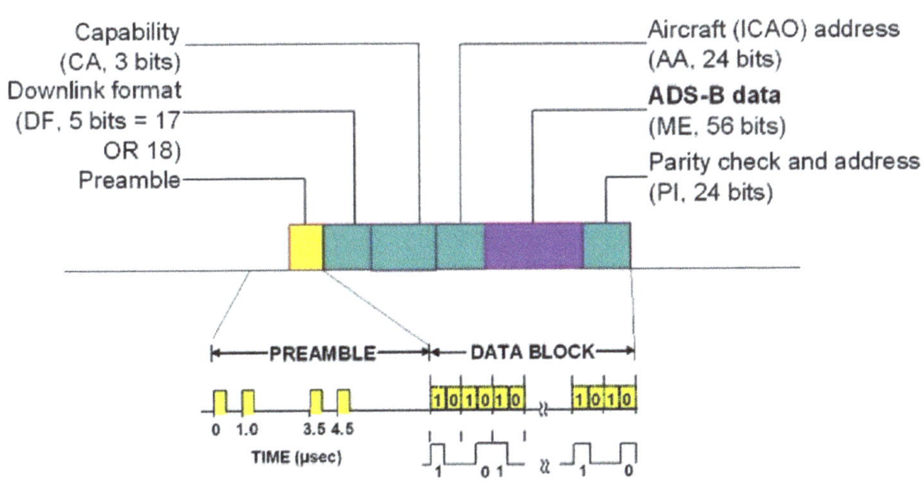

Total duration = 120μs (8μs preamble and 112 μs data block)

Data block transmitted using Pulse Position Modulation

4. **Capability (CA Field):** This 3 bit field indicates the capability of the Mode S transponder;
5. **Aircraft (ICAO) Address (AA - Address Announced - Field):** This is the 24 bit address of the airframe;
6. **ADS-B Data (ME Field):** This is 56 bits long and contains the of ADS-B data. Its contents depend on which Extended Squitter is being transmitted;
7. **Parity Check (PI Field):** This 24 bit field is an error detection code to help a receiver determine errors in the received message.

The data is transmitted using a modulation scheme called Pulse Position Modulation (PPM). This is a relatively simple modulation scheme for a 1090 MHz receiver to decode in the presence of non-overlapping (in time) replies.

Structure of ADS-B System on a Base of Satellite Communication System Iridium

Based on the analysis of existing technologies and literature, a generalized structure of the ADS-B system was created based on the Iridium satellite system (Figures 4-7).

Figure 4. Scheme for determining the coordinates by airplanes with the help of a satellite system

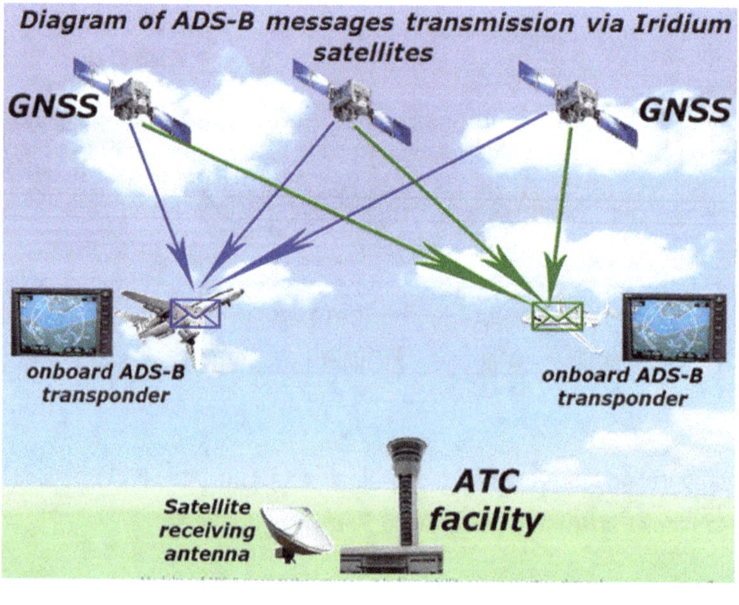

Figure 5. Scheme of the message transmission by the right aircraft with its coordinates through satellites to another aircraft and to the Earth

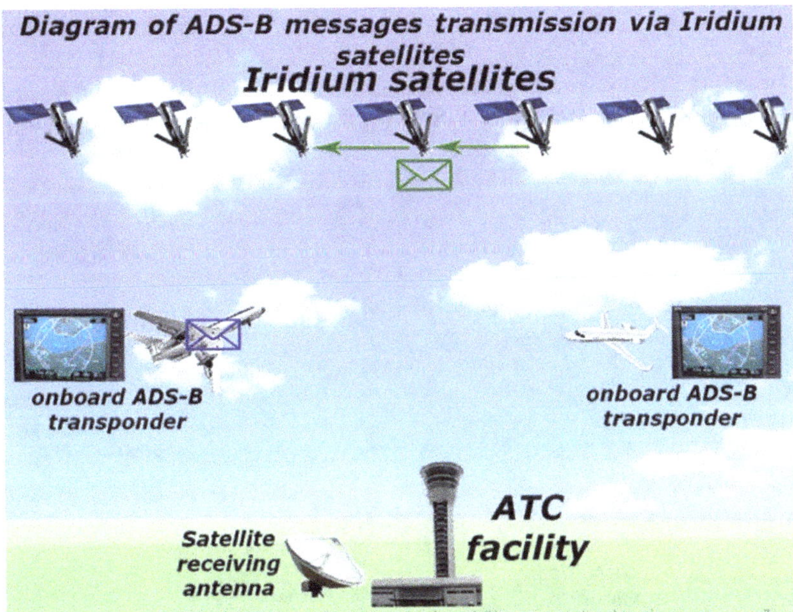

Figure 6. Scheme of the message transmission by the left aircraft through the satellites with its coordinates to another aircraft and to the Earth

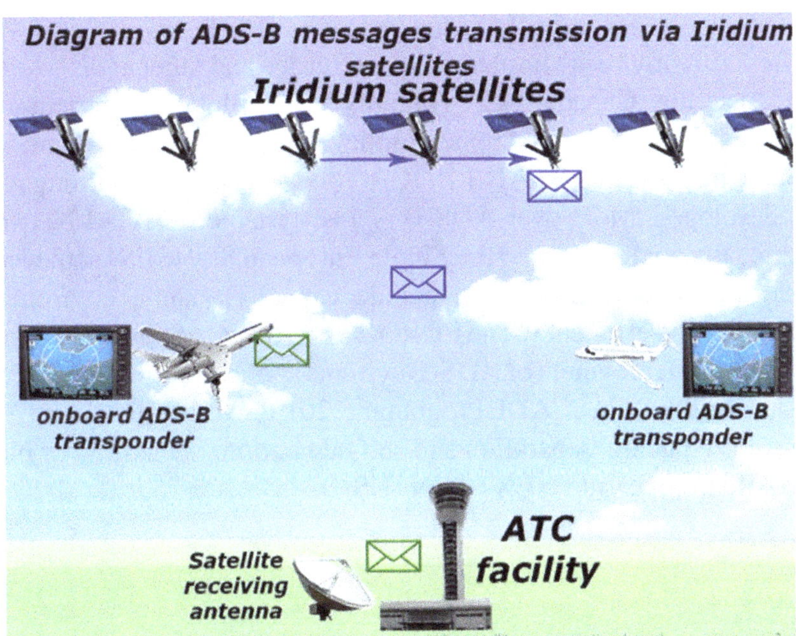

Figure 7. Scheme for exchanging of information about the location between planes and ground services via satellites

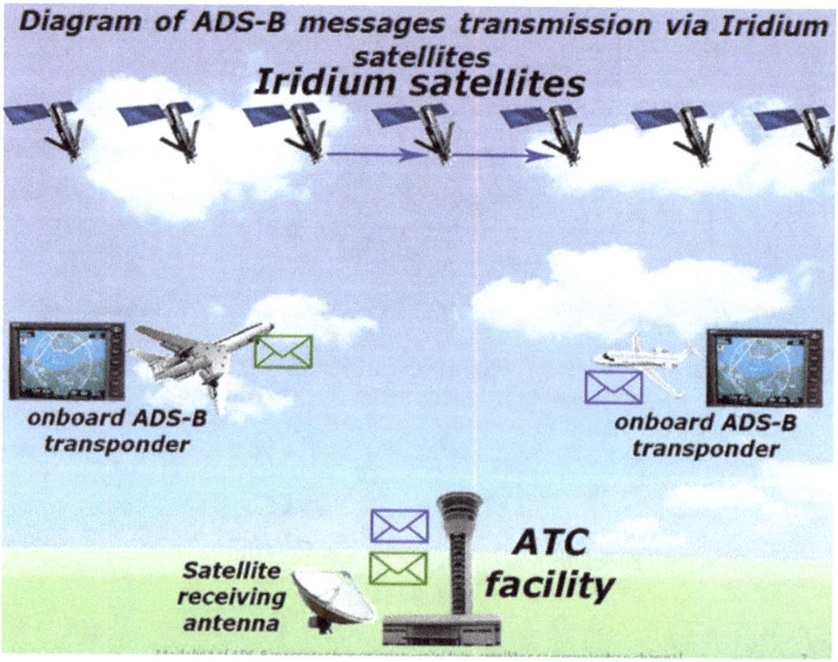

ADS-B Deployment Worldwide

ADS-B is currently being implemented in Europe and other areas worldwide - Asia, Australia, Canada, USA (Figures 8, 9). Global interoperability is ensured at system level and application level. Both EUROCAE and RTCA developed the standards for ADS-B. ICAO documentation regarding ADS-B is also developed. More over ADS-B is a key enabler of ATM Network of the future, as a contribution to the Single European Sky (SES) achievements in performance objectives that include capacity, safety, environmental sustainability and efficiency. The EUROCONTROL CASCADE Programme co-ordinate the deployment of ADS-B systems application all over the Europe (EUROCONTROL CASCADE Programme, 2013). Work on the future ADS-B applications (spacing, separation and self-separation) is ongoing or planned by SESAR (Europe) and NEXTGEN (USA).

Figure 8. FlightAware ADS-B coverage map over the Europe (https://ru.flightaware.com/adsb/coverage#data-coverage). The actual coverage depends on many factors (antenna/receiver quality, equipment location, obstructions, etc.). This map currently only reflects space-based ADS-B from high-altitude (e.g., > 30000 ft) flights and was generated with only a partial constellation. Full, global coverage will be available in late 2018

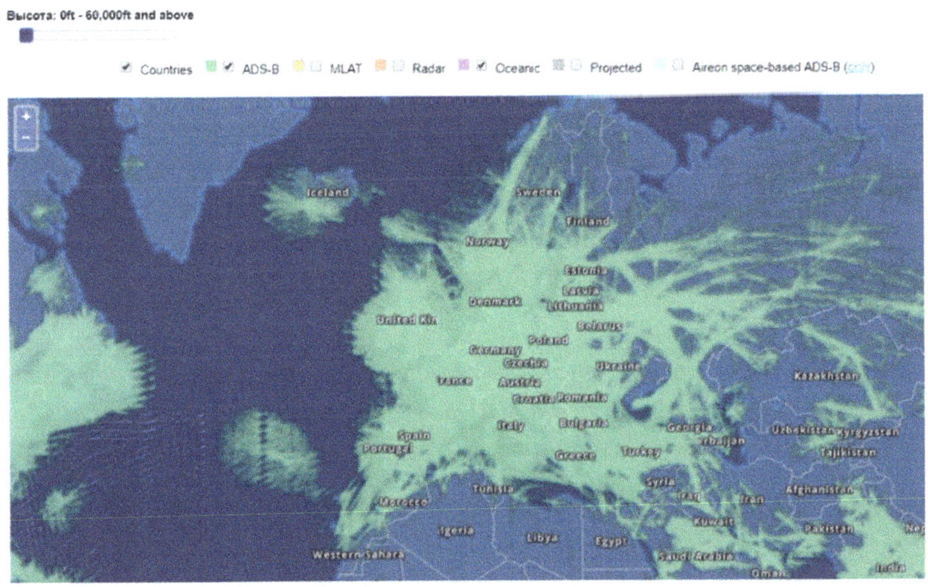

MONITORING OF AIR TRAFFIC USING ADS-B SYSTEM

Equipped with ADS-B transceiver aircraft during all flight transfers in real time their exact position, speed, altitude, course and other information. Access to ADS-B information is free for all. ADS-B signal can be received on earth for the purpose of surveillance (ADS-B-OUT) or other aircraft for information about the surrounding traffic (ADS-B-IN) and avoid collisions (Mode S and ACAS Program, 2008). ADS-B-OUT system began operating in 2008, ADS-B-IN - in 2011. System ADS-B-OUT can be used for purposes of self-observation, as well as with radar systems and MLAT (multilateration). For transmition of ADS-B messages transponder Mode S Extended Squitter is used (Guidance for the Operational Introduction of SSR Mode, 2012). On-board transponders transmit data at a frequency of 1090 MHz. To receive these messages we must have the receiver at that frequency - ADS-B-receiver. The frequency of 1090 MHz belongs to the L-band. Electromagnetic waves with such high frequencies are distributed in the atmosphere straight without being reflected from the ionosphere. Receiving of these signals can only be possible if between aircraft and the receiving antenna is no obstructions. Receiving antenna must be installed as high as possible and have free space in all directions. The greater is the height of the aircraft, the greater is distance that signal may be accepted. For example, the signal from the aircraft, which is at an altitude of 30000 feet (10 km), may be received up to 350 km. This requires a sensitive antenna and receiver. However, if we consider the distance to the aircraft within 100 miles we can receive ADS-B signals even using simple equipment. There are exist some schemes for simple ADS-B systems (Manual for the adsbScope Software V1.9 & adsbPIC Decoder V2 Fw6, 2010).

The aim of this study is to build a simple system for receiving ADS-B signals and to use it for monitoring of air traffic, which was described in paper (Kharchenko, Barabanov, Grekhov, Ghaznyuk, & Kolchev, 2012).

Figure 9. ADS-B coverage map over the USA (https://www.faa.gov/nextgen/programs/adsb/ICM/). This tool identifies where ADS-B services are generally available. Not all services are available at all altitudes

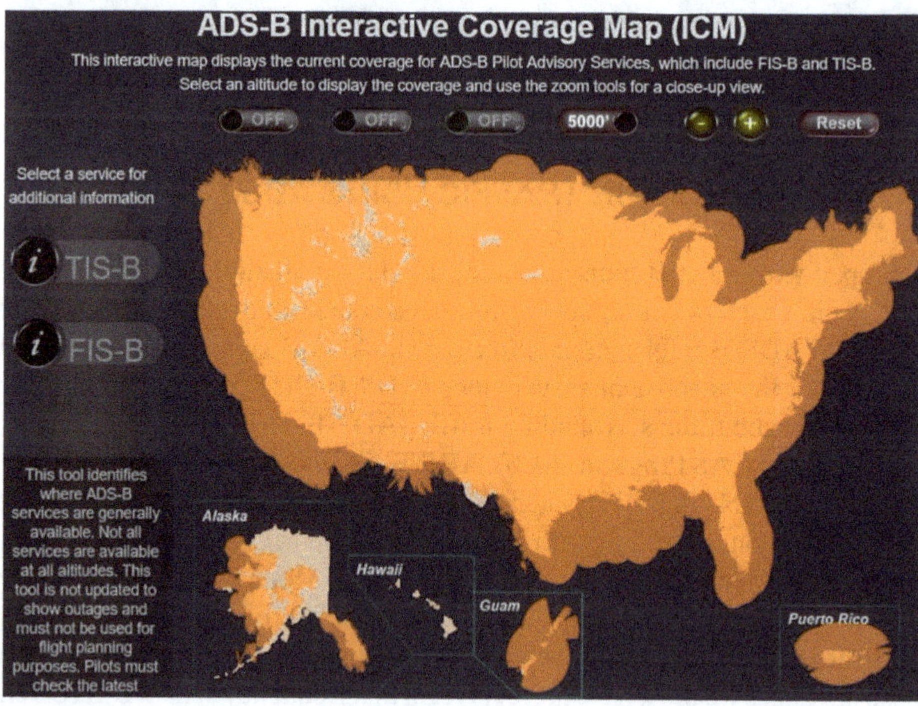

Structural Scheme of ADS-B System

A system that generates ADS-B signals can be used as virtual radar to create real-time picture of air traffic and consists of four components (Manual for the adsbScope Software V1.9 & adsbPIC Decoder V2 Fw6, 2010): an antenna, receiver decoder and computer software (Figure 10).

The antenna receives the signal at a frequency of 1090 MHz and converts it into high-frequency electrical signal. The receiver selects, amplifies, demodulates received signals and generates an analog video signal. Decoder converts analog video to digital and detects ADS-B message. ADS-B signals are then transmitted via the USB port on your computer. Using computer software, we can decode ADS-B information and generate virtual display radar. In this system miniadsb receiver and adsbPi decoder software adsbScope (Receiver and Decoder - miniADSB 1090 MHz Receiver, 2010) are implemented.

Antenna

Transponder signals are polarized in the vertical plane and they need to be received by an antenna with vertical polarization at a frequency of 1090 MHz. The simplest antenna is a vertical wire with length of 0,13 m, which corresponds to half the wavelength of the frequency of 1090 MHz. This is an electric dipole antenna, which can receive the signal from all directions. To increase the sensitivity of the antenna several electric dipoles need to be applied. Nevertheless, if they are placed beside one another then the antenna will cease to be non-directional. Therefore, individual dipoles placed one above the other. In addition, the dipoles must be connected. The top and bottom edges of the dipole oscillate in phase shift of 180 degrees. We have connected dipoles by means of horizontal coils of the length λ / 2 (0.13 m), which serve as devices for the phase shift of 180 degrees (Figure 11).

Figure 10. The system for receiving ADS-B signals

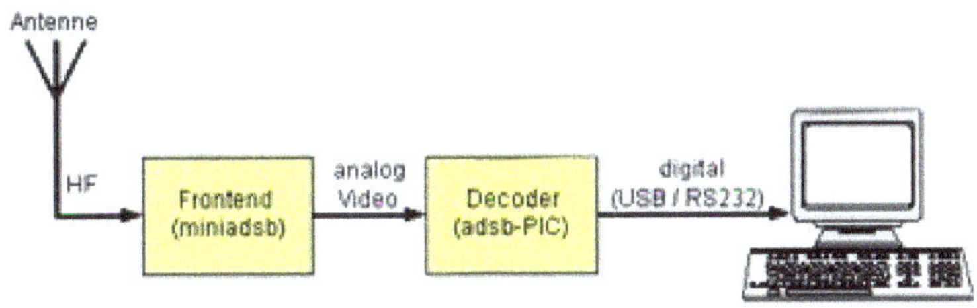

Figure 11. Antenna for receiving ADS-B signals

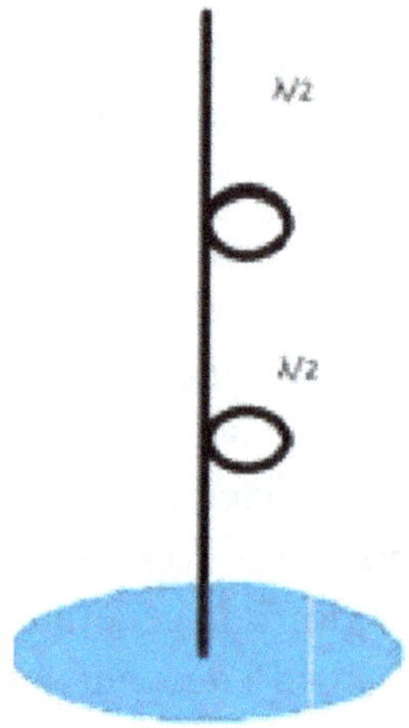

Dipoles and coils by itself must be made from one long piece of wire. Wire diameter should provide the necessary rigidity of the antenna. The bottom end of this set of dipoles must be connected to the middle wire of 50-Ohm coaxial cable. For grounding can be used round metal plate with a diameter of $\lambda/2$ (0,13 m) located radially or pieces of wire (at least 4 wires at 90 degrees to each other). Antenna impedance is not consistent with the cable impedance but it can be changed by bending the wires down.

Figure 12. Schematic diagram of the ADS-B receiver

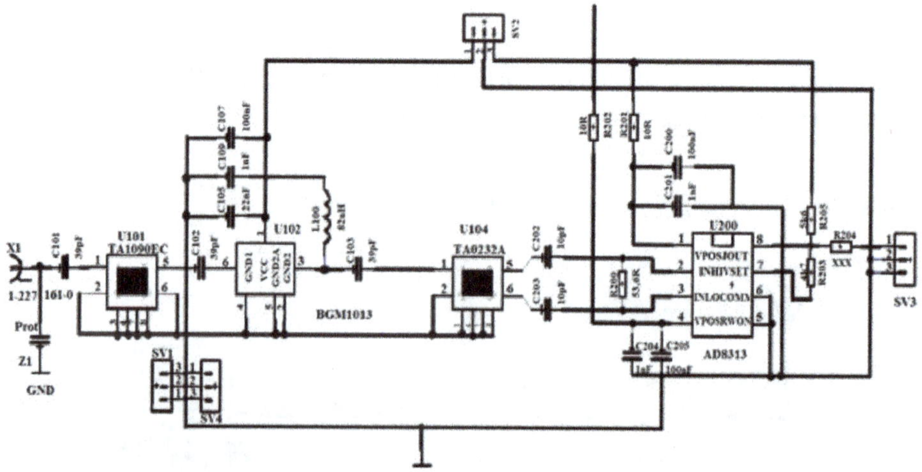

ADS-B Receiver

ADS-B receiver was assembled using the scheme shown in Figure 12. Antenna is connected via 50-Ohm coaxial cable.

ADS-B Decoder

Decoder scheme is given in Figure 13. The main element is the decoder microcontroller PIC18F2550. It converts analog video to digital using the internal comparator. These ADS-B messages are transferred via the USB port at a PC. The decoder is designed for use with miniadsb receiver and software adsbScope or Planeplotter (PlanePlotter 6.4.5.1. 2016).

adsbScope Software

Software adsbScope contains outlines of continents, list of cities, airports list, and a list of aircraft with ICAO numbers of these aircraft information to determine the nationality of aircraft, the location and length of runways, information about air routes, the location and areas of ILS systems, data about ground radar location. When you run the software AdsbScope opens the program, tests subdirectories and loads data files. The program window contains a text field for the decoded data, table for the detected aircraft, decoder control panel and information field (Figure 14). The data with revealed aircraft are shown in the table between the two text boxes. Figure on the left

Figure 13. Scheme of a decoder adsbPIC

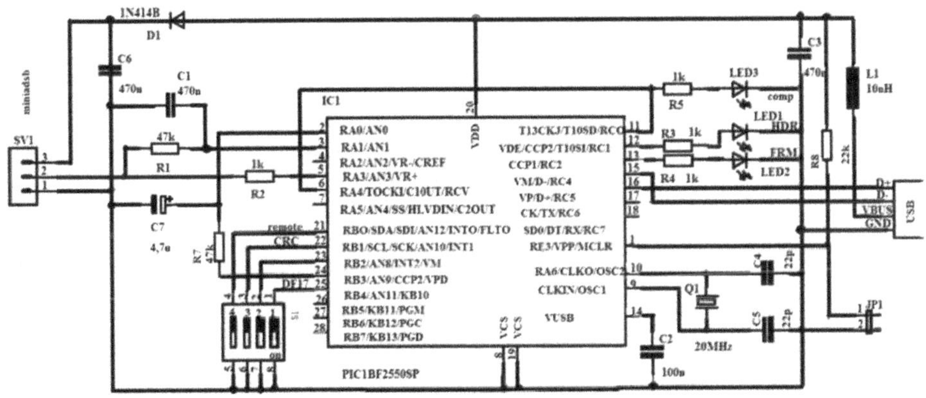

Figure 14. Main window of adsbScope

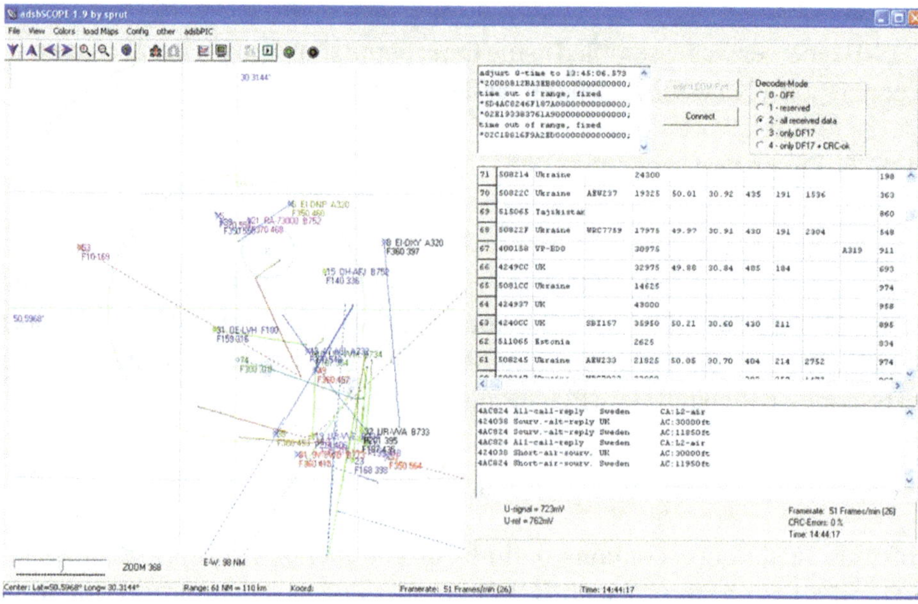

side of window shows the position, track and additional information about the aircraft. The software counts the number of ADS-B frames (packets) per minute and displays it in the bottom right corner. In addition, the average number of frames received from an aircraft per a minute is displayed. Besides the software checks the CRC-checksum for each frame.

Results of Air Traffic Monitoring

The graphic display shows a part of the globe, the size and location of which can be changed from a site of 2 NM (sea miles) to the mapping of the entire globe with the mouse. As the scale increases, the names of the airports, the ILS system and the radar may be displayed on the display.

The user can choose which information to display and which colors are used for different types of objects. The display is updated at least once per second (usually 4 times per second).

The detected aircraft with its location is displayed on the display. The software creates lines (flight paths), from the place where the aircraft was first detected. The first location of the aircraft is depicted as a small circle. Next is a track number, aircraft ID, flight altitude (echelon) and speed at the nodes. If the aircraft is unknown (not on the list of aircraft with ICAO numbers of these aircraft), then only the track number is indicated. The track number corresponds to the first column of the table in Figure 14.

If there is no information about the aircraft at this time, then its position is calculated based on the last known location, speed and direction of travel. The flight path from the last known position in the predicted trajectory is depicted by a dotted line. User can ban position prediction.

The software selects a random color for each aircraft. The user can select a mode in which the color of the flight trajectory represents the altitude of the aircraft. In this mode, 0 ft. Appears in red, while 20000 ft. In green and 40000 ft in blue.

The antenna for receiving ADS-B signals was mounted on the roof of National Aviation University Building number 11 and can monitor aircraft at a distance of 160 km and heights up to 12000 m equipped with the appropriate transceiver. The number of aircraft on Flightradar24 site display coincides with the number of aircraft that "sees" our system.

FlightAware and ADS-B

FlightAware (FlightAware, 2018) receives data from more than 45 air traffic control services and private data channels from different countries, as well as through the global network of ADS-B and Mode S receivers, which receive data from aircraft transmitting ADS-B or Mode S. Aircraft transmitting ADS-B signals are able to track their exact location, and the aircraft transmitting Mode S signals can be tracked using MLAT when the signal is received by three or more receivers. FlightAware owns such receivers at many airports around the world and uses them in conjunction with airport operators.

FlightFeeder Receiver

FlightFeeder is a network ADS-B receiver. In "FlightAware Major Events and New Features" (FlightFeeder, 2014) was mentioned that "FlightFeeder - FlightAware ADS-B receivers begin shipping to over a dozen countries" in January 2014 (information about our system was published in 2012). It receives ADS-B data directly from airplane transponders via a small antenna. FlightFeeder then makes the real-time data available on the local network to users and sends it to FlightAware over the Internet. FlightFeeder only requires an external antenna (included), Internet connection over Ethernet, and power (via a USB cable into a wall adapter).

FlightFeeder is compatible with virtually all existing ADS-B software packages like PlanePlotter and BaseStation. Users can connect to their FlightFeeder over their LAN to receive all data. Users can also choose to view flight data on the FlightAware.com web site.

FlightFeeder is roughly the size of a can of soda and uses less power than a night light. There are no moving parts or temperature restrictions.

FlightFeeder Features

FlightFeeder is self-configuring.

FlightFeeder is remotely managed and upgraded automatically by FlightAware.

Real-time ADS-B data feeding to FlightAware.com.

Real-time Mode S MLAT participation with nearby FlightAware receivers.

Automatic FlightAware Software Upgrades.

Easy Web Interface.

Real-time Flight Tracking Map.

ADS-B Coverage Statistics on FlightAware.com.

ADS-B/Mode-S Data Output for Optional Application Integration and Data Feeds.

TCP port 30002 for raw/unparsed messages in AVR format.

TCP port 30003 for parsed messages in BaseStation format.

TCP port 30005 for raw/unparsed messages in Beast binary format.

TCP port 30105 for multilateration results (only) in Beast binary format (for FlightFeeders, the device must be running 7.x or newer software).

TCP port 30106 for multilateration results (only) in extended BaseStation format (for FlightFeeders, the device must be running 7.x or newer software).

HDMI TV/Display Interface.

View Network Configuration and Real-Time Statistics.

Reliable and always-on (24/7) Internet connection (broadband not required).

Available uplink bandwidth of <10Kbps for ADS-B or <50Kbps for ADS-B+MLAT.

Standard RJ45 Ethernet connection.

WiFi (hardware version 8 or newer; or hardware version 7 with user-provided USB WiFi adapter).

Standard router/modem with DHCP auto-configure network setup.

NAT compatible.

Reliable electrical power (battery back-up preferred but not required).

Standard residential/office 108-240V (usage: 1A @ 5V).

Minimum 200 joules surge protector, 800 joules if the area has known power issues.

ADS-B Antenna Site:

Antenna should have a clear line of sight to the sky with no obstructions.

Antenna must be located within 15m/50ft from FlightFeeder (shorter is preferable).

FlightFeeder Site

Must be located within 15m/50ft from the antenna.

Must be located within 100m/330ft from the Internet connection.

Must be installed indoors (hangar or garage is acceptable).

Must be installed in a dry environment and not be subject to rain or moisture.

FlightFeeder Kit Contents

FlightFeeder device.

External Antenna (1 ft./300 mm antenna, total size 2 ft./600 mm with mounting bracket and antenna cable).

Antenna cable, low-loss with SMA connectors (length choice of 5m, 10m or 15m).

Ethernet (length of 10 ft./3 m).

USB power cable (with local plug or adapter).

1090 MHz bandpass filter (provided with FlightFeeder Orange hardware only).

FlightFeeder Specifications

Weight: 225g/0,5lbs

Dimensions: 115mm x 77mm x 64mm / 4.5” x 3” x 2.5”

Antenna Port: SMA male (requires SMA N-female antenna cable)

Power: 1A via microUSB

Range: Over 250nm depending on installation quality

Performance: Over 1500 Mode S frames per second

Computer: Does not require a computer

Operating Temperature: 0C to 40C

Storage Temperature: -20C to 60C

Obtaining a FlightFeeder

MODELING OF RPAS COLLISION AVOIDANCE SYSTEM OPERATION USING MATLAB SOFTWARE

Airborne Collision Avoidance System

In January 2005 the requirement of ICAO concerning installation of Airborne Collision Avoidance System ACAS-II came into force for all civil fixed-wing aircraft with the maximum certificated take-off mass exceeding 5700 kg or authorized to carry more than 19 passengers (ICAO, 2007; PANS-OPS, 2006; PANS-ATM, 2007; ACAS, 2017). As a result, the number of aircraft equipped with ACAS-II significantly increased.

The problem of aircraft collisions in the air became palpable in the 50's. When there began to be fatalities with massive loss of life, the ICAO took up this issue. The concept was developed and then the international standards of the Airborne Collision Avoidance System (ACAS) was accepted. Of all the developments under this concept, the main distribution was the TCAS (Traffic alert and Collision Avoidance System) system. This system (in its latest versions and modifications) surveys the airspace around the aircraft, detects other aircraft, analyzes the information received, issues it to the crew, and in case of collision danger, warns pilots about this and issues the necessary recommendations for immediate action. Now the latest version of the system is TCAS II. Only TCAS II system is fully compliant with ACAS standards now. It is currently installed on most commercial aircraft and is produced by its firm Rockwell Collins, Honeywell and ACSS.

The TCAS equipment set includes: a computer unit that calculates the development options and determines the issued commands, two transceiver antennas installed at the top and bottom of the fuselage (one directed (from above), the other omnidirectional), individual antennas for S-transponders Next) and the display indicator in the cab. This display-indicator provides information on the movement of close-up aircraft, as well as visual commands to prevent collisions.

WHAT IS ACAS

ACAS is also known as TCAS. It is a transponder based interrogation system capable of displaying conflicting traffic and providing resolution advice. ACAS provides an extra level of protection against mid-air collision.

ADVISORIES

Traffic Advisory (TA): Help the pilots in the visual search for the intruder aircraft, and alert them to be ready for a potential resolution advisory.

Resolution Advisory (RA): Avoidance manoeuvres recommended to the pilot; when the intruder aircraft is Mode S equipped both systems will co-ordinate their RA's through datalink.

HOW IT WORKS

ACAS consists of aircraft mounted antennas, computer processor, flight deck display and controls, and Mode S transponder. Altimeter and radar altimeter inputs are used to determine relative altitude and inhibit RA's close to the ground.

ACAS is capable of displaying targets for other transponder equipped aircraft. The ACAS set transmits an interrogation signal, which other aircraft transponders reply to. The ACAS computer determines rate of distance change. TA's and RA's are based on this closure rate.

Targets are displayed on the ACAS screen as different symbols, which depict level of threat.

The visual cues will be accompanied by an aural warning. Example: TA "Traffic, Traffic" or RA "Climb, Climb". Relative altitude will be displayed in conjunction with a trend arrow. Relative altitude is the targets altitude in relation to the aircraft.

Example: 03↑ depicts a target 300 feet below and climbing at 600 fpm or more. If the target is climbing or descending at less than 600 fpm the arrow will not appear.

ACAS is incapable of displaying aircraft without transponders.

PROTECTION THRESHOLDS are shown in Figure 15.

TYPES

There are three main types of ACAS in different stages of use or development:

ACAS I: will provide a warning (TA traffic advisory) of proximate traffic without guidance to avoid potential collisions. A traffic advisory is given anytime a target enters the TA range and altitude limits of the ACAS. (The

Figure 15. Protected zones

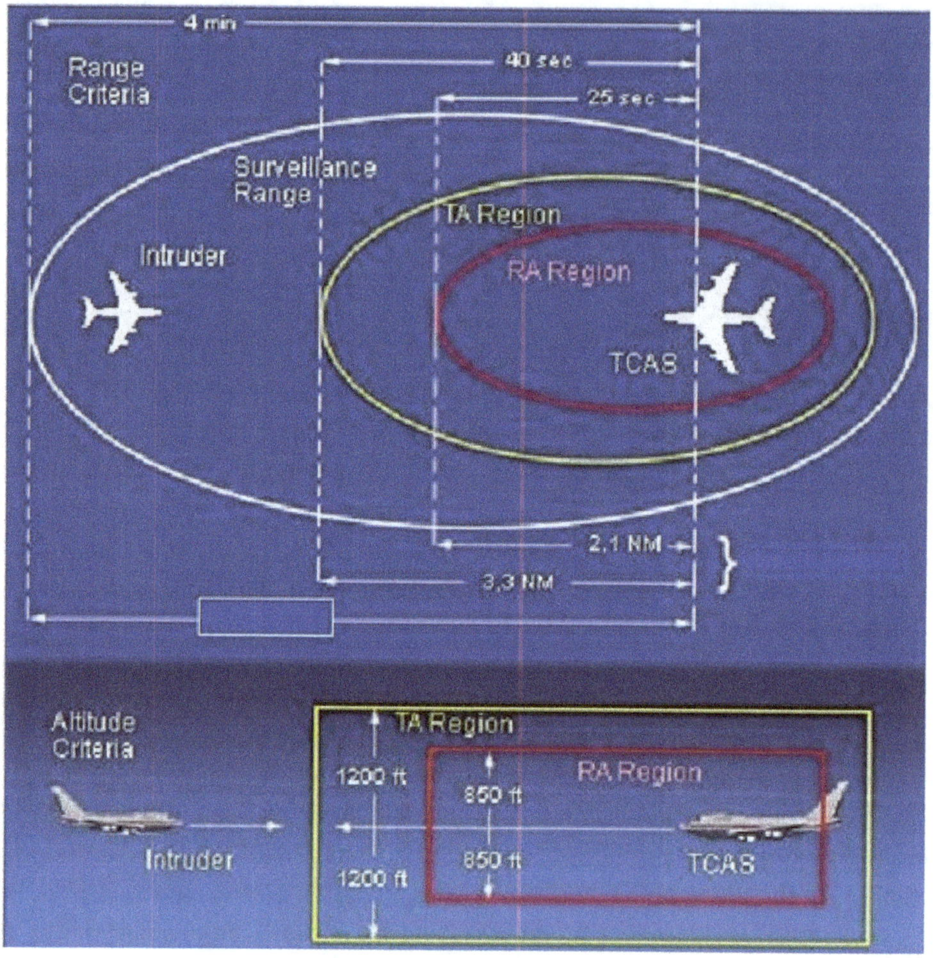

TA will be an aural "Traffic, Traffic). Traffic with Mode A transponders will be depicted with position information only. Traffic with Mode C transponders will be depicted with position and relative altitude information. Traffic with Mode S transponders will be depicted with position and relative altitude information.

ACAS II: Will provide warning (TA traffic advisory) and vertical plane guidance (RA resolution advisories) to avoid potential collisions. ACAS 11 is capable of providing resolution advisories anytime an aircraft enters the RA range and altitude limits of the ACAS. The RA will be a combination of aural and visual cues. Traffic with Mode A transponders do not allow the

ACAS to compute an RA. Traffic with Mode C transponders allow the ACAS to determine and display a RA. Traffic with Mode S transponders allow the two aircraft to communicate via datalink and provide complimentary RA's.

Example: one aircraft climbs while the other descends.

ACAS lll: The next generation of ACAS still under development will provide TA's and RA's which include vertical plane and horizontal plane guidance. ACAS lll will have the added capability of providing lateral and vertical RA's. No equipment currently exists due to technical and operational difficulties. Safe and accurate RA's in the vertical and horizontal plane are difficult to program especially when dealing with multiple targets.

ACAS SYMBOLOGY

Hollow blue or white diamond - non-intruding traffic.

Solid blue or white diamond - intruding traffic.

Solid amber circle - intruder that triggers a TA.

Solid red square - threats which trigger a RA.

RESOLUTION ADVISORY

The RA's are displayed on an electronic Instantaneous Vertical Speed Indicator.

Red arcs depict rate of climb/descent that must be avoided.

Green arcs depict required climb/descent rates to resolve conflict.

The visual resolution will be accompanied by an aural message.

"Descend, descend"

"Maintain vertical speed, maintain"

"Clear of conflict"

PILOT/CONTROLLER ACTIONS

Pilots shall not manoeuvre their aircraft in response to TA's only.

In the event of an RA to alter flight path the alteration should be the minimum required and ATC must be notified as soon as possible.

Pilots who deviate from an ATC instruction or clearance in response to an RA shall promptly return to the terms of that instruction or clearance when the conflict is resolved.

What Is TCAS?

TCAS: Traffic alert and Collision Avoidance System. TCAS development is an American initiative, preliminary studies in 1955. TCAS is the only equipment complying with the ICAO ACAS standards:

TCAS I is an ACAS I
TCAS II complies with ACAS II standards
The TCAS III/IV projects were not pursued.
Currently two TCAS II versions in service:

- TCAS II v6.04a introduced in 1994
- • TCAS II v7.0: introduced in 1999/2000, fully complying with ICAO ACAS II standards.

TCAS II PRINCIPLES

- An airborne equipment that interrogates adjacent SSR transponders.
- Collision avoidance criterion based on time.
- **Independent From:** All external systems (e.g. ground radars, satellites), ATC, FMS and autopilot, on board NAV systems (GPS, INS).
- Designed for collision avoidance only.

TCAS II system is an autonomous onboard collision avoidance system. TCAS II system should be switched on in areas of airspace in which the use of TCAS is required. In activated state TCAS system provides additional resource for collision avoidance with another aircraft, which have switched on transponders with Mode S. This defense is not ensured when transponders do not meet ICAO standards. The system is designed to provide safe separation of aircraft, trajectories of which indicate the probability of a collision, while minimizing the deviation from the proposed air traffic control flight course.

TCAS II system realizes continuous surveillance of airspace around its own aircraft transmitting active interrogations and receiving responses from transponders of the other aircraft in nearest vicinity.

TCAS II system recognizes responses from the aircraft transponders with ICAO Mode A, C, S and determines the range to the other aircraft, their relative bearing and relative altitude if the height indication is functioning. Using this information, the system predicts the trajectory, estimates distance in areas of nearest approaching and determines if there is a potential threat of a collision. In case of such threat the system issues instructions for performance of the optimal maneuver in the vertical plane to avoid the collision.

TCAS system receives also input signals from systems of its own aircraft and has information about their state. This information is taken into account during analysis of conflicts. TCAS system creates around its own aircraft defended airspace, which is called "collision zone".

Because TCAS system is designed in accordance with the principle of a calculation the time before a possible collision, than sizes and a shape of such zone is changing in accordance with approaching velocity of conflicting aircraft and its relative bearing. If a trajectory of conflicting aircraft will cross the collision zone, which surrounds a certain aircraft with TCAS II system, than this system will inform pilots about this event using visual and speech means of a warning.

TA and RA are aimed at maintaining or increasing vertical separation required for preventing the collision with conflicting aircraft. With the objective of collision avoidance TCAS II monitors up to 50 aircraft, which are expected to enter the zone of collision.

On 30 March 2017 information appeared (Northrop, 2017; ACSS, 2017) that "Northrop Firebird tests collision avoidance system for UAVs. Northrop Grumman's Firebird optionally manned demonstrator has completed an early shakedown of the anti-collision system intended to make the skies safe for unmanned and manned aircraft.

The flight tests over Mojave, California, will provide the data that shows how well the version of the ACAS designed for unmanned air vehicles (UAVs) conform to simulation models. Any discrepancies will be used to refine the models, which shoulder the load of the development activity for the new system. "The flight tests collected data necessary to validate simulation models and help inform the continued development of ACAS Xu," says Greg Boerwinkle, UAV programme manager for ACSS, a joint venture of L-3 Communications and Thales.

A mixture of government agencies, national laboratories and private contractors is developing ACAS. The US Federal Aviation Administration is spearheading the development of ACAS, which is intended to replace the TCAS for all aircraft, whether manned or unmanned. Working for the FAA, the Massachusetts Institute of Technology's Lincoln Laboratory and the Johns Hopkins Applied Physics Laboratory **developed the software for ACAS Xu**, the version dedicated for UAVs. The laboratories delivered that software to ACSS, which integrated the code into its TCAS hardware, which includes functions for traffic and collision avoidance, terrain awareness and warning and a transponder. The result, ACAS Xu, was installed in the Firebird demonstrator, a twin-boomed, optionally manned aircraft designed by Scaled Composites engineer Cory Bird.

ACSS and Northrop supplied additional aircraft for the demonstration, which flew a series of collision trajectories with the Firebird. The ACAS Xu system on the Firebird is supposed to detect the impending collision using an onboard sense-and-avoid sensor package, and then perform a series of steps. It first sends "resolution advisories" to the host aircraft and the "intruders", then coordinates manoeuvres with the TCAS on board the other aircraft. Ideally, ACAS Xu will make the UAV safer to operate in national airspace than if a pilot was in the cockpit, Boerwinkle says.

"The capabilities under development for large UASs to detect and avoid collisions will be substantially better than the ability of a pilot to see with the human eye and avoid collisions," he says. In addition to ACAS Xu, the project is also developing ACAS Xa and Xo for manned aircraft equipped with TCAS already".

Structure of ACAS II/TCAS II Model

In this section, original software for RPAS on-board collision avoidance system is described (Kharchenko, Barabanov, Grekhov, & Shishkov, 2013). TCAS II consists of a set of antennas on the aircraft, computing unit, Mode S transponder, indicators and control panel installed in the cockpit. In Figure 16, a scheme of the system is shown.

The computing unit of TCAS II receives data from the aircraft instruments (radio altitude, barometric altitude, if the plane stays in the air or on the ground, as well as if landing gears are released). The computing unit interrogates transponders of other aircraft, computes their location, tracks their trajectory, indicates various warnings and recommendations on the display, and transmits voice messages to the pilot through voice warning systems located in the cockpit. The computing unit is the basis of TCAS II.

TCAS II system has Mode S transponder, which responses on interrogations of air traffic control radars with Mode A, C and Mode S, as well as interrogations of other aircraft equipped with TCAS. Each aircraft equipped with the Mode S transponder has its own unique ICAO address, allowing direct exchange of data with compatible systems.

Figure 16. TCAS system

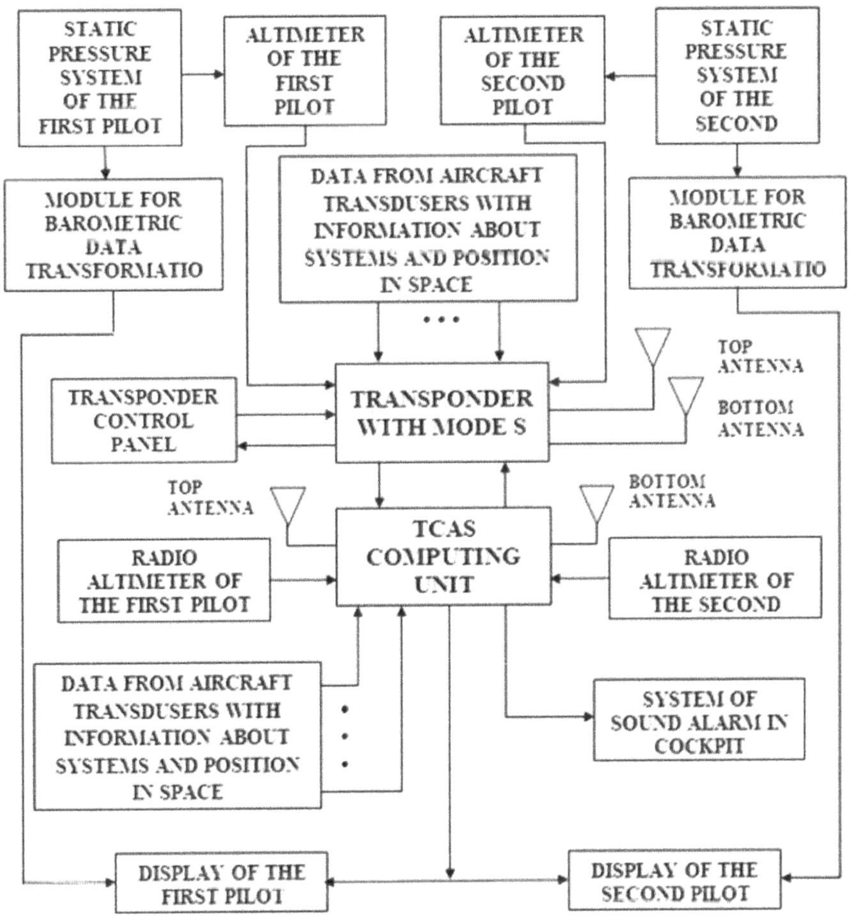

An operation of TCAS II is simulated in a paper (Livadasy, Lygerosz, & Lynch, 1999) and conditions are determined, under which the system guarantees adequate altitude separation for aircraft at risk of collision.

The aim of this study is to develop software for modeling real-time operation of on-board collision avoidance system and to create an experimental model with data exchange between on-board systems.

To simulate the real time work of the on-board collision avoidance system programming environment MATLAB was used. Model of TCAS system is shown in Figure 17.

Figure 17. Model of TCAS system

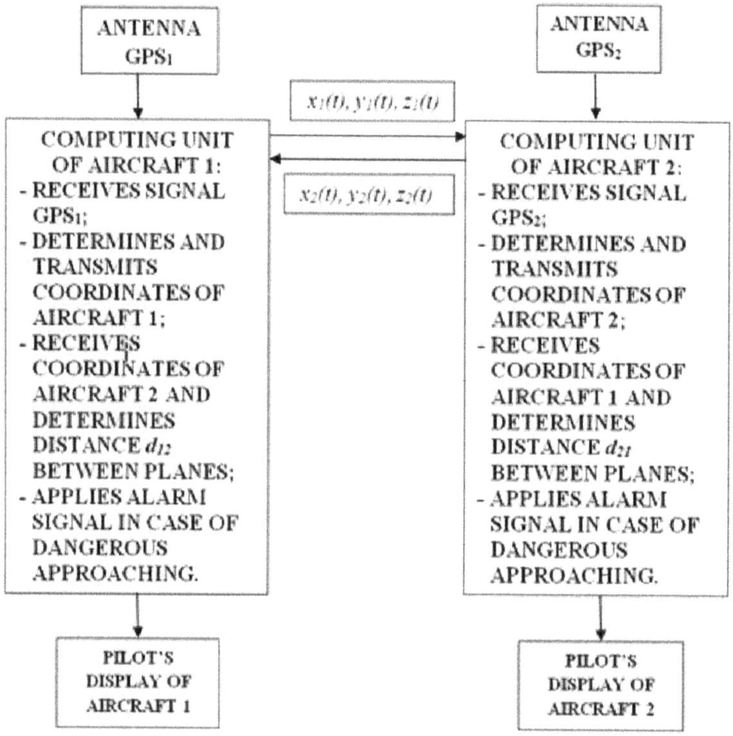

The position of aircraft is determined in the computing unit using data from GPS-receiver signals. During the relative motion of aircraft their computing units estimate continuously in real time the relative distance between them and compare it with a given minimum acceptable value.

National Marine Electronics Association (NMEA data, 2018) data format was used for writing the algorithm of the computing unit, in which the most important information about 3D aircraft position is contained in GGA sentence. The example of GGA sentence:

```
$GPGGA,123519,4807.038,N,01131.000,E,1,08,
0.9,545.4,M,46.9,M,,*47
where:
GGA           Global Positioning System Fix Data
123519        Fix taken at 12:35:19 UTC
4807.038,N    Latitude 48 deg 07.038' N
01131.000,E   Longitude 11 deg 31.000' E 1
Fix quality: 0 = invalid
1 = GPS fix (SPS)
```

```
2 = DGPS fix
3 = PPS fix
4 = Real Time Kinematic
5 = Float RTK
6 = estimated (dead reckoning) (2.3 feature)
7 = Manual input
8 = Simulation mode
08  Number of satellites being tracked
0.9 Horizontal dilution of position
545.4,M Altitude, Meters, above mean sea level
46.9,M eight of geoid (mean sea level) above WGS84 ellipsoid
(empty field) time in seconds since last DGPS update
(empty field) DGPS station ID number
*47  the checksum data, always begins with *.
```

The distance between aircraft is calculated using the Pythagorean Theorem, which takes into account the height of each aircraft and the distance between them on the ground. Distance to the Earth's surface was calculated using the 'haversine' formula:

$$a = \sin^2(\Delta lat/2) + \cos(lat_1)\cdot\cos(lat_2)\cdot\sin^2(\Delta long/2)$$

$$c = 2\cdot \text{atan2}(\sqrt{a}, \sqrt{(1-a)}),\ d = R\cdot c,$$

where R is Earth radius (6371 km) and ellipticity of the Earth is not considered. In the result aircraft computing units work out the displayed warning about critical distance achievement (Figure 18) when the relative distance is less than the given minimal acceptable value.

Simulation of Aircraft Approaching

Model of ACAS/TCAS (Figure 17) has been implemented in a lab class on two computers that perform data exchange in real time according to pre-programmed algorithms in MATLAB environment using the Wi-Fi network.

For simulation the relative motion of aircraft the model was modified (Figure 19). For this GPS-navigator Garmin.eTrex was connected to one of the computers using a COM port.

From this port data in the NMEA format arrived from the navigator into a computing unit 1, which determined its own coordinates x_1, y_1, z_1. These coordinates were transmitted in real time to the second computer, where a computing unit of aircraft 2 generated its coordinates in accordance with a given law, imitating the motion of the aircraft 2 relative to aircraft 1. Variable

Figure 18. Interface of the computing unit

COORDINATES OF AIRCRAFT 1

GPS RECEIVER TIME

DECIMAL LATITUDE

DECIMAL LONGTITUDE

ALTITUDE

COORDINATES OF AIRCRAFT 2

GPS RECEIVER TIME

DECIMAL LATITUDE

DECIMAL LONGTITUDE

ALTITUDE

DISTANCCE BETWEEN AIRCRAFT

WARNING!
CRITICAL DISTANCE IS REACHED!

Figure 19. Scheme of simulation of aircraft approaching

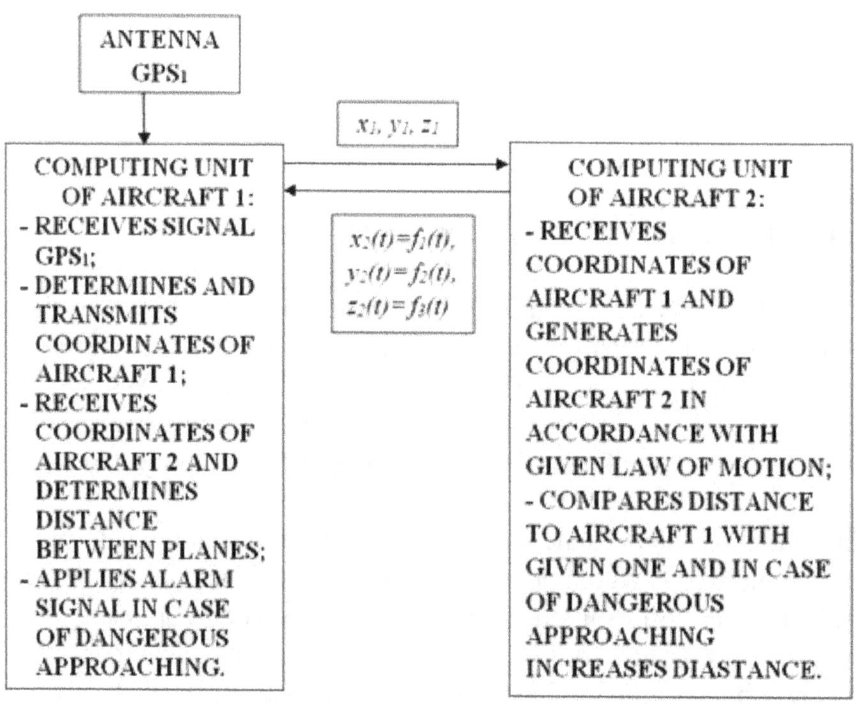

in time coordinates of aircraft 2 were continuously transmitted via Wi-Fi in the computing unit of the aircraft 1. Both computing units were continuously estimating the relative distance between the aircraft and issued a warning when reaching given minimum acceptable value that can be set arbitrarily. After an alarm the law of a motion of the aircraft 2 is changing and planes began to "diverge" in space avoiding collisions.

Results of simulation for aircraft approaching at different speeds are presented in Figure 20.

Two aircraft from a distance of 2900 m begin to approach uniformly along the shortest path. The horizontal line shows the distance at which the computing unit triggers the alarm about the danger.

At lower speed of approaching the system "separates" aircraft without appearance of a warning and at higher speed - issues warning to pilots and aircraft begin to move away from each other.

Figure 20. Dependence of the distance between airplanes on time: curve, marked by circles corresponds to the speed of approaching 412 m/s (1484 km/hour), and marked with crosses – 526 m/s (1893 km/hour)

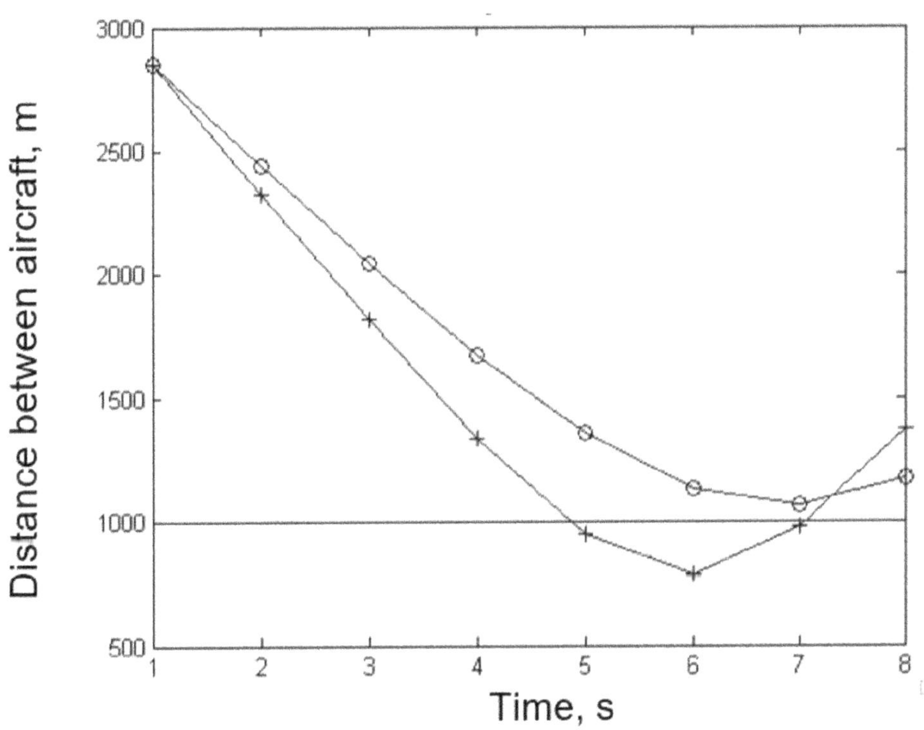

If we assume that the coordinates are not processed every second, but every ten seconds, the curves in Figure 20 will correspond to the speeds of approaching ten times lower. This case meet situation with Unmanned Air Vehicles.

CONCLUSION

1. Original software for modeling of real-time TCAS operation was developed using MATLAB.
2. The experimental model with data exchange between onboard systems via Wi-Fi network was created.
3. This model was used for modeling of aircraft/RPAS approaching.
4. This model can be used as a base for creation a collision avoidance system of RPASs.

REFERENCES

ACAS. (2017). *Airborne Collision Avoidance System (ACAS) Guide*. Retrieved from https://www.eurocontrol.int/publications/airborne-collision-avoidance-system-acas-guide

ACAS Guide Airborne Collision Avoidance. (2017). Retrieved from https://www.eurocontrol.int/sites/default/files/content/documents/nm/safety/ACAS/safety-acas-II-guide.pdf

ACSS. (2017). *ACSS, NASA plan UAS collision avoidance flight tests this summer*. Retrieved from http://www.uasmagazine.com/articles/1674/

EUROCONTROL CASCADE Programme. (2013). Retrieved from https://www.eurocontrol.int/sites/default/files/publication/files/2013-ads-b-wam-leaflet.pdf

FlightAware. (2018). Retrieved from https://ru.flightaware.com/adsb/

FlightFeeder. (2014). Retrieved from https://ru.flightaware.com/about/history

Guidance for the Operational Introduction of SSR Mode. (2012). Retrieved from http://www.eurocontrol.int/articles/eurocontrol-documents-ssr-mode-s

ICAO. (2007). *Annex 10. Vol. IV. Aeronautical Telecommunications - Surveillance and Collision Avoidance Systems, Fourth Edition*. Retrieved from http://www.eurocontrol.int/msa/public/standard_page/ACAS_ICAO_Provisions.html#manual

Kharchenko, V. P., Barabanov, Y. M., Grekhov, A. M., Ghaznyuk, M. O., & Kolchev, M. V. (2012). Monitoring of air traffic using ADS-B system. *Proceedings of the National Aviation University*, *53*(4), 13–19.

Kharchenko, V. P., Barabanov, Y. M., Grekhov, A. M., & Shishkov, F. O. (2013). Modeling of Airborne Collision Avoidance System operation using MATLAB software. *Proceedings of the National Aviation University*, *56*(3), 3–17. doi:10.18372/2306-1472.56.5407

Livadasy, C., Lygerosz, J., & Lynch, N. A. (1999). *High-level modeling and analysis of the traffic alert and collision avoidance system (TCAS)*. Retrieved from http://people.csail.mit.edu/ clivadas/pubs/IEEE00.pdf

Manual for the adsbScope Software V1.9 & adsbPIC Decoder V2 Fw6. (2010). Retrieved from http://adsbradar.ru/sites/default/files/i/adsbScope_en.pdf

Mode S and ACAS Program. (2008). Retrieved from https://www.eurocontrol.int/eec/public/standard_page/WP_Mode_S.html

NMEA data. (2018). Retrieved from http://www.gpsinformation.org/dale/nmea.htm

Northrop. (2017). *Northrop Firebird tests collision avoidance system for UAVs*. Retrieved from https://www.flightglobal.com/news/articles/northrop-firebird-tests-collision-avoidance-system-f-435730/

PANS-ATM. (2007). *Procedures for Air Navigation Services - ICAO Doc. 4444 Fifteenth Edition 2007-ATM/501*. Retrieved from http://www.eurocontrol.int/msa/public/standard_page/ACAS_ICAO_Provisions.html#manual

PANS-OPS. (2006). *Procedures for Air Navigation Services - Aircraft Operations* (Vol. 1). Flight.

PlanePlotter 6.4.5.1. (2016). Retrieved from https://planeplotter.en.lo4d.com/

Procedures - ICAO Doc. 8168 OPS/611, Fifth edition plus Amendment 3. (n.d.). Retrieved from http://www.eurocontrol.int/ msa/public/ standard_page/ACAS_ICAO_Provisions.html#manual

Receiver and Decoder - miniADSB 1090 MHz Receiver. (2010). Retrieved from miniadsb.web99.de/homepage/index.php

Richards, W. R., O'Brien, K., & Miller, D. C. (2010). *New air traffic surveillance technology.* Academic Press.

Wi-Fi Research Center. (n.d.). Retrieved from http://www.networkworld.com/topics/wi-fi.html

Appendix

ABBREVIATIONS

3GPP E-UTRA: Third Generation Partnership Program Evolved UMTS Terrestrial Radio Access
AAS: Adaptive Antenna System
ACARS: Aircraft Communications Addressing and Reporting System
ACAS: Airborne Collision Avoidance System
ACSS: Aviation Communication and Surveillance Systems
ADS-B: Automatic Dependent Surveillance – Broadcast
ADS-R: Automatic Dependent Surveillance – Rebroadcast
AFTN: Aeronautical Fixed Telecommunications Network
AIS: Automatic Identification System
AM: Amplitude modulation
AM/AM: Amplitude Modulation/Amplitude Modulation
AM/PM: Amplitude Modulation/Phase Modulation
AMSS: Aeronautical Mobile Satellite Service
ANSP: Air Navigation Service Provider
ARINC: Aeronautical Radio Incorporated
ATC: Air Traffic Control
ATIS: Automatic Terminal Information Service
ATM: Air Traffic Management
ATN: Aeronautical Telecommunication Network
ATS: Air Traffic Services
AU: Average Utilization
AW: Average Workload
AWGN: Additive White Gaussian Noise
BER: Bit Error Rate
BLAST: Bell Labs Layered Space-Time
BPSK: Binary Phase Shift Keying

BRLOS: Beyond Radio Line-of-Sight
C2: Command and Control
C3: Command, Control and Communication
CASCADE: Cooperative ATS through Surveillance and Communication Applications Deployed in ECAC
CC: Convolutional Code
CDTI: Cockpit Display of Traffic Information
CIDIN: Common ICAO Data Interchange Network
CIR: Channel Impulse Response
CNS: Communication Navigation Surveillance
CPDLC: Controller-Pilot Data Link Communication
CRC: Cyclic Redundancy Code
CSMA: Carrier Selecting Multiple Access
D8PSK: Differential 8 Phase Shift Keying
DC: Direct Current
DECnet: Digital Equipment Corporation
DF: Downlink Format
DLC: Data Loss Coefficient
DME: Distance Measuring Equipment
DVB: Digital Video Broadcast
E_b/N_0: Ratio of bit energy to noise power spectral density
EFIE: Electric-Field Integral Equation
ES: Extended Squitter
EUROCAE: European Organization for Civil Aviation Equipment
EUROCONTROL: European Organization for the Safety of Air Navigation
FAA: Federal Aviation Administration
FCC: Flight Control Computer
FDMA: Frequency-Division Multiple Access
FEC: Forward Error Correction
FFT: Fast Fourier Transform
FIR: Flight Information Region
FIS-B: Flight Information Service–Broadcast
FM: Frequency Modulation
FMS: Flight Management System
FSPL: Free Space Path Loss
GEO: Geostationary Earth Orbit
GES: Ground Earth Station
GI: Guard Interval
GNSS: Global Navigation Satellite System

GPS: Global Positioning System
GSO: Geostationary Orbit Satellite
GVA: Geometric Vertical Accuracy
H-ARQ: Hybrid-Automatic Repeat Request
HF: High Frequency
HHO: Network-Optimized Hard Handoff
ICAO: International Civil Aviation Organization
IEEE: Institute of Electrical and Electronics Engineers
IFFT: Inverse Fast Fourier Transform
IFR: Instrument Flight Rules
ILS: Instrument Landing System
IP: Internet Protocol
IPX/SPX: Internetwork Packet Exchange/Sequenced Packet Exchange
IQ: Inphase and Quadrature
IRS: Inertial Reference System
ISI: Inter Symbol Interference
ISO: International Organization for Standardization
LAN: Local Area Network
LEO: Low-Earth Orbit
LNA: Low-Noise Amplifier
LOS: Line Of Sight
LSTC: Layered Space-Time Codes
LTE: Long-Term Evolution
MA: Manouevre Advisories
MAC: Medium Access Control
MAN: Metropolitan Area Network.
MBS: Multicast and Broadcast Service
MCP/FCU: Mode Control Panel/Flight Control Unit
MediaFLO: Media Forward Link Only
MEO: Medium Earth Orbit
MFP: Management Frame Protection
MIMO: Multiple Input – Multiple Output
MISO: Multiple Input – Single Output
MLAT: Multilateration
MoM: Method of Moments
MPLS VLANs: Multi-Protocol Switching Over Labels of a Virtual LAN
M-PSK: M-ary Phase Shift Keying
M-QAM: M-ary Quadrature Amplitude Modulation
MSK: Minimal Shift Keying

MSS: Mobile Satellite Services
NCBPS: Number of Carrier Bits per OFDM-symbol
NextGen: Next Generation Air Transportation System
NM: Nautical Mile
NMEA: National Marine Electronics Association
NOTAM: Notice to Airmen
OFDM: Orthogonal Frequency Division Multiplexing
OSTBC: Orthogonal Space-Time Block Code
PANS: Procedures for Air Navigation Services
PAPR: Peak to Average Power Ratio
PBN: Performance-Based Navigation
PDF: Probability Density Function
PDH: Plesiochronous Digital Hierarchy
PER: Packet Error Rate
PFC: Packet Fail Chance
PHY: Physical layer
PIC: Pilot in Command
PLCP: Physical Layer Convergence Protocol
PM: Phase Modulation
PN: Pseudorandom Noise
PPM: Pulse Position Modulation
PRF: Pulse Repetition Frequency
PSDN: Public Switched Data Network
PSK: Phase-Shift Keying
PSR: Primary Surveillance Radar
PSTN: Public Switched Telecommunication Network
QAM: Quadrature Amplitude Modulation
QoS: Quality of Service
QPSK: Quadrature Phase-Shift Keying
RA: Resolution Advisory
RCP: Required Communication Performance
RF: Radio Frequency
RLOS: Radio Line of Sight
RNP: Required Navigation Performance
RPAS: Remotely Piloted Air System
RS: Reed-Solomon
RTCA: Radio Technical Commission for Aeronautics
SARPs: Standards and Recommended Practices
SC: Single Carrier

SC-FDMA: Single-carrier Frequency Division Multiple Access
SES: Single European Sky
SESAR: Single European Sky ATM Research
SINR: Signal to Interference plus Noise Ratio
SISO: Single Input - Single Output
SM: Spatial Multiplexing
SMGCS: Surveillance Moving Ground Control System
SNA IBM: Systems Network Architecture
SNOC: Satellite Network Operating Center
SNR: Signal to Noise Ratio
SOFDMA: Scalable OFDM Access
SPI: Special Position Indication
SSPA: Solid State Power Amplifier
SSR: Secondary Surveillance Radar
STBC: Space-Time Block Codes
STC: Space-Time Coding
STDMA: Self-Organizing Time Division Multiple Access
STTC: Space-Time Trellis Codes
STTTC: Space-Time Turbo Trellis Codes
TA: Traffic Advisory
TBT: Time between Transactions
TCAS: Traffic Alert and Collision Avoidance System
TCP/IP: Transmission Control Protocol/Internet Protocol
TDD: Time Division Duplex
TDMA: Time Division Multiple Access
TG: Protective Interval
TIS-B: Traffic Information Service–Broadcast
TS: Transaction Size
TT: Travel Time
TTAC: Tracking, Telemetry and Control
TTI: Transmission Time Interval
UAS: Unmanned Aircraft Systems
UAT: Universal Access Transceiver
UHF: Ultra High Frequency
VDES: VHF Data Exchange System
VDL: Very-High Frequency Digital Link
VHF: Very High Frequency
VLOS: Visual Line of Sight
VOR: Very High Frequency Omni directional Range

VSAT: Very Small Aperture Terminal
WAN: Wide Area Network
WAVE: Wireless Access in Vehicular Environment Standard
WiMAX: Worldwide Interoperability Microwave Systems
WLAN: Wireless Local Area Network
XNS: Xerox Network Systems
X.25: ITU-T Standard Protocol for Packet Switched WAN

About the Author

Andrii Mikhailovich Grekhov earned a BSc, MSc (1968–1973) in Theoretical Physics at Physics Faculty and Commander of Radar Set at Military Department of Kyiv State T. Shevchenko University, and PhD (1974–1980) in Institute of Semiconductors at Ukrainian Academy of Sciences, Doctor of Physical and Mathematical Sciences (1990) in Moscow Institute of Steel and Alloys. He has worked as a junior researcher (1974-1980) in Institute of Semiconductors at Ukrainian Academy of Sciences, senior research in Institute of Physics at Ukrainian Academy of Sciences and Head of the Department of Mathematics and Physics in Kyiv Antiaircraft Missile Higher Military Educational Institution (1981–1991), Head of the Physics Department at National University of Food Technologies (1991–1996), diplomat and Acting Representative of Ukraine to ICAO (1997-2000), Vice-Rector for R&D in Kyiv European University (2001-2006), Director of ICAO Institute in National Aviation University, Professor at Air Navigation Systems Department (2006 - to present). He is a EUROCONTROL expert in ADS-B systems. His research interests include surveillance, ADS-B systems, telecommunications, computer modeling. He is the author of about 190 scientific papers, textbooks, training aids, and proceedings.

Index

N

O

R

S

www.ingramcontent.com/pod-product-compliance
Lightning Source LLC
LaVergne TN
LVHW060306011026
849161LV00011B/2782